CW01067176

The Lost Age Of High Knowledge

Evidence of an Advanced Civilisation Prior to Recorded History

Keith M. Hunter

Published by Keith M. Hunter

ISBN: 978-0-9564563-1-1

Printed in Great Britain by the MPG Books Group, Bodmin and King's Lynn

Acknowledgements

This author would like to acknowledge the work of **Bruce Cathie**, as quite simply inspirational as a guide to understanding the deeper mysteries of the Earth and its subtle forces.

Special thanks also go to:

Ed Williams, for his kind permission to use his *Great Circle Calculator* program, which can be found on the Internet at:
http://williams.best.vwh.net/gccalc.htm

Robert Johnston, for allowing this author to make use of his extensive database of nuclear tests, and also for his invaluable help through correspondence, whose own website is located at:
http://www.johnstonsarchive.net/index.html

U.S. Naval Observatory, whose work has been of aid to this author in the realm of astronomical modelling. The website of this particular (U.S.) agency can be found at: http://www.usno.navy.mil/

Contents

Official Website of Keith M. Hunter

http://www.ancient-world-mysteries.com

For the latest research and updates

Introduction

Human beings are like no other life form on this planet. They are unique from all the rest, in that only a human can discover the universal principles of nature that underlie the existence of the universe, and which govern the movement and transformation of all things that physically manifest within its confines. The very desire to discover is innate to the human condition, and as a result, human knowledge has advanced through the ages. For the most part, upward progression has been slow over much of recorded history. And yet, quite remarkably, the last few centuries of this present age have seen man acquire knowledge of the physical world at a rate unmatched by any known previous generation. Even with this said however, one should never assume that human knowledge always advances throughout the ages, uninterrupted. Indeed, there are moments in history when knowledge has been lost; replaced by ignorance. When the whole of a society is thrown into deep crisis, when excepted customs cease to be; regardless of the cause, if the ability to recover quickly is lacking, then long term decline is assured. One notable example of this is the fall of the Roman Empire in Europe during the 5th century AD. With its collapse the continent suffered a dark age for the next thousand years. During this period knowledge itself was indeed actually lost, as oppressive regimes and authorities emerged and the whole region became fragmented. Many discoveries that had been made during antiquity disappeared. Once, it was known in ancient times that the Earth orbited the sun [1] and that its form was that of a sphere [2]. The very size of the Earth was also known to a fair degree of accuracy [3]. All of these things however, including other facts of a scientific nature, had to actually be re-discovered during the Renaissance, when the Dark Age finally came to an end.

In the present age it is most unfortunate that those who study history do not pay nearly as much attention to man's loss of knowledge as they do his acquisition of it. Indeed, as a general rule, if not a general prejudice, it is tacitly assumed by many researchers today that the further back in time one goes, the less knowledge man possessed. However, this would presuppose that man only ever acquires knowledge, successfully transmitting it from generation to generation without ever losing it. But this of course, is quite simply not how it works. As history itself confirms, every civilisation that has ever arisen has had occasions when they have advanced in their knowledge over successive generations, and occasions where their knowledge has been lost by their failure to successfully transmit it to future generations. It is due indeed to this very fact – that the level of knowledge of past civilisations periodically fluctuates over time – that those who study them, are sometimes led to encounter what can only be described as strange anomalies; essentially, evidence of advanced knowledge amongst those thought too primitive to be able to possess it, at least judging from the outward appearance of the physical infrastructure of their society, and their 'way of life' in general. Such indeed is indicative of the fact that the acquired knowledge of any civilisation up to a societal highpoint is never completely lost following any subsequent cultural-physical-political collapse. Remnants of high knowledge can still survive amongst a people following such a catastrophe, even though they may live in the midst of a dark age, where the very pre-conditions for the actual discovery of knowledge itself no longer exist as they once did.

In the modern age, many studies have been conducted of ancient civilisations from all over the world. A large proportion of them have been carried out by highly qualified people well trained in the scientific method. However, it is fair to say that most of the studies produced have focused mainly upon the more 'primitive' aspects of such societies. The conclusions reached in such works have tended to be overly conservative, and in almost all cases thoroughly in line with the axiomatic assumption that past civilisations are less advanced than those that follow on from them, essentially in all fields of endeavour. Thus, most mainstream scientists today see the people of prior ages as being both primitive in a physical-technological sense, and primitive in their beliefs. One may consider a few notable examples to illustrate this point, which include studies of such as the Great Pyramid of Egypt, Stonehenge in England, or the calendars and systems of measure employed in ancient times by the Babylonians, Egyptians, or the Maya of Central America. Briefly stated, the conclusions drawn from such studies, which indeed form current orthodox opinion today, are that the Great Pyramid was built as a tomb for a notable Egyptian Pharaoh, that Stonehenge was a site of pagan 'mystical rituals', and that the units of measure employed by the ancients were established somewhat arbitrarily. However, it is not true to say that all scientists and researchers of the present age subscribe to these views, though they are in truth the most dominant. Indeed, over the past few decades, due to the efforts of a small minority of alternative thinkers, a series of radical new theories have been developed to explain the purpose of these gigantic monuments, including the systems of measure employed by the ancients. The solutions put forward fly in the face of established ideas.

As incredible as it may seem, evidence has been uncovered by some to suggest that the Great Pyramid of Egypt may in actuality have been built in some previous age as a power plant of some sort [4], and that Stonehenge may have been constructed as a giant 'crystal set' for receiving and transmitting signals from space [5], and moreover, that the units of measure employed by the ancients including the time cycles built in to their calendars, were chosen to represent the physical energy configuration of the Earth itself, and also, to denote the time intervals of various celestial cycles whose periodic completion was thought to be marked by sudden physical planetary upheaval. As a result of the development of these theories, many alternative researchers consider it very likely that prior to established recorded history, a highly advanced civilisation did once exist upon the face of the Earth, whose achievements were comparable to those of this present age, if not greater, but that for some unknown reason this civilisation suffered catastrophic failure. If such indeed is true, then it is highly conceivable that various ancient stone monuments such as include the Great pyramid and Stonehenge, in addition to the systems of measure of the peoples of the past, may all be remnants of such an advanced civilisation; the last vestiges of a lost technology and a lost cosmology.

Of course, mainstream academics and researchers tend to be highly dismissive of the radical ideas generated by some of their colleagues. Their own rather conservative studies, in conjunction with their axiomatic belief in the primitive nature of the peoples of the past, have forced them to accept the conclusion that all of the accomplishments of those living in ancient times do not and could not in any way whatsoever demonstrate an advanced understanding of the principles of the natural world. In essence, they claim that though such as the Great Pyramid or Stonehenge are indeed large structures, quite impressive to behold, they are also basic structures, and the reasons underlying their construction are thus unremarkable. In a similar vein, the units of measure and the mathematics of the ancients are also held to be of no special significance; their choice of values not being governed by any particularly profound reasoning. Indeed, almost all mainstream researchers today who have studied these ancient civilisations reject out of hand the very notion that they could possibly have constructed such monuments based upon an advanced knowledge of engineering and material sciences, or that their systems of measure were established due to a sophisticated knowledge of physics and astronomy. Consequently, it is the position of such researchers that there could not possibly have been a highly advanced civilisation existent upon the Earth prior to recorded history. Indeed, all of the great stone monuments of the past and the systems of measure employed during ancient times are all considered to have been created by 'known' civilisations, according to orthodox opinion. And yet, if this is not true, and the more radical theories are correct, then the people who lived in what historians today call ancient Egypt, Babylon, or England for that matter, may not in actuality have been the same people responsible for constructing the Great Pyramid or Stonehenge, or establishing the systems of measures found to be in use during this period. Rather, they may have simply inherited them all from a much earlier civilisation that flourished in a far more remote era; one that indeed they may be the direct descendants of. In which case, the stone structures would have been ancient even to them during their time, as would also the systems of measure they employed.

Over the recent period many books have been produced examining the possibility of a once highly advanced civilisation existent upon the Earth in some forgotten age. Quite a number have focused upon the massive stone structures built by ancient cultures from all over the world. The result of such efforts has led to a great deal of support in favour of the idea that they were indeed built by an advanced people. However, much less attention has been given to the units of measure that were employed in the past, including the calendrical systems established by the ancients. Though they have been studied by many, the analyses that have been conducted to date have invariably failed to reveal decisive evidence that the measures and time cycles chosen by the ancients reflected a most profound understanding of the natural-physical world. What is lacking from most of these studies is a critical breakthrough, conclusively demonstrating the true foundation of these ancient systems of measure, which simultaneously confirms also the genius and high knowledge of those responsible for devising them. In light of this fact, it is therefore the intention of this present work to seek to accomplish exactly this.

By conducting an in-depth examination of the mathematics and the units of measure actively employed in ancient times by such as the Sumerians, Babylonians, and Egyptians, including others, this work shall seek to uncover the original basis of their established systems of measure; determining the very foundation upon which the essential components as a whole rest. For if the true basis of the system can be uncovered, there exists the possibility that by its very nature, it may in and of itself constitute decisive proof that the forbears of these ancient civilisations did indeed possess an advanced understanding of the physical world; one actively reflected by their choice of measures. Such would demonstrate if true, that their whole system of mathematics and of measure would constitute a valid remnant of high knowledge, once held by an extremely sophisticated civilisation that in a previous age was active upon the face of the Earth.

Part 1:

Transformation Of the Celestial Realm

Chapter 1

The Basis of Ancient Measure

Over the last few years several notable works have been produced detailing a number of unusual discoveries that relate to the geo-physical structure of the Earth and its surrounding fields of energy. Of those researchers who have been inspired to study such things from a non-conventional standpoint, a significant number have over the recent period uncovered evidence to suggest a close affinity between the physical form of the Earth, its known energy fields, and certain units of measure actively employed by various ancient civilisations of the past. A brief account of some of this work is highly appropriate, as indeed it forms the very foundation of this present study.

The Energy Fields of the Earth

It is well established among scientists who have investigated the properties of the Earth, that as a body, it possesses a number of different fields. Most common amongst them are its magnetic and gravitational fields. Concerning the former, it is generally acknowledged that it is roughly aligned along the north-south axis of the planet, with a slight angular deviation to the true (physical) axis. The gravitational field by contrast would seem to be more in line with the physical symmetry of the outward planetary form. Both fields appear though to permeate the planet and extend well into space. Although initially thought to be rather uniform and somewhat featureless, a number of researchers have over the recent period uncovered evidence of a definite structure to the Earth's energy fields. Indeed, certain revealing studies would appear to suggest that the Earth possesses a highly intricate system or matrix of energy lines spread out over the surface of the planet in ordered geometrical patterns. In essence, the Earth's energy fields are highly organised, being embedded with energetic lines of force that would seem to be active conduits of power [1]. Several distinct patterns or line formations have been uncovered by a variety of individuals and scientific groups, though all are undoubtedly facets of one unified system. Most remarkably, the points at which numerous lines have been found to crossover one another, establishing energy nodes, would appear to coincide with the precise locations not only of prominent cities and various stone monuments that have existed since ancient times, but also points upon the globe wherein strange magnetic anomalies have been found to manifest. Moreover, there is also evidence to suggest that engineered technological devices may actively be used to tap the energy from the various grid matrixes for highly practical purposes.

Of the known energy grid formations that have been uncovered over the past few decades, there is one in particular that is deserving of special note, due to the fact that the matrix identified was found to possess a strong affinity to the units of measure employed by a well known civilisation from the remote past. The grid pattern in question was discovered by a New Zealand researcher named Bruce Cathie, its existence being validated in a most unusual way: from his early studies of the UFO phenomenon in New Zealand during the 1950's through to the 1960's. During these years many reports of UFO sightings were made, and not just in New Zealand. They came from all over the world from numerous eyewitnesses, a significant portion being highly credible. Such sightings continue to this day even [2]. As a consequence of these early sightings, many organisations were formed by people interested in this phenomenon, which set about the task of trying to collect reports of as many encounters and sightings of UFO's as possible for further analysis. Usually such organisations were privately established with no official backing. Due to his own interest in the subject, Cathie set out to evaluate this information for himself by reading the reports collected by various UFO organisations to try to find a pattern to many of the sightings in the area around New Zealand. If there was a definite reason as to why these objects appeared where they did, then this would be indicative of some sort of active intelligence at work controlling these objects or guiding them. If this were so, then there may be an ordered pattern to their appearances [3]. This is exactly what Cathie found. After separating the good reports from the bad, he was able to plot on a map to a fair degree of accuracy the positions at which various UFO's were spotted. It was a long process, but after some time of careful analysis of the various reports from witnesses; as a result of many different people viewing the same objects from several different locations, he was able to plot the sighted positions of various UFO's and in some cases their line of movement, upon a map. From this, a remarkable discovery emerged. It was found that they appeared to move in a very ordered pattern over the surface of the Earth. The pattern itself was that of a grid like matrix composed of a series of parallel lines orientated approximately north-south and east–west. Interestingly though, the grid lines were found to be spaced at intervals of 7.5 minutes of arc covering the surface of the planet both in respect of latitude and longitude [4]. Such is indeed a most intriguing finding when one considers that minutes of arc are units of measure inherited from Babylon, a civilisation of the ancient world whose origin dates back to 1900 BC, and which only came to an end in 600 BC [5]. Indeed, the very system of angular geometry employed in mathematics to this day, of which minutes of arc are a component part, is widely thought to have originated with the Babylonians, who are also held responsible for establishing the very convention of splitting up the circle into 360 equal parts, known as degrees. Within their system, the minute of arc itself is simply a further refinement of this value, being 1/60th of a degree; essentially a more acute angle. Employing such measures, exactly 15 fractional halves (7.5 / 0.5) of one minute of arc were thus found to mark out the intervals separating the lines of movement of the observed UFO's. In light of this, Cathie was led to conclude that there must be an important affinity between these units of angular measure and the actual flight patterns of the UFO's, and that in some way the uncovered grid matrix was possibly an aid to the navigation or propulsion of such craft [6].

Although initially the grid lines were established by Cathie through observations of UFO's solely in and around the New Zealand area, a certain Frenchman who had himself been investigating the flight of UFO's over Europe, had uncovered evidence of exactly the same ordered patterns [7]. It appeared that in Europe too these strange objects seemed to confine their movement to certain lines criss-crossing the continent, just as they did in New Zealand. Indeed, this led to the belief that the system was global in nature, but still presented a problem: how was it precisely aligned to the Earth? The answer to this was finally revealed when Cathie discovered the presence of yet another geometrical pattern of energy lines crisscrossing the Earth - a second energy grid; one that appeared to possess a direct link to the first one that he had uncovered. He was able as a result of this to align the matrix of lines from his first grid on to various key points upon this newly discovered grid formation. Interestingly, this alternate grid was not found to be in symmetry to the geophysical axis of the planet, but slightly offset, being aligned instead to the Earth's magnetic field [8]. Moreover, the structure of the grid was far more elaborate than that of the first. It was found to be composed of a series of 'great circles' and 'small circles' intersecting one another in both the southern and the northern hemispheres. The manner in which the circles interacted established the presence of 4 nodal points in each hemisphere; both sets being in joint symmetry. According to Cathie, these nodal points are highly significant energy points in the Earth's grid network. Intriguingly, what led initially to their discovery was an undersea survey conducted by an American vessel called the Eltanin in 1964. Quite by accident, at a very deep ocean depth, a photo was taken of what appeared to be some sort of metal aerial possessing a series of cross-shaped shafts placed along it at 90 degrees to the central pole, each of which looked to be staggered to the next so that they radiated out at intervals of 15 degrees with respect to one another [9]. It was the very placement of this artefact that actually revealed the existence of one of the 8 focal points, from which Cathie was then able to deduce the overall system pattern and the positions of the other points. The location of the aerial itself was given at the time as being just off the coast of South America at 59 degrees, 8 minutes south, and 105 degrees west [10]. The true purpose of the device is not presently known. However, its existence does seem to indicate though that certain parties, whose identity currently remains a mystery, do have an interest in the Earth's energy grid and are actively involved in either monitoring it or manipulating it.

When Cathie first revealed the energy grid system that he had uncovered, it was heavily criticised, purely as a result of the fact that it rested upon the units of measure of a very ancient culture. His critics pointed out that such ancient measures were chosen quite arbitrarily and thus were not special in any way with regard to the surface of the Earth. Consequently, they held that his entire grid system was in essence nothing more than a mathematical fiction. However, some of these critics were effectively silenced at a later date when Cathie established that the Babylonian units of measure were actually related to the Earth and its fields, and that they were indeed based upon the natural field lines that he had identified. Before proceeding to detail how this was achieved, a brief introduction to the Babylonian system of angular measure as it is commonly understood today is most definitely in order, along with a description of certain units of length intimately associated with the system, some of which were used also in ancient Egypt, and are still in use even to this day in many parts of the world.

Angular Geometry

Most students who are taught the mathematics of geometry and angular measure are readily informed that it is customary to split-up the circle into 360 equal parts, each known as a degree, and moreover, that 1/60th of a degree is designated as a minute of arc, and that 1/60th of a minute of arc is a second of arc. Consequently, a full circle may be expressed in an angular sense using either of these measures:

360 degrees
21600 minutes of arc
1296000 seconds of arc

Combinations of these three units are often used to more precisely specify an angle when required. A good example would be the measure of the angle between the equatorial plane of the Earth and the orbital plane of the Earth, which is currently (1988) set at about 23 degrees, 26 minutes, and 27 seconds of arc (it changes slightly every year) [11]; a deviation that is responsible for the four seasons.

Angular measures are often employed in such fields as astronomy or to aid in the mathematical modelling of systems rotating about a fixed point. It is well to note also that they are used to map the Earth and its surface, and not just the stars. The minute of arc unit of measure in particular is very important in this regard, with its 'angular sweep' being actively used to derive curved units of distance over the surface of the Earth in a variety of ways. Some of the more prominent are worthy of mention.

The Geographical Mile

The geographical mile is a unit of length upon the surface of the Earth derived from a single minute of arc angle swept out from the centre of the planet upon the plane of the equator. It is thus equal to a curved arc-length measure of precisely 1/21600th of the Earth's (circular) equatorial circumference.

The actual value of the geographical mile is calculated to be about 6087.25 British feet, making it a slightly longer unit of measure than the more commonly used statute mile, whose value by comparison is set at 5280 ft:

Earth Equatorial Circumference = 24902.4 statute miles [12]

$24902.4 \times 5280 = 131484672$ feet

$131484672 \times (1 / 21600) = 6087.25333$ feet

The Nautical Mile

Whilst the geographical mile is quite an uncommon unit of length, and extremely specific to the plane of the equator, there is another unit of measure whose method of derivation is similar which is very much in use, being widely employed for the practical purposes of navigation at sea. It is known as the Nautical Mile, and is established upon a different plane to that of the equator. By 'slicing' the Earth in half from pole to pole – a plane cut that is 90 degrees to the equatorial plane - the cross section of the Earth produced takes the form of an ellipse, a slightly 'squashed circle'. Over one quarter of such an ellipse from either geographical pole to a point on the equator, 5400 (90 × 60) minutes of arc radiate out from the centre of the planet to the surface. Because however the circumference is elliptical in this plane, the arc length on the surface of the Earth subtending each minute of arc angle is different. The first minute of arc swept out from a point on the equator for example, subtends a surface arc length equal to about 6046 ft, whereas the last minute of arc that connects to either north or south geographic pole, subtends a surface arc of approximately 6108 ft [13]. Given this fact, in order to establish a uniform distance measure, one must wilfully split up the Earth's *surface* into 5400 equal parts; making the angular units non-uniform in their place. In doing so, a unit measure is thus defined as 1/5400th of the full surface arc length of an elliptical quadrant, which is an average of the two above noted extremes; being about 6077 ft [14]. It is this unit of length that is the nautical mile. By international agreement, many countries have fixed the nautical mile at 6076.11549 feet; a unit of measure known as the International Nautical Mile, established in accordance with a specific yard to metre Conversion [15].

The Fathom

The fathom is a much shorter distance unit than either the nautical or geographical mile. In relation to the standard foot, one single fathom is defined as being equal to precisely 6 feet. And, were the total circumference of the Earth at the plane of the equator to be expressed in terms of this unit, it would be equal to about 21914112 fathoms (131484672 / 6). It may further be noted that 1 minute of arc swept out – also upon this plane – is itself equal to 1014.54222 fathoms.

The fathom unit will play an important part in this current work as it proceeds, and thus its brief introduction here was a necessity. Much more shall be said of this unit of measure in the following chapter.

The Origin of Babylon

With a brief introduction given of the system of angular measure that this present age has inherited from the Babylonians, it is most appropriate that something be said of this ancient civilisation itself. To this end, a brief account of the history of this culture will be provided, along with a more in-depth look at the basic organisation of their overall system of mathematics.

In the modern age, much of what is known of ancient Babylon essentially derives from major archaeological excavations that were conducted throughout the 19th and early 20th centuries. From them, it was established that the Babylonian Empire arose circa 1900 BC in ancient Mesopotamia (modern-day Iraq), and that it did so alongside a competing rival empire called Assyria. The length of both was quite considerable, each lasting for well over a thousand years. It was only during the 7th and 6th centuries BC that they finally came to an end. Through a study of the ruins of Babylonian and Assyrian cities throughout Mesopotamia it was found that both empires had much in common, including a shared language, known as Akkadian. From the many sites that were excavated, thousands of records were uncovered in the form of clay tablets detailing innumerable aspects of the life of these two empires. They were written in a script that is known today as cuneiform [16]. The writings discussed such topics as the gods, astronomy, cosmology, history and mathematics. Included also were records of such things as marriage, divorce, and business contracts. Taken as a whole they are indicative of a highly evolved civilisation. However, yet further studies revealed that the language and customs of both the Babylonians and Assyrians were not invented by them at all, but instead originated with an even earlier Mesopotamian civilisation, known as Sumer, which began itself circa 3800 BC, almost 2000 years prior to the rise of Babylon and Assyria. Many of the actual writings (on clay tablets) uncovered in Babylonian and Assyrian cities were found to be copies of older documents that went back to the Sumerian era. Indeed, Akkadian inscriptions in cuneiform were established by scholars to have been predated by a Sumerian method of writing using pictographs, similar to Egyptian hieroglyphs [17]. From the detailed study of such writings, much has been learnt of ancient Sumer. However, there are still a great many mysteries about this civilisation that remain. Most notably, there is its sudden appearance. It apparently sprung up from nowhere; already complete and possessing all the aspects of a highly advanced culture. Even today it is still not known who the Sumerians were, their origin, or the means by which their civilisation appeared so suddenly around 3800 BC [18].

Setting aside for the moment the deep mystery surrounding the actual emergence of the Sumerian civilisation, it is the mathematics of this ancient culture that is of greatest importance to this present work. Indeed, it is worthy of note that the knowledge held by the Sumerians in this field is that which later on the Babylonians themselves came to adopt. In point of fact, the conventions of angular measure did not originate with the Babylonians at all, but rather were simply passed down to them from the Sumerians, who indeed may also have received them from a civilisation that preceded their own; one that at present has yet to be identified. Thus, though one may speak of the mathematics employed by both these ancient cultures, the knowledge they each had of this subject may not have originated with either of them. In truth, the answer to this puzzle still remains open. Even so, it has not prevented present day scholars from studying the mathematics itself.

Sumerian Mathematics

Nobody would question the fact that all civilisations by their very nature possess some significant measure of mathematical knowledge. The use of numbers for such as counting or forming the basis of a system of physical measures is a feature practically inherent to any civilisation. That being said, not all civilisations treat numbers in the same way or even understand them in the same way. Indeed, detailed studies of the primary mathematical system employed by the Sumerians have led scholars to conclude that they most certainly viewed numbers in a very different light to those of this present age. Such is evident from the unusual features inherent to their system.

In the current age, the mathematical system of the Sumerians is known as the sexagesimal or 'base 60' system; a result of the fact that special importance was placed by them upon the numbers 10 and 6, which were the primary values employed to generate a particularly important numerical set. To achieve this, the numbers were used in an alternating sequence in combination with one another. A primary value could be transformed, for example, by multiplying it by 10, then 6, then 10, then 6, and so on [19]. This has the effect of generating a most peculiar series [20]:

10
60
600
3600
36000
216000
2160000
12960000...

Such a system of numbers may seem quite strange when compared to the more familiar decimal system used by most countries today; a strict 'base 10' system, where numbers are increased or decreased to a higher or lower magnitude by simple multiplication or division by the number 10. Under this sort of scheme, a numerical progression of the following form is generated:

10
100
1000
10000
100000
1000000
10000000
100000000…

In comparing the base 10 system to the sexagesimal system, it is quite obvious that the latter demonstrates a greater level of complexity. Indeed, the very simplicity inherent to the former system and its natural affinity to 'counting numbers' and their manipulation, explains why it has tended to be employed most often throughout the ages for highly practical purposes i.e. basic numerical calculations. By contrast, were the sexagesimal system employed for such as this it would no doubt prove to be somewhat awkward and cumbersome. Consequently, it is highly unlikely that an accomplished civilisation would ever have designed such a system with the intention of handling rather low level mathematical operations. Indeed, this is confirmed due to the fact that the true foundation of the sexagesimal system is not as shrouded in mystery as the people who first devised it. Scholars who have studied the civilisations of ancient Mesopotamia have indeed been able to adduce the primary basis of the system, including its essential mathematical progressions. In point of fact, there is no real secret at all as to its foundation, for the Babylonians themselves indicate quite clearly that the primary values of the system are based upon astronomical cycles; most especially, those that relate to the Earth itself. One may indeed cite the degree unit as evidence of this.

According to Babylonian conventions of angular measure, as noted previously, the circle is divided into 360 equal parts; so called degrees. And the very reason for this one may find contained within many an introductory textbook of mathematics. The circle was so split, due to a belief amongst the ancient Babylonians that at some point in the past the Earth itself possessed exactly this number of days to one complete orbit about the sun [21].

Accordingly, it was decided that the Earth should be used as the basis of their system of angular measure, with each standard solar day (24 hours) taken to represent a single degree unit. With 360 days completing one full orbital period about the sun, it therefore seemed 'natural' to use solar days as the basis for dividing or sweeping out a full circle. To allow for greater refinement of the system though, it was decided that each degree should itself be further subdivided into 60 equal parts called minutes, and each minute into 60 still even smaller parts called seconds. The units of angular measure currently employed in the modern age are thus thought then, to have been originally based upon an ancient belief in a 360-day year; and, that such a year, serving as a primary standard, was itself further sub-divided through use of the number 60, in order that the system as a whole might achieve a yet greater level of refinement. One should of course note in view of this the currently observed measure of the Earth year, which stands at about 365.2421897 solar days [22]; a significant deviation from an exact value of 360 days.

360 Days per Year: Fact or Fiction?

Although most scholars today do accept that the Babylonians (and Sumerians) established their primary units of measure upon their belief that the Earth once possessed 360 days per year, very few actually believe themselves that such an orbital period has ever truly existed. Indeed, present-day academics are of the mind that these ancient civilisations were in error to think that a 360-day Earth year was ever a physical reality, for if it were, then the Earth must have significantly shifted its orbit in some past age in order to establish the length of the current year, of 365.2421897 days. For their part, they strongly argue against this possibility, stating that there is no evidence of such a transformation ever taking place. And thus consequently, the Babylonians must have been mistaken to think that the Earth ever held 360 days to one full year. The acceptance of this has led researchers to further conclude that the units of measure they employed could never have been established upon a true physical standard. They must have been chosen either arbitrarily, or at best in accordance with a vague approximation of the Earth year. In sharp contrast though, the very people of ancient Mesopotamia themselves, including neighbouring Egypt, were truly adamant that the Earth had indeed undergone a profound shift in its regular course about the sun at some point in the past; one that had directly resulted in the transformation of its orbital period. Such indeed is explicitly stated in their writings (as shall be examined in some detail in a later chapter). Upon the issue of the past reality of a 360-day Earth year, there is thus total disagreement between modern day scholars and those of ancient Mesopotamia. Somebody therefore must be in error.

In order to resolve the conflict between the views of the ancients and those of modern scholars, a thorough examination of the primary values of the sexagesimal system is required. If it is the case that the system was arbitrarily devised as the latter group maintains, then no evidence should be uncovered of a link between the values themselves and the Earth year. So far, with the work of Bruce Cathie considered, there is at least support for the idea that the minute of arc unit has some sort of real physical basis: the structure of the Earth's magnetic field. It is not therefore so totally inconceivable that degrees themselves - closely related to minutes of arc - also possess a real physical basis: an 'ideal' Earth year of 360 days, as suggested by the ancients. In attempting to test the truth of this, it would seem appropriate to conduct a more thorough analysis of Cathie's grid system, to see if one is able to uncover evidence of a valid physical link between the noted intervals of the energy lines, and the number of days contained within the Earth year; a link that may indeed decisively confirm the truth of a once existent 360-day orbital period.

Chapter 2

The Earth Orbit & Form Transformed

Geometry & the Earth Energy Grid Matrix

Previously it was stated that the New Zealand researcher Bruce Cathie had uncovered evidence of the existence of a natural grid of 'energy lines' covering the Earth, initially from the study of UFO sightings in and around New Zealand. The fundamental units of the grid based upon the movements of the UFOs themselves were shown to be 7.5 minutes of arc covering the surface of the Earth with respect to both latitude and longitude, with the whole matrix itself aligned to the axis of the Earth's magnetic field. In several of his books, in order to validate the existence of the proposed energy grid, Cathie offers up a certain geometrical figure for consideration as proof of the physical reality of his grid lines. The figure in question is that of a triangle whose proportions are exactly 3, 4, and 5 [1]. One may ask of course what is so special about this particular triangle. All that can really be said in response to this is that it does possess a certain mathematical significance, in that it is the smallest triangle to contain a perfect right angle within itself that may be constructed using the simplest of whole numbers. Through a careful examination of this triangle in an extended form, Cathie was able to uncover a series of quite remarkable associations between the length of its sides, the lines of his matrix, and the fundamental units of the Babylonian system of measure.

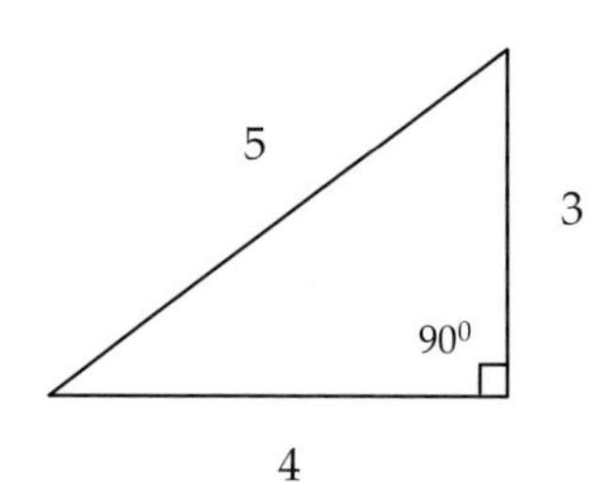

Starting with the basic 3-4-5 triangle, Cathie points out that if each of the side lengths are multiplied by 72, then the triangle is extended such that its side lengths become 360, 288 and 216 [2]:

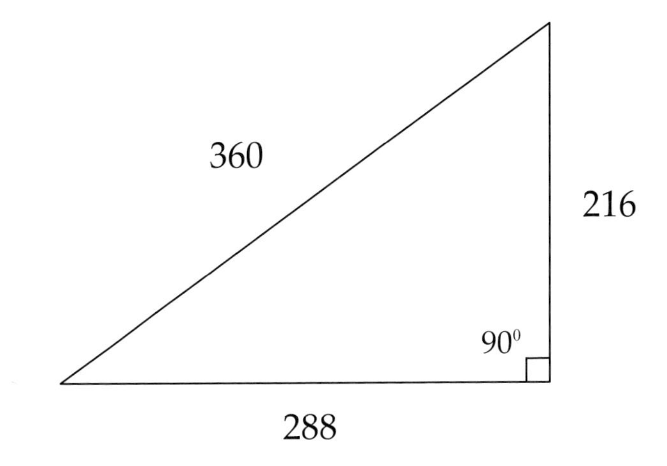

From this new extension of the basic triangle, Cathie notes that if the shortest side were extended yet further to 21600, and treated as an angular measure expressed in minutes of arc, then when divided by 7.5 – the interval between successive grid lines upon the Earth's surface – the obtained answer is 2880 [3]; identical to one of the side lengths of the triangle, but with an added zero. Indeed, with regard to his willingness to add extra zeros to establish such associations, Cathie is simply conforming to the essential principles that govern the entire Sumerian / Babylonian numerical system, doing nothing other than transforming numbers by multiplication or division using combinations of 10 or 6.

The mathematical associations uncovered by Cathie linking the sides of the extended 3-4-5 triangle to his grid line intervals were but the first stage of the true physical validation of the energy matrix, the key to which would appear to be the very significance of the number 72, used to extend the 3 sides of the basic triangle itself. Indeed, it was through this very number that Cathie was able to show that the angular separation of his energy lines possessed an important affinity to a well known physical magnitude; namely, the speed of light. However, the demonstration of a valid connection between the speed of light and the energy grid intervals becomes possible only when the magnitude of light speed is expressed using a special combination of units. The use of kilometres or statute miles to express distance, or even seconds to mark time, all fail to reveal the decisive connection. Instead, the link is only established when light is measured in terms of nautical miles, in conjunction with a unit of time based upon one 8/9th of a standard second, which Cathie himself refers to as a grid second [4]. When light is converted into these specific units, the sequence of numbers denoting its measure proves most interesting indeed:

Speed of light (in vacuum)
= 299792.458 kilometres per second [5]

Multiplied by 0.6213711922 [6]
= 186282.397 statute miles per second

Multiplied by (5280 / 6076.11549)
= 161874.977 nautical miles per second

Multiplied by (8 / 9)
= 143888.868 nautical miles per grid second

Following the conversion from kilometres per second to nautical miles per grid second, one can see that the value for light speed propagation through a vacuum differs only slightly from an exact value of 144000. Indeed, upon this very point, Cathie theorises that such is in point of fact the true maximum speed of light, a value that he himself associates directly with the equatorial surface of the Earth [7]. Moreover, he also seems to consider light speed expressed in nautical miles per grid second as not just being a simple measure of its magnitude, but also indicative of the 'fundamental pulsation rate' of the light itself. His work in this area has led him to directly conclude that 'reality' is in effect intermittent in its manifestation, and that it is essentially based upon the very pulsation rate of light [8]. Without diverging too deeply into these aspects of his work at the moment, it is enough to state here that the key to understanding why the number 72 'works' to extend the basic 3-4-5 triangle, lies with the fact that light itself actually manifests in terms of a frequency, and that some frequencies are of special significance compared to others. This would seem to be what Cathie appears to imply throughout much of his work. And if there is truth to this, then light must possess an affinity even to musical theory.

A theoretical maximum light speed value of 144000 nautical miles per grid second, understood also as the rate at which the light physically pulsates, may thus be considered in a similar manner to the sounds generated by a musical instrument. In the case of a string instrument for example, each individual string can be played open, or be held down and played at various points to produce a variety of different harmonies e.g. ½, 2/3, ¾ etc. The open length of string can be regarded as the fundamental wavelength. In the case of light, a fundamental magnitude of 144000 cycles per grid second can be halved to 72000, which when divided by 1000 gives simply 72. Thus, a half harmonic of light is produced, in a musical sense, but also in a true physical sense. Moreover, just as the fundamental frequency of light may be halved and reduced to its constituent numbers to give a value of 72, one may also simply double the light harmonic to produce 288; the actual measure of one of the side lengths of the 3-4-5 extended triangle. With such associations uncovered, not only does Cathie generate a significant measure of support for the true physical existence of his energy grid matrix, but also for his use of both nautical miles and grid seconds to express the speed of light. Indeed, his discovery that these particular units in contrast to others seem to provide a much deeper insight into the fundamental frequency base of pulsating light, may be regarded as one of his most significant accomplishments.

Of Abstract Angles and True Physical Magnitudes

In light of the above analysis of the extended 3-4-5 triangle it would appear that the side lengths of the figure do possess an affinity to the basic mathematical values of the Babylonian system, viewed as a set of angular measures. The longest side length of 360 is the exact number of degrees in a circle, whereas the shortest length multiplied by 100, produces the total number of minutes of arc that also sweep out a full circle, 21600. One may note that the simple multiplication of 360 by 60 produces the exact same answer. Even with such connections evident though, can one be absolutely certain that the values specified are truly meant to be representative of a set of *angular* measures?

Although the work of Cathie linking the extended 3-4-5 triangle to the speed of light does appear to offer some support to the existence of an Earth energy grid matrix, over and above relying solely upon the flight patterns of UFO's, this does not necessarily imply that the grid lines themselves, based upon the actual values of the sexagesimal system, specifically denote a set of *angular intervals* overlaid upon the planet. For indeed, the true significance of the intervals between successive parallel grid lines may not rest upon the idea of (abstract) angular separation at all, but something far more substantial. The suggestion is thus hereby made, that over the course of time, actual knowledge as to the real foundation of the sexagesimal numerical system may have been lost at some point, and as a result, future generations working with mere fragments of ancient knowledge, took to using certain key numbers of the system to signify angular intervals, when such was never the original intention of the founders. In investigating this very possibility, it would seem appropriate to begin with a complete re-evaluation of the 3-4-5 extended triangle, to determine if there is any evidence to support the idea that its side length values are linked to certain real physical properties of the Earth in some way.

A Modified Extension of the 3-4-5 Triangle

In beginning the newly proposed analysis, a certain intuitive leap is required. As the ancients were of the view that the Earth once possessed a 360-day year, and there is the suspicion that perhaps the longest side of the 3-4-5 triangle is in some way representative of such an ideal, would it not seem appropriate to replace it with none other than the value for the current Earth tropical year, and use this as a base-line measure for the extension of the other side length values? Indeed, if the three sides of this figure truly are associated with the Earth, would not a set of values actually representative of the real measure of the planet form its dimensions?

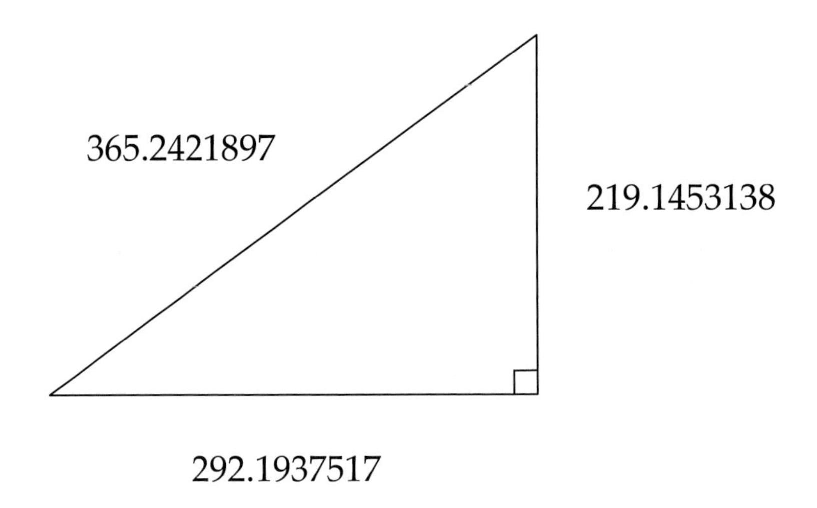

With this new extension, the longest side is clearly representative then of an actual physical characteristic of the Earth i.e. its number of days per year. Moreover, multiplication of this figure by 60 now produces a value of

21914.53138, whereas in the earlier extension this would have been 21600; a numerical value that would have been viewed as an angular measure denoting the full sweep of a circle. However, in angular measure, *all* circles sweep out or contain 21600 minutes of arc; even those that possess very different real circumferences. And yet, if the choice is made to discard degrees for days, further progress in understanding the value 21914.53138 may well rest upon discarding minute of arc angular measures; also for something far more substantial.

From careful consideration, it would seem fitting that none other than the concept of actual arc length measures swept out over the surface of the Earth should replace the idea of abstract minute of arc angular sweeps. But is there any support for this however? Most assuredly. Indeed, direct support is to be found when the unit length of the Fathom (6 feet) is considered in relation to the size of the Earth. As was noted previously, the total circumference of the Earth at the plane of the equator is equal to approximately 21914112 fathoms. When the placement of the decimal point here is set aside though, it is easily seen that this number sequence is extremely close to that obtained by simply multiplying the Earth year by 60, giving 21914.53138. This is highly suggestive of a link between the Earth year given in solar days (time-based element), and a circular arc length measure (spatial distance) of the planet's total equatorial circumference, as given in fathoms. Simple multiplication of the number of days per year by the numerical value 60000 would thus appear to generate a true answer for the equatorial circumference of the present Earth in terms of precisely this unit: 21914531.382 fathoms.

An Ideal Year, an Ideal Equatorial Circumference?

Consideration of the above facts must lead then to an almost total re-evaluation of the true significance of the 3-4-5 extended triangle. It would seem that over and above being merely a mathematical curiosity, the triangle is in fact a geometrical figure that hints at a profound physical relationship between the length of the Earth tropical year and the circumference of the Earth at the plane of the equator, as measured by the fathom unit. For indeed, the very ratio between the current Earth year and an ideal of 360 days, were it applied to 'back-transform' the current equatorial circumference of the planet (expressed in fathoms), would return it to an almost exact value of 21600000, and with such a small discrepancy, that one is bound to suspect *a principled* connection:

365.2421897 / 360 = 1.0145616380…

21914112 / 1.0145616380… = 21599586.63 fathoms

21600000 – 21599586.63 = 413.36 fathoms

In order to realise the significance of this result one must understand something of just how units of distance per se are ultimately derived; a true sense of history being of critical importance. The creation of the metre unit hereby serves as a decisive example. Devised in the late 18th century in France, the metre was established upon the standard of one quarter of the elliptical (pole to pole) circumference of the earth; the result of conducting a survey between Dunkirk and Barcelona to measure the earth to a new and exacting standard as had not been achieved previously. Following the survey it was decided that the full measure of one quadrant of the Earth would be split up into precisely 10,000,000 equal parts; each part being designated as a metre unit. The full elliptical circumference of the planet was thus *defined* as being 40,000,000 metres. In the very creation of such a unit there was thus an almost inherent desire by man to select a 'natural standard', i.e. the Earth, and then split it up most harmoniously and in a basic fractional sense, as a means to establish a new unit of measure.

In light of the known facts thus surrounding the creation of the metre unit, and also the evaluation of the fathom in relation to the Earth equator, it is almost inconceivable that such a unit as the fathom would actually have been established upon the standard of the current equatorial circumference. For who would deliberately choose to split up the Earth into 21914531 equal parts to derive a new unit of measure? Such a peculiar, almost random sort of number would appear to stand against reason itself. However, it is well within the bounds of reason to suggest that the fathom was originally established – and intentionally so – in accordance with the Earth at a time when there were precisely 21600000 such unit intervals to the full equatorial circumference of the planet, *when it was in real physical terms, smaller than at present*. And this indeed, is exactly what the above analysis would appear to strongly hint at.

It is therefore suggested, that the very *reasoning* at work that established the fathom, would have stood as identical to that which governed the minds of the French when they established the metre unit – with only very subtle differences. Essentially, whereas the French employed a 'base 10' harmonic of 10000000 with which to derive the metre, and chose the elliptical circumference of the Earth as a primary standard, those who established the fathom employed a 'base 60' harmonic of 21600000, and instead of the elliptical Earth circumference, chose the equatorial circumference of the Earth as the primary standard.

A Law of Physical Transformation

Taking the above points to their ultimate conclusion, and accepting that the ancient Sumerians and Babylonians were right that the Earth year had increased in some remote era; the very fact that the ratio of change between the present year and an ideal of 360 days, when applied to 'back-transform' the equator of the Earth is shown to harmonise the total number of fathoms within its equator to a key numeric value within the sexagesimal progression; it is thus possible that there may exist a *natural physical law of proportion* applicable to the Earth, that actively governs the mutual transformation of both of these noted features over time. Expressed mathematically, the form of such a law would be as follows:

$$TY \propto PE_{(e)}$$

Where: TY & PE (e) are ratios:

TY = Tropical Year (present) / Tropical Year (past)

PE (e) = Physical Equator, Earth (present) / Physical Equator, Earth (past)

$\propto$ = Symbol for: Proportional to...

In the form of an equation, the law would be of the following type:

$$\frac{TY}{ty} = \frac{PE_{(e)}}{pe_{(e)}}$$

Where: TY = Tropical Year (Present)

PE (e) = Physical Equator, Earth (Present)

ty = Tropical Year (past)

pe (e) = Physical Equator, Earth (Past)

What the law hereby stated implies then, is that *any given change to the total orbital period of the Earth, is directly proportional to a change to its equatorial circumference.* Thus, an increase to the Earth year (in some remote age) to bring it to its current level from an ideal of 360 days, was directly accompanied by a physical increase in the size of the Earth at the plane of the equator. Indeed, using the ratio of increase for the Earth year as the primary standard (of change), the value generated by multiplying it by 21600000, which yields 21914531.382 (fathoms), may be accepted in principle as the true measure of the current Earth equatorial circumference. Moreover, the very reality of the stated law, if true, validates as a whole the entire set of unit measures of the Imperial System, including such as the foot, being 1/6th of a fathom, and the inch, being 1/12th of a foot. In addition to this, in following through on the analysis of the minute of arc, understood to be a true arc measure in relation to the Earth and not an abstract angular sweep; under an Earth circumference of 21600000 fathoms, 1/21600th part of this is equal to precisely 6000 feet:

$$21600000 \times 6 = 129600000 \text{ feet}$$

$$129600000 \times (1 / 21600) = 6000 \text{ feet}$$

And this unit, one may suggest, was in fact the true original basis of the minute of arc: an *arc length* measure equal to precisely 6000 feet. A unit of distance that shall hereafter be referred to as an *Ideal Geographical Mile* (IGM), in contrast to the standard geographical mile of 6087.25333 feet, given previously. Consequently, the simple multiplication by 60 of the current Earth year in solar days, as formerly evaluated, would produce a real physical measure for the true equatorial circumference of the Earth given in terms of ideal geographical miles, shown to be equal to 21914.531382 IGM.

In Support of Harmony

With the law formally stated, it is a bold claim at this juncture to say that the evidence so far presented is enough to confirm its truth. For it is fair to say that the actual validity of the proposed law rests at least for the moment almost entirely upon indirect support via the ancient belief in a past 360-day year, by way of demonstrating that were the physical circumference of the planet at the equator to be reduced by the same ratio as that which transformed the tropical year itself, then it would conform most harmoniously to the primary values of the sexagesimal progression under the Imperial System of measures. One is bound not to be content therefore with just this one point supporting indirectly the validity of the proposed law. Further supporting evidence must be obtained, involving nothing less than the study of the very notion of proportional laws per se, and of their presence within the universe as a whole.

To this end, a review of certain relevant discoveries within the realm of astronomy over the past few centuries is most definitely in order. For it is here, that the decisive breakthroughs of one man led ultimately to true confirmation of the existence of principled laws of proportion actually operative within the universe, the very establishment of which over-turned ideas that had dominated astronomical theory since the time of antiquity. And it is precisely the discoveries of this one man that lend strong support to the law of proportion as put forth above.

Chapter 3

The Geometry of Astronomy

The Emergence of Modern Astronomy

From the viewpoint of history, modern astronomy began with the work of one man: Johannes Kepler (1571-1630 AD). Prior to Kepler astronomy as a natural science was in the dark ages, resting as it did upon the flawed theories and false methods of such as Ptolemy, Copernicus, and Brahe, that had dominated since early antiquity. As a result, there existed no true knowledge at all of the physical principles that governed planetary motion. The work of these early thinkers had led only to the development of a variety of false mathematical models of the heavens. And what defined such models, whether heliocentric or geocentric, was the thought that all of the planets in their orbits were engaged in uniform circular motion. Of course, actual observations did not themselves appear to support this. However, instead of questioning their fundamental belief that all motion within the celestial realm was uniform and circular, they merely devised ever more complicated models, overlaying circles upon circles upon circles, in the vain effort to force their models – whose fundamental axioms they refused to abandon – to fit the observations.

The Revolutionary Discoveries of Johannes Kepler

What marked out Kepler from those that had gone before him was that he allowed the actual observations themselves to completely determine the mathematical structure of the planetary orbits *without interference* – the hallmark of any true scientist. In doing so he made the discovery that the true character of planetary orbits was elliptical in nature. That is, a single ellipse shape defined the orbit of each planet about the sun, with the sun itself located at one of the focal points of the ellipse. Moreover, the internal area of each planet's ellipse could be used to determine a universal principle of action that governed the continuously changing speed of motion of a given planet – a fact that confirmed once and for all that their orbital motion was *inherently* non-uniform in nature.

Going beyond even these initial discoveries though, Kepler went further still, and uncovered the existence of a general mathematical law that was found to relate the orbital periods of the planets to their actual mean distance from the sun. Consequently, as a result of such new and revolutionary discoveries, Kepler completely destroyed the models of Ptolemy, Copernicus and Brahe, as were developed before him. And indeed, with the eventual widespread acceptance of Kepler's new theories of planetary motion, science was much advanced. For one to thus develop an understanding of the true nature of celestial motion, one must be familiar with Kepler's basic findings, and most especially the character of the ellipse shape itself.

The Ellipse

The general appearance of an ellipse is that of a somewhat squashed circle. The main difference being that an ellipse has two centres about which two radii rotate, whereas a circle has the appearance of possessing only one of each of these components. More commonly referred to as focal points, the two centres of the ellipse are always separated by some discreet distance, and not merged into one, as with the circle. This gives rise to a certain added level of complexity in the formation of an ellipse, as is detailed in the diagram below:

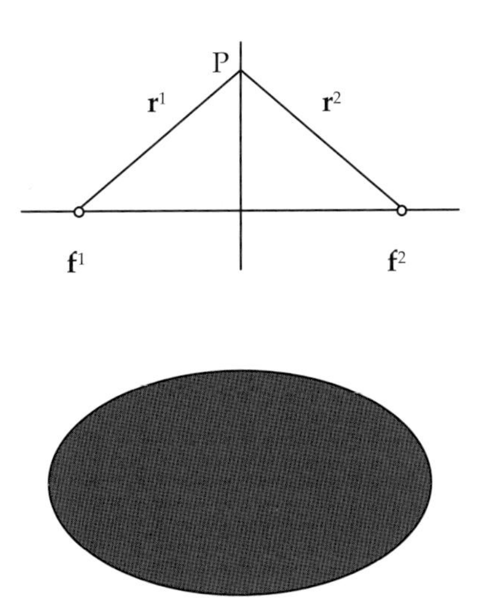

*Above: From an initial starting point where the ellipse is as yet unformed (Top), the two radii are shown to be of equal length in their extension from the noted focal points (**f**[1] and **f**[2]), meeting up at point **P**. The actual formation of the shape begins with a clockwise rotation of both radii about their respective focal points, so that the maximum area permissible is covered to generate the shape. The actual perimeter of the ellipse is 'drawn' from the point of contact of each radius as they rotate. As the process unfolds, there is a transfer of length from each radius to the other. For one half of the formation of the shape* **r**[1] *increases its length whilst* **r**[2] *decreases. In the second half, the process is reversed with* **r**[1] *decreasing its length as* **r**[2] *increases. However, the total length of both radii when added together remains constant throughout the entire process of actually generating the full shape (bottom).*

With the above account given as to how an ellipse shape is actually generated, it is also important to know just how some of its internal components relate to one another. The example ellipse as shown below helps to illustrate this:

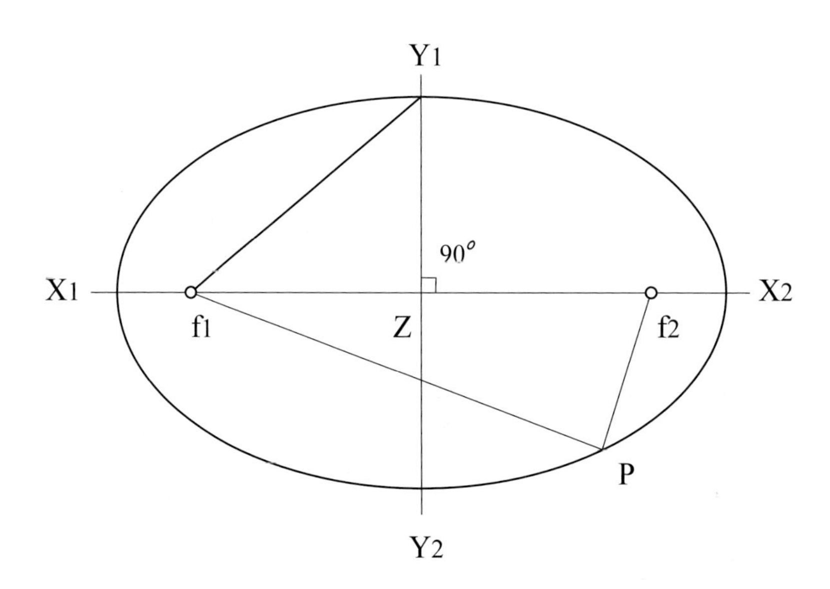

<u>The Main Components</u>

Major Axis = X1 – X2
Minor Axis = Y1 – Y2
Semi-Minor Axis = Z to either point Y1 or Y2
Semi-Major Axis = Z to either point X1 or X2, and also either focal point f1 or f2, to either point Y1 or Y2. E.g. f1 – Y1 as detailed

Notable Properties of the Ellipse:

1) A set of straight line distance measures from any point on the perimeter of an ellipse to each of the two focal points is equal to the full length of the Major Axis component of the ellipse. E.g. f1 to P + f2 to P = Major Axis, as detailed above.

2) *Eccentricity.* All elliptical figures possess a unique curvature, the measure of which is expressed mathematically as a number between 0 and 1 which denotes the eccentricity of the shape. The eccentricity value is calculated by dividing the distance from either focal point to the centre of the figure (e.g. f1 to Z, detailed above), by the semi-major axis component of the form (e.g. X1 to Z, above). From a careful study of the relations in question, it may be noted that a 'perfectly squashed ellipse' will have an eccentricity value of 1, and a perfect circle will possess an eccentricity value of 0. Of the planets within the solar system one may note that all of them deviate only slightly from an exacting circular orbit, and thus their eccentricity values are all close to 0. As an example one may cite the eccentricity of the Earth itself: 0.016708617 [1].

The Mean Point of Approach

As was noted briefly above, each planet possesses a unique elliptical orbit, wherein the sun itself is positioned at one of the focal points of the ellipse, the other being merely empty, as such. As a result of the inherent properties of the ellipse shape, when a planet reaches the point in its orbit wherein the semi-minor axis touches the perimeter of its ellipse, e.g. at either points Y1 or Y2 (as pictured above), at that given moment, the distance between the planet and the sun is the arithmetical mean of the entire orbit. The straight line distance connecting the planet to the sun at this point is identified as the semi-major axis of the orbit. Moreover, and of great importance to this work, is the fact that the length of the semi-major axis of an ellipse shape is identical to that of the distance between the exact centre of an ellipse and either of the two most extreme radius values of the shape, specifically Z to X1 or Z to X2, as pictured above.

The Harmonic Law

Following his correct identification of the ellipse as the shape which defines a planets' orbit about the sun, Kepler further discovered a very specific relationship that appeared to exist between the semi-major axis values of the planets and their orbital periods; one that could be expressed as a precise mathematical law. The very character of the law seemed to indicate that the mean distance of a planet from the sun *determined* the time it took to complete one orbital period, or vice versa. Commonly referred to as Kepler's 'Harmonic Law', it may be expressed as follows:

$$\mathrm{p}^2 \propto \mathrm{a}^3$$

Where: **p** = Orbital Period, **a** = Semi-Major Axis

2 and 3 are powers

In the form of a mathematical equation:

$$(p \,/\, \mathrm{p})^2 = (a \,/\, \mathrm{a})^3$$

Where:

p = Orbital Period (Alternate)
a = Semi-Major Axis (Alternate)

p = Orbital Period
a = Semi-Major Axis

What Kepler's law reveals is that were a planet to change its orbital period by some proportional measure to an alternate value, then this would also cause a direct change to its mean distance from the sun, giving it a new (alternate) semi-major axis, as would be required to maintain a stable orbit. To illustrate this one may consider a theoretical change to the orbit of Mars, involving an increase to its orbital period from its current value, to one of 800 days:

As per observations the orbit of Mars as it currently exists possesses the following characteristics [2]:

Orbital Period = 686.9297110 Solar days (24 hours)
Semi-Major Axis = 124638661.47 IGM

If Mars were to increase its orbital period to 800 days, Kepler's law (in its equation form) allows one to determine the new semi-major axis (***a***) of the planet as would be required under such a new orbit:

$$(800 / 686.9297110)^2 = (a / 124638661.47)^3$$

$$800^2 / 686.9297110^2 = a^3 / 124638661.47^3$$

$$(800^2 / 686.9297110^2) \times 124638661.47^3 = a^3$$

$$a = \sqrt[3]{[(800^2 / 686.9297110^2) \times 124638661.47^3]}$$

$$a = 137965733.3 \text{ IGM}$$

One can see then that Mars is forced to extend its mean distance from the sun to maintain an orbital period of 800 days. Of course, the example as given here is of a theoretical shift. However, the full truth of Kepler's law can indeed be had from the actual examination of any two planets within the solar system; an exercise which readily confirms its validity. In effect the law is thus sun centred, uniting the orbits of all of the planets about the central solar body.

An Extension of Kepler's Findings

Although the basic geometry of the orbits of the planets as given in Kepler's work is shown to be that of an ellipse, the full truth of the matter is slightly more complex. For though the planets do each appear to possess a flat elliptical orbit about the sun; when they are measured to an extreme level of accuracy, it is found that the true curvature of their orbits is not in fact that of a perfect 2-dimensional ellipse at all, but rather a 3-dimensional ellipsoid. The actual orbital planes of the planets are continually transformed at an exceptionally fine rate, with each planet never truly returning exactly to where it once was at the beginning of a given orbital period. This indeed is a most profound finding and of crucial importance to this present work. For what this point in fact reveals is that the *orbital geometry* of the planets is identical to that of their *physical geometry* i.e. *both are ellipsoid in nature.*

Indeed, concerning the outward form of the Earth, this fact was first mentioned in chapter 1 when the nautical mile unit was introduced; a unit shown to be derived from the pole to pole, or elliptical circumference of the planet. Thus, from a full 'side on' view does the Earth possess an apparent 2-dimensional outline, with the complete physical form of the planet being further generated by the rotation of such a shape about the Minor Axis to produce a 3-dimensional ellipsoid form, indicative of the material 'shell' of the planet. In exactly this same way so too is the full orbit of the Earth – or any celestial body about the sun, generated. Mathematically therefore, the very components that make up the Earth form have their equivalent in its orbit:

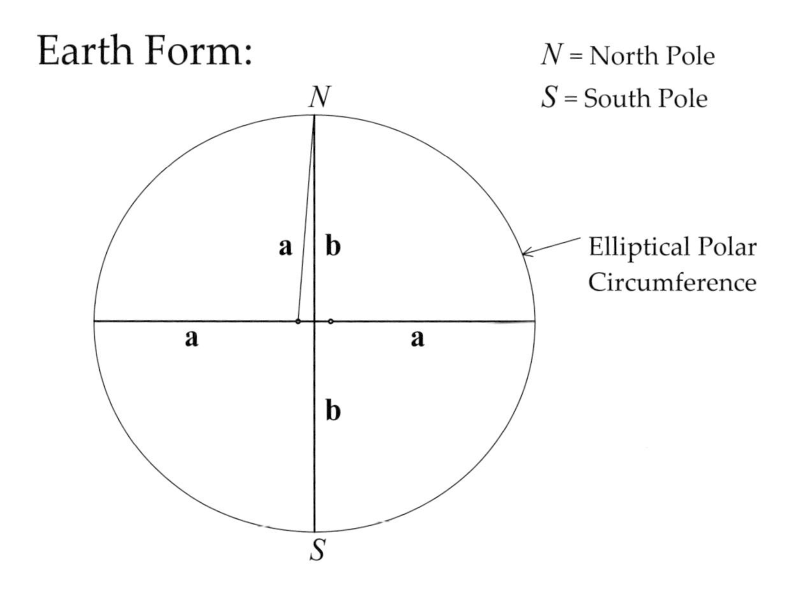

Side View: Apparent 2-Dimensional ellipse

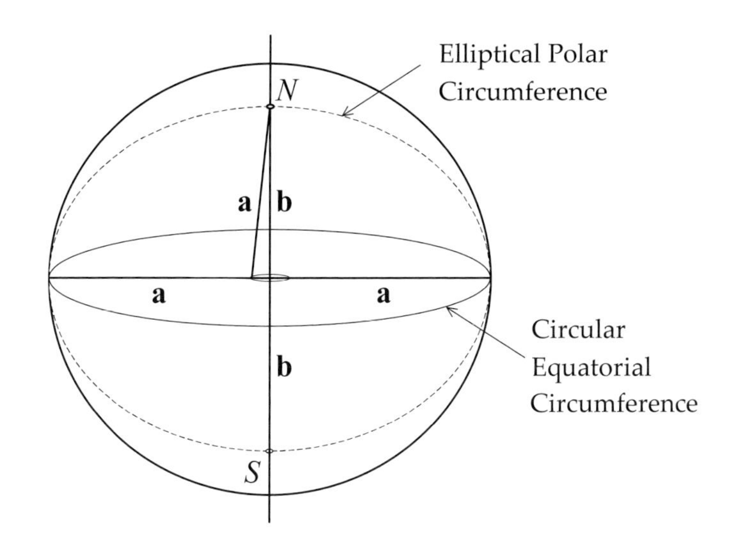

Angled View: True 3-Dimensional Ellipsoid Form

Earth Orbit:

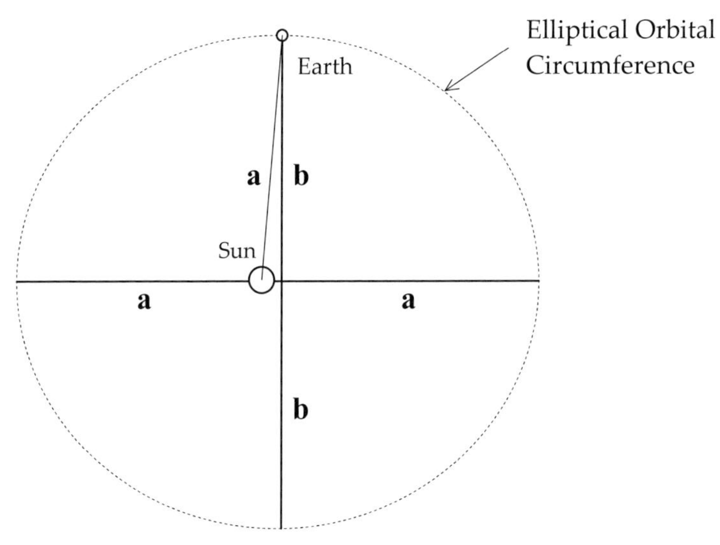

Plan View: Apparent 2-Dimensional ellipse

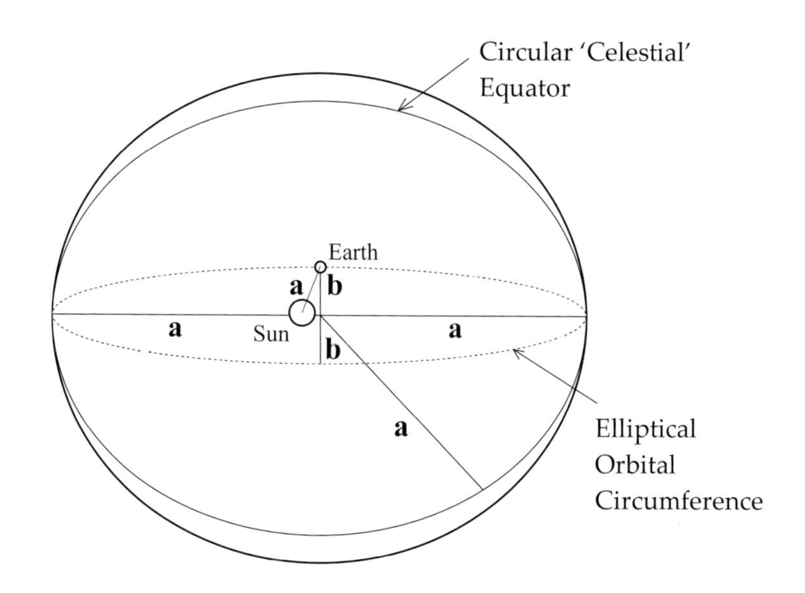

Angled View: True 3-Dimensional Ellipsoid Form

From the above diagrams one can see then the actual equivalence of the mathematical-physical components of both the Earth form and the Earth orbit:

For the Earth Physical Form:

a = Equatorial radius or semi-major axis of the planet
b = Polar radius or semi-minor axis of the planet

For the Earth Orbit about the Sun:

a = Distance between the Earth and the sun at their mean point of approach i.e. the Earth's orbital semi-major axis
a = Radius of celestial equator of the Earth's orbit
b = Orbital semi-minor axis

In addition to these straight-line components, one can see also that the full elliptical circumference of the orbit of the earth is a mathematical match to the elliptical circumference of the Earth form. Moreover, the Earth's actual circular equatorial circumference has its equivalent in the orbit of the planet, in what may be regarded as the circular 'celestial' equator of the orbit, as is detailed above.

An Energetically Manifest Orbital Form

Of course, in direct defiance of the above suggested mathematical-physical equivalence between the orbit of a planet and its material form, there are those, known to history as empirical scientists, who argue (as they did indeed throughout much of the late 17th and early 18th centuries, with some tragically still doing so even today) that the orbit of any given celestial body has no 'true' existence at all. And that all that one can say concerning any orbit is that it is nothing more than a 'tracing out' of the path of a celestial object, established by 'connecting up the dots' of its various point-like positions over time such as would tend to produce some sort of curve. To these people the orbit of a planet has in effect: *no efficient history*. However, this flawed position was totally destroyed in 1801 by the brilliance of Carl Fredrick Gauss, in his groundbreaking evaluation of the minor planet Ceres, located in the asteroid belt. Being the first individual to employ infinitesimal calculus for the practical purposes of astronomy, combined with Kepler's laws of planetary motion, Gauss was able to correctly determine the character of the orbit of Ceres, and in doing so the character of the orbits of all celestial bodies.

Being a most exceptional mathematician as well as physicist, Gauss realised that the smallest section of the elliptical orbit of Ceres was irreducibly a curve. And indeed, that such is a critical property of any ellipse i.e. that it is not in fact composed of points at all (as the empiricists themselves believe), but that the entire shape is of such an irreducibly curved nature, that the smallest section determines the total curvature of the whole shape. Consequently, the smallest movement of any celestial body in orbit is itself irreducibly curved in nature. What this means in real physical terms, is that *the geometry of an orbit is pre-existent with respect to an actual orbital body, and that any given orbital ellipsoid constitutes a real,* ***physically energetic structure****, rarefied in nature and invisible to the naked eye, but no less substantial than the obvious and more 'solid' physical ellipsoid form of a material body.* In effect, it is this invisible geometry that determines the true course of a planet, as opposed to the view of the empiricists that its orbit has no real existence, being a mere incidental tracing out of its path produced by the 'blind' movement of the body as it flies through space.

Harmony Supported

From careful consideration of the above it should be quite clear then just how the discoveries of Kepler in an extended form offer significant support to the law of proportion put forth in the previous chapter. Kepler's work establishes the actual physical validity of laws of proportion per se within the universe; with the extended evaluation demonstrating that such laws apply not just to the specific geometry of orbital ellipsoid forms, but also the outward material structure (form) of such celestial bodies themselves. The proportional law given previously then linking a transformation of the Earth tropical year to a change in the physical circumference of the Earth at its equator, is thus shown to be a law specific to the Earth, and yet operative simultaneously with Kepler's more general law, which itself holds for all of the planets within the solar system, though solely in terms of the geometry of their orbits it would seem.

Chapter 4

The Earth-Moon System: The Physical Basis of Angular Measure

The Physical Foundation of the Second of Arc

With initial principled support thus uncovered favouring the reality of the stated law that the Earth year is indeed transformed in direct proportion to its equatorial circumference, as via the discoveries of Kepler, one would of course hope that further additional supporting evidence may yet be had that would decisively confirm its validity. Extending the investigation to consider the original basis of the second of arc unit of measure would seem a natural way to proceed towards this end, as it is quite evident there is a close relationship between this unit and the minute of arc.

So far, at least from an angular perspective, it has been stated that the total number of seconds of arc sweeping out a full circle is 60 times the total number of minute of arc units required to complete a circle; the former held simply to be more acute angles. However, the actual mathematical operation of converting from minutes to seconds of arc involves multiplication. Thus is the number 21600 multiplied by 60 to obtain the value 1296000 – an expression of the measure of a full circle in seconds of arc. And yet, if the minute of arc was originally based upon a true arc length distance of 6000 ft, this operation may rightly be viewed as one wherein the value 21600 IGM, understood to denote the equatorial circumference of the Earth in some past age, is multiplied by 60 so as to produce simply a larger circle of 1296000 IGM; one extended into space from the very surface of the planet. Such thinking would seem wholly reasonable. However, this would only be so if one could demonstrate that such an extended circle were physically meaningful, and not simply a mere mathematical abstraction.

Remarkably, it would seem that just such a demonstration is possible, as the circle so generated in the specified manner does indeed appear to be possessed of true physical significance; a fact that is evident when one considers the actual dimensions of none other than the orbit of the moon itself about the Earth. For it so transpires that the current orbital circumference of this very satellite is close to being almost exactly 60 times the physical circumference of the Earth form; the actual ratio between them currently being just slightly over this figure, at about 60.26 [1]. However, when this fact is viewed in light of the research already presented in this work, the existence of such a discrepancy would seem most intriguing, leading one to consider that perhaps at some time in the past the ratio between the total circumference of the moon orbit and the physical circumference of the Earth was once exactly 60 to 1. Necessitated by harmony itself, the time in question could have been none other than that when the Earth was possessed of an equatorial circumference of precisely 21600 ideal geographical miles, in conjunction with an orbital period of exactly 360 solar days. Consequently then, this would imply that the numerical value of 1296000 at such a time would have been representative of the actual orbital circumference of the moon in ideal geographical miles; exactly 60 times that of the Earth's true physical equator.

In considering these very facts, the idea is bound to present itself that this very celestial arrangement may well have formed the true physical basis of the entire Babylonian system of angular (so called in this present age) measure. A rigorous proof would of course be required to establish this. One would need to demonstrate that at such a time when the Earth appears to have possessed an orbital period of 360 days with an accompanying equator of 21600 IGM, the moon itself simultaneously possessed also a full orbital circumference of precisely 1296000 IGM at exactly the same time. Given this, the way forward in this matter would seem clear. The proof, if it exists, will doubtless be in the form of a further law of proportion, in addition to the one already presented in the previous chapter. Indeed, from the realisation that the geometry of the orbit of a celestial body is physically identical to that of the form of such a body – both being ellipsoid in nature – one would be seeking specifically a law that would relate the ellipsoid form of the Moon orbit to that of the physical form of the earth, which would also tie in to the noted Earth tropical year law. A careful study therefore of the exact physical form of the earth including the properties of the Moon orbital ellipsoid is thus required.

The Earth Form

From the analysis provided in previous chapters it was shown how a refined value for the physical equator of the Earth could be derived, based upon the proposed tropical year ratio of increase. Combining this value with a recognised flattening or eccentricity measure for the Earth based upon observations, allows one to further derive all other measures that define the physical form of the Earth [2]:

Equatorial Circumference = 21914.531382 IGM
Elliptical Polar Circumference = 21877.793744 IGM

And:

Semi-major axis / Equatorial radius (**a**)
= 3487.8059949 IGM

Semi-minor axis / Polar radius (**b**)
= 3476.1120418 IGM

IGM = *Ideal Geographical Mile,* (1 IGM = 6000 British feet)

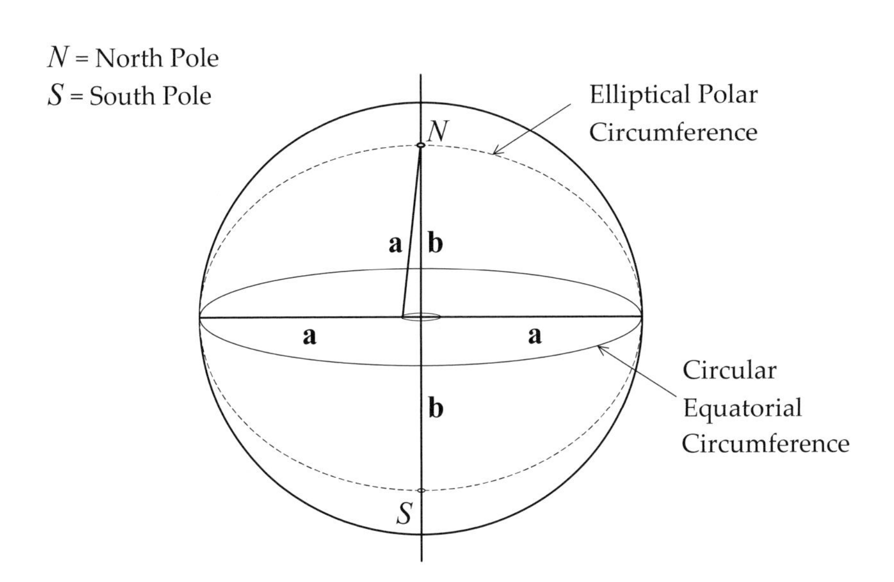

Earth Ellipsoid Form (Angled View)

NB: Of the specified magnitudes above, the Semi-major axis (**a**) *component is identical to the radius of the Earth on the plane of its equator. Also, as can be seen, in a full ellipsoid form the focal points are not points as such, but constitute an 'inner ring' or circle on the same plane as the equatorial plane. The distance from any point on this ring to either north or south geographic pole is also equal to the Semi-major axis. The Semi-minor axis* (**b**) *is the distance from the centre of the planet to either north or south geographical pole on the line of the Earth's axis of spin. Indeed, the full minor axis is the rotational axis of the planet.*

The Moon Orbit

In comparison to the obvious material ellipsoid form of the Earth, there is then the orbit of the moon. The geometry of both are of course identical; the true orbital configuration of the moon merely being that of an 'invisible' ellipsoid so to speak. The path of the moon as it travels through space about the Earth does though reveal the dimensions of its component parts. In this particular instance one may note that it is the Earth itself, as the primary body, which is located at one of the focal points of the moon's orbital form. The magnitudes of the component parts of the moon orbital ellipsoid are known to a high degree of accuracy, and are as follows [3]:

Celestial Equator = 1320692.019179 IGM
Elliptical Orbit = 1319696.123465 IGM

Semi-major axis (**a**) = 210194.6631544 IGM
Semi-minor axis (**b**) = 209877.6597029 IGM

A 2-dimensional overview of the Moon's orbit, with a visual representation of its full 3-dimensional ellipsoid form, is as follows:

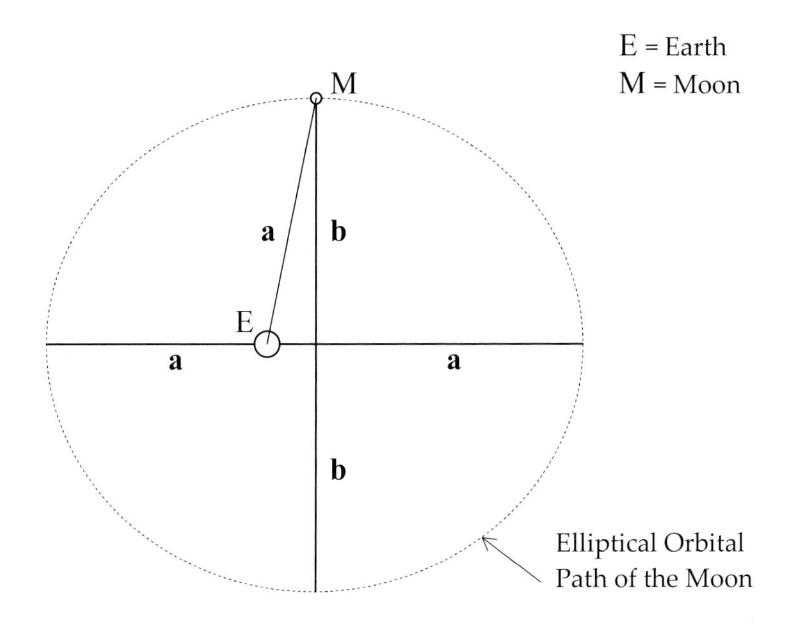

Moon Orbit (Plan View)

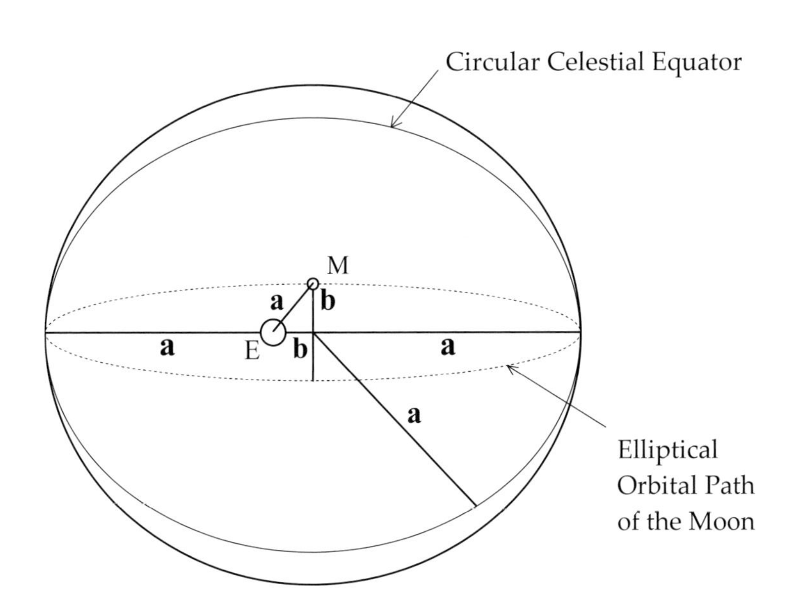

Moon Orbit, Ellipsoid (Angled View)

A Comparison of the Earth Form & Moon Orbit

Through a careful comparison of the physical ellipsoid form of the Earth and the invisible ellipsoid form of the moon (orbit), it may be perceived that the actual orbital path of the moon equates in principle to the physical pole to pole elliptical circumference of the Earth, whereas the Earth's circular equatorial circumference is geometrically equivalent to the circular 'celestial' equator of the moon orbit, perpendicular to the plane of its orbital ellipse. From this one may infer then that were the value 21600 (IGM), linked specifically to the ideal equator of the Earth, to be physically associated with the number 1296000, then this latter value must itself be representative of the circular celestial equator of the moon, and not the measure of its elliptical journey about the Earth as such. Consequently, any meaningful evaluation of the orbit of the moon must involve a comparison of the respective magnitudes of both its present and proposed ideal celestial equator.

A Harmonic-Type Law for the Moon

Previously it was shown that the ratio between the current Earth year and the theoretical ideal of 360 days was identical to that between the current equatorial circumference of the planet and an ideal of 21600 IGM. How does this compare though with the ratio between the present circular celestial equator of the moon orbit, and a proposed ideal of 1296000 IGM? An evaluation of the relevant values of both the earth ellipsoid form and the moon ellipsoid orbit, as detailed above, reveal the following:

Ratio between current Earth Equator and proposed ideal:

21914.531382 / 21600 = **1.014561638055555...**

Ratio between current Celestial Equator of Moon orbit and ideal:

Semi-major axis = 210194.6631544 IGM

Celestial Circumference = 210194.6631544 × 2 × PI
= 1320692.019179

And: 1320692.019179 / 1296000 = **1.019052483934999**

In comparison, the above given ratios readily demonstrate that any connection between an increase in the physical Earth equator and an increase in the celestial equator of the moon orbit, were it to exist, could not be based upon a 1 to 1 power relationship i.e. they would not transform in direct proportion to one another over time, as was shown to be the case with respect to the Earth year and Earth equatorial circumference. However, as was seen in Kepler's work, celestial elements are not always related in such a simplistic manner, for he himself demonstrated that a power relationship of 3 to 2 is existent between the semi-major axis of a planet's orbit, and its orbital period (about the sun). In light of this fact, it is quite conceivable that the ratios of increase for both the physical Earth equator and moon celestial orbital equator may be linked by a power relationship unique to these variables, the reality of which would suggest the presence of another harmonic-type law governing their transformation.

In the search for a likely candidate, a power relationship of 4 to 3 linking these celestial features would appear most promising. By expressing such a relationship in the form of an equation, and by using the Earth physical equator increase as a basis, a value for the ratio of increase of the moon circular celestial equator from an ideal of 1296000 to its present magnitude, can be attained that is quite close to the value derived above, based upon the known semi-major axis of the moon. Such an equation is of the following form:

$$(PE_{(e)} / pe_{(e)})^4 = (CE_{(m)} / ce_{(m)})^3$$

Where:

PE (e) = Physical Equator, Earth (present)

CE (m) = Celestial Equator, moon (present)

pe (e) = Physical Equator, Earth (past)

ce (m) = Celestial Equator, moon (past)

Inputting values into the equation produces the following:

$$(21914.531382 / 21600)^4 = (CE / 1296000)^3$$

$$\sqrt[3]{(1.01456163805^4)} = CE / 1296000$$

$$1.01946248613 = CE / 1296000$$

$$1.01946248613 \times 1296000 = CE$$

CE = 1321223.382034 Ideal geographical miles

From the above it can be seen how a value of 1.01946248613 is obtained for the ratio between the current moon celestial equator (CE) and proposed ideal of 1296000; and moreover, how the current moon celestial equator is determined itself to be the product of both the expressed ratio and 1296000. As a result of this, one is therefore able to compare the values so generated to those derived earlier for the moon celestial equator and ratio of increase, each of which were based upon the actual observed value of the current semi-major axis of the moon orbit:

Earth Year based values (using above equation):

Celestial Equator of Moon Orbit = **1321223.382034** IGM
Ratio of increase of Celestial Equator = **1.01946248613**

Moon Celestial Equator based on Observed Semi-Major Axis value:

Celestial Equator of Moon Orbit = **1320692.019179** IGM
Ratio of increase of Celestial Equator = **1.01905248393**

Although the respective values may look very similar, upon comparison it must be conceded that the measure of difference between them is too significant to be simply ignored. However, to dismiss the possibility outright of such an existent power relationship between these variables would at this initial stage appear far too hasty, for indeed, it so transpires that with one critical alteration to the way in which the current circumference of the moon celestial equator is determined, the margin of difference is reduced to almost nothing. In order for this to be achieved, what must be insisted upon, is that instead of PI being used for the determination of the circumference of the celestial equator of the moon, the ratio 22/7 is used. Compared side by side, the values of each are presented below:

PI = 3.1415926535897932384626433832795…
22/7 = 3.1428571428571428571428571428571…

Although both values do appear to be close numerically, it is important to realise that they are fundamentally different in a qualitative sense. The value PI is known to be an 'irrational' number, for it does not appear to be composed of any ordered or predictable sequence of values, whereas 22/7 is an exact fraction expressed as a ratio of two whole numbers. By comparison it is thus ordered, and therefore considered rational. (For a more in-depth consideration of the nature of PI as an irrational number see Appendix D).

By employing the use of 22/7 instead of PI as the numerical link between the circumference of the moon celestial equator and the full major axis of its orbit, an alternate value for the equator may be generated based upon the known magnitude of its semi-major axis; one that may then be compared with the value of the equator derived from the Earth year ratio of increase, using the specified formula:

Alternate Moon CE based upon known value of Moon semi-major axis employing 22/7 in place of PI:

210194.6631544749 × 2 × (22 / 7) = **1321223.596970985** IGM

Moon CE based upon proposed formula (as calculated above):

= **1321223.3820341711** IGM

Difference = 0.2149368139 IGM, or 1289.6208834 feet

It can be seen then that when 22/7 is used instead of PI, the actual difference between the two values is reduced to an extremely low level; a result that would appear to support the validity of the specified 3 to 4 based law of proportion. However, true confirmation of such support is hardly established simply with a more favourable match between the relevant numbers. Indeed, the very decision made to substitute PI for the fraction 22/7 is one that requires considerable justification, for certainly it is a most unusual decision. In addition to which, there is little doubt that the mere suggestion that PI is somehow an inappropriate measure of the ratio between the circumference of a circle and its diameter is certainly something that mathematicians the world over are bound to dispute. Highly persuasive evidence is therefore required in order to account for this decision, to fully validate the use of 22/7 in opposition to PI, and thus explain why the former should be preferred over the latter. Only then could one have confidence in the proposed harmonic-type law relating the Earth equator to the moon celestial equator.

In addressing this issue, it is necessary for one to consider nothing less than the very nature of the fundamental relationship existing between physics and mathematics. Indeed, the statement that most comprehensively sums up the precise nature of their association is: *the physics determines the mathematics*. It is a simple truth, universal in scope; and one known to many of the greatest thinkers throughout the ages. To truly comprehend why one is justified in using 22/7 over PI in relation to the geometry of the Earth-moon system, a critical understanding of this statement is essential. It does though require a great deal of thought, and it would not be wrong to say that an entire book could be written upon this subject alone. A truly in-depth study of this matter is thus not possible within the confines of this present work. However, it must at least be addressed in brief in order that this work may lawfully proceed. Consequently therefore, a brief statement of explanation as to the falsity of PI and the validity of 22/7 in relation to the Earth-moon system may be given here; though a slightly more detailed evaluation of the issues can be found in Appendix E. For the immediate present then, the argument stated in brief, is that PI is inappropriate because it is a *mathematical* value generated within the confines of an artificial framework of axiomatic assumptions based upon certain sense-perceived notions of spatial extension. As a result it is incapable of being validated physically outside of the mathematical domain of apparent static form based entities. However, in the case of the fractional value 22/7, evidence does exist, albeit of a most subtle nature, which *experimentally* confirms its valid incorporation as a significant measure within the Earth-moon system. Indeed, when the *natural physical* magnitudes of both of these bodies are evaluated, the actual *unity* of the system is existent *only* when 22/7 is actively recognised as the connecting ratio between certain of its critical elements.

Actual evidence revealing the intimate association of 22/7 to the celestial equator of the moon orbit, including the Earth tropical year and equatorial circumference, presents itself when one is taken to study a certain rather simple geometrical configuration, wherein a sphere is placed inside a cube, such that the surface of the sphere exactly touches each of the six sides of the cube without crossing. With such an arrangement, the diameter of the sphere is identical to that of the side length of the cube containing it. The formulas for determining the volumes of each object are as follows:

$$\text{Volume of Sphere} = \text{Diameter}^3 \times \text{PI} \times (1 / 6)$$

$$\text{Volume of Cube} = \text{Side Length} \times \text{Side Length} \times \text{Side Length}$$

Substituting into both equations any given value to represent simultaneously the diameter of the sphere and the side length of the cube, the ratio between the cube and the sphere (volumes) may thus be determined:

$$\frac{Cube}{Sphere} = 1.9098593171027440292266051604702$$

However, if PI is not employed in the formula for the sphere, but instead, the rational value 22/7 is used, the answer obtained is quite different. Of course the very decision to do this breaks completely with the entire framework of common sense notions concerning spatial extension, such that one indeed is no longer even able to visualise the very form of the sphere within the confines of the cube. Be that as it may, this does not invalidate the decision at all. In reality, it simply transforms the geometrical domain within which one is working. One departs from the domain of static form and enters into that of physical action. In using therefore 22/7 in place of PI for the formula of the sphere, the following alternate ratio is obtained:

$$\frac{Cube}{Sphere} = 1.90909090909090909... \qquad \text{As an exact fraction} : 21 / 11$$

With this new value, a simple set of ordered manipulations reveal its strong association to the current Earth-moon system:

$$1.909090909090\ldots / 100 = 0.019090909090\ldots$$

$$1 - 0.019090909090\ldots = 0.980909090909\ldots \ (1079 / 1100)$$

$$1 / 0.980909090909\ldots$$

$$= 1.01946246524\ldots \ (1100 / 1079)$$

One may compare this value with that derived previously from the 3 to 4 power equation, for the proposed increase to the moon celestial equator:

$$= 1.01946248613\ldots$$

The level of accuracy is great indeed, exact to 7 places after the decimal point, and close to eight, suggestive of a highly significant association. In fact, it would seem most proper to consider the fraction 1100 / 1079 to be the most exact expression of the deviation of the moon celestial equator from a magnitude of 1296000 IGM to its present value. If this is then combined with an acceptance of 22/7 itself as a valid component integral to the physical geometry of the Earth-moon system, then yet another most incredible association is revealed; one between the true magnitude of the current semi-major axis of the moon, and its very ratio of increase:

$$1296000 \times (1100 / 1079) = 1321223.3549582947\ldots \text{IGM}$$

$$\frac{1321223.3549582947}{(2 \times (22/7))} = 210194.62465245597\ldots \text{IGM}$$

If the actual number sequence of this magnitude, representative of the moon semi-major axis, is examined carefully, it can be seen that when the number 2 is set aside, the sequence that remains is an exact match to the sequence of numbers in the ratio of its increase. Compare the following:

2 10194.624652455977757182576459685...

1.0194624652455977757182576459685...

This is powerful support indeed for the idea that 22/7 is an active ratio associated with the physical geometry of the Earth-moon system. The system values appear by necessity to indicate that this is so. By contrast, PI on the other hand is completely falsified, and shown to have no real connection to the geometry of the system at all. As a result of this, it would appear that a true physical link has indeed been identified between the Earth physical equator and the moon celestial equator; each of which are transformed over time directly in accordance with a power relationship of 4 to 3, as detailed. In truth, the very validation of 22/7 *is* the validation also of this physical connection, the two being inextricably linked. In light of such facts, the presence of a general harmonic-type law of proportion applicable to the Earth-moon system based upon the indicated powers does suggest itself. Expressed mathematically, the law is as follows:

$$PE^{4}_{(e)} \propto CE^{3}_{(m)}$$

Where: PE (e) & CE (m) are ratios:

PE (e) = Physical Equator, Earth (present) / Physical Equator, Earth (past)

CE (m) = Celestial Equator, moon (present) / Celestial Equator, moon (past)

The law of proportion hereby proposed would of course be linked to that given earlier relating the tropical Earth year to the actual equatorial circumference of the planet; the implication being that they are both *physically synchronised*. Thus, an increase from a 360-day year to the present value would *require by necessity* both a direct proportional increase in the physical equatorial circumference of the Earth to match, *and* an increase in the celestial equator of the moon orbit in accordance with a 3 to 4 power association to the Earth year.

Such a connection between the two laws may be more easily perceived with a re-statement of the first law of proportion, taken from chapter 2:

$$TY \propto PE_{(e)}$$

Where: TY & PE (e) are ratios:

TY = Tropical Year (present) / Tropical Year (past)

PE (e) = Physical Equator, Earth (present) / Physical Equator, Earth (past)

Moreover, by combining both laws, a direct connection is evident between the Earth tropical year and moon celestial equator, giving rise to a further law of proportion:

$$TY^4 \propto CE^3_{(m)}$$

Where: TY & CE (m) are ratios:

TY = Tropical Year (present) / Tropical Year (past)

CE (m) = Celestial Equator, moon (present) / Celestial Equator, moon (past)

The Physical-Celestial Basis of Angular Units of Measure

From the most exacting mathematical relations uncovered, one would feel hard pressed to deny at his juncture the validity of the laws of proportion as set forth. The evidence when carefully examined would appear most strongly to affirm their truthfulness; that they are indeed physically operative within the universe, and do govern the real synchronous transformation of certain key elements of the Earth-moon system. Taken together as a full set they force one to conclude that the primary units of the Babylonian system were never originally established as a series of angular measures at all, but instead as a set of actual physical magnitudes characteristic of these bodies, as were genuinely manifest in some harmonious, past age:

360 Days = Earth Tropical Year
21600 IGM = Earth Physical Equatorial Circumference
1296000 IGM = Moon Orbital Celestial Equator

Days = Solar days (86400 seconds)
IGM = Ideal Geographical Miles (6000 feet)

In what previous age and for what length of time the Earth and moon possessed such an idealised orbital configuration, nothing can be said at present. All that one is able to say is that as time passed, an ordered transformative change to the system did occur. It may have been sudden or it may have been gradual. It may have been due to the influence of some unknown disruptive agency, or perhaps due to the natural unfolding of the universe. However it occurred, and however long it took, the principle magnitudes of the original system were all raised to their present levels, a process governed by the specified laws of proportion. Thus was there an ordered synchronised increase in the length of the Earth year, physical size of the planet, and the distance separating the Earth and moon.

With the laws of proportion now fully specified one is able then to use them to actually generate an exacting figure for each of the various physical magnitudes as are manifest in this present age. In generating the relevant values, it would seem most fitting that the key ratio that should be used to derive the whole set should be none other than the exact fraction 1100/1079 which itself is representative of the increase of the moon semi-major axis. Using this as a base then, the primary values of the Earth-moon system derived from the laws of proportion are as follows:

365.2421840 Days = Earth Tropical Year
21914.53104 IGM = Earth Physical Equatorial Circumference
1321223.354 IGM = Moon Orbital Celestial Equator

One can easily perceive a close match between them and the values cited already from other known sources:

365.2421897 Days = Earth Tropical Year
21914.11200 IGM = Earth Physical Equatorial Circumference
1321223.596 IGM = Moon Orbital Celestial Equator

The Frequency Base of the Sexagesimal System

From the above analysis, considerable support then would appear to be had for the idea that the ancients originally based the degree unit of measure upon a standard solar day, whilst the minute and second of arc were each based upon a geographical mile of precisely 6000 ft; all three types of unit being component parts of one unified system. However, upon more careful reflection, it is difficult to reconcile the idea that both degrees on the one hand, and minutes and seconds of arc on the other, could truly be a part of the same system when clearly they each appear to rest upon completely different physical foundations.

The concept of a day, which underlies the degree unit, is a concept of frequency. In essence, one asks oneself how many times the Earth rotates on its own axis to complete one full year; or more simply, how many times does one event occur within the space of another event? The idea of spatial extension does not apply at all in such instances. Thus, expressing the Earth year in terms of solar days, nothing is said about the apparent distance of the Earth from the sun or the circumference of the orbit as a whole. Yet by contrast, in the case of seconds and minutes of arc, it is well intimated in the above analysis that the physical base of them both would appear to be a curved arc length distance of 6000 ft; most definitely a spatially extended measure. But just how can this be? How can the primary values denoting degrees, minutes and seconds, be a part of the same unified system of measures if they each possess a fundamentally different physical base? A study of waveform propagation would seem to reveal the answer.

It so transpires that both spatially extended distances and pulsating cycles existent in nature can be united quite easily as component parts of a propagating waveform. This is evident when one considers that wave-speed is the measure of the wavelength of a propagating disturbance multiplied by its frequency, the latter of which is itself often expressed in terms of cycles per second (Hertz):

$$f \times w = c$$

f = Frequency (Hertz) w = Wavelength (IGM)
c = Wave-speed (IGM / Second)

Given the form of this equation, it is not unreasonable to suggest that the primary values of the whole sexagesimal system, generally expressed as a pure number series, are in fact a set of 'wave-speed' measures. Consequently, the solar day underlying the degree as a frequency component, must be accompanied by a valid wavelength measure, both of which combine to express the wave-speed magnitude of the entire Earth orbit. In view of this it is easy to see then that the Earth itself in its regular journey about the sun whilst rotating on its axis is none other than one giant propagating cyclical disturbance ploughing through space.

Extending this analysis, it must follow also that the ideal geographical mile underlying minutes and seconds of arc, is itself clearly a wavelength measure; one that must be combined with a frequency component just as with the Earth year. In this particular instance the wave-speed disturbances associated with this unit are localised to the physical and celestial equators respectively of both the Earth form and the moon orbit. However, unlike with the actual Earth year which is a clearly visible phenomenon, the wave speed disturbances here must constitute 'invisible cycles'. They are efficiently real and manifest, but are not obvious to behold in a basic visual sense. Properly understood then, the ideal geographical mile is not merely an arbitrary measure of spatial extension, but a unit length inextricably combined with a frequency base actively forming a part of the overall energy configuration associated with the Earth and moon. One must therefore rise above the notion that both the body of the Earth and the orbit of the moon are 'captured as static forms' by the set measure of the geographical mile, but instead realise that a valid description of their 'dimensions' with respect to this unit is ultimately frequency based and energetic in nature. Thus, in the case of the ideal Earth, the number 360 is the frequency pulsation of its orbit; 21600 the frequency linked to the physical equator of the planet, whilst 1296000 is a frequency associated with the orbit of the moon; the latter two manifesting as invisible cycles. Of course, with the actual physical transformation of the noted bodies, the frequency measures associated with them have suffered alteration, and currently stand at 365.2421840, 21914.53104 and 1321223.354.

Once there is the realisation that the primary values of the sexagesimal system actively express the energetic structure of the Earth and moon in the form of characteristic sets of propagating wave-forms, one is much more able to grasp the precise manner in which the ratio 22/7 is linked to both of these bodies. The essential point to comprehend is that the validity of 22/7 does not rest at all within the Euclidian domain of static, strictly form-based geometry. Instead, the true validity of 22/7 is within the domain of physical action which indeed is exactly the domain within which propagating waveforms associated with physical bodies are to be found. The ratio 22/7 may thus be regarded as an actual link between the frequencies associated with an apparent 'circular event' e.g. the orbit of the Earth about the sun, the physical equator of the Earth, or the moon celestial equator, and the frequency measures 'built in' to the straight-line major axis values of such circular events. Using the ideal orbit of the moon by way of example, when the frequency base of its semi-major axis 206181 & 9/11 is multiplied by 2 × 22/7 to give the celestial equator of the body, the resultant value 1296000 must be expressive of a physically existent though invisible wave-speed measure associated with its orbit, and not the measure of a static circular curve at all.

The Absolute Supremacy of Imperial Measures

In final summary of this chapter, there is but one further point to note. And that is the incredible accuracy of the full range of British Imperial Measures. Indeed, from the above analysis it should be readily apparent that *the true standard of the primary distance units of the British Imperial System can be none other than the Earth tropical year itself,* whose orbital frequency can be seen to possess a distinct association to the numerical-fractional division of the mean distance between the Earth and moon.

By using the ratio of change for the Earth tropical year (from the ideal of 360 days) in the 3 to 4 power law to generate a value for the current moon semi-major axis, the numerical figure as returned is shown to be only very slightly removed from the observed measure given in units of IGM, without thought to the tropical year at all, as such:

210194.6631544 = Observed value (converted from 384404 km)
210194.6246524 = Derived via proportional (3 & 4 power) law

A simple division:

$$210194.6631544 / 210194.6246524 = 1.00000018317...$$

What this sum demonstrates is that the unit measure of 6000 British feet (IGM) currently recognised, differs only very slightly from an IGM unit derived from the proportional law (above) held to govern the real transformation of the Earth tropical year from an ideal standard of 360 days. In very real terms then, British Imperial Units are in no way set to an arbitrary standard. The true natural beat that is the pulsation rate of the Earth itself is the standard.

Chapter 5

Of Ancient Myths & Astronomy

As strong as the evidence appears to be in the form of exacting proportional laws that the Babylonian system of measure was originally established upon certain key values of an ideal Earth-moon system, one must acknowledge that the nature of the evidence supporting this contention as detailed so far, is purely mathematical. Everything rests essentially upon the strength of the mathematical relations as identified; their exactness. And yet, if one is to truly establish the reality of the proposed laws to the highest possible level, one should hope that additional supporting evidence of a non-mathematical nature would also exist to confirm the celestial basis of the primary system values; evidence essentially in the form of historical records of some sort. Indeed, if the Sumerian/Babylonian sexagesimal system was ultimately based upon the magnitudes and ratios of an ideal Earth-moon configuration, it is highly reasonable to suppose that ancient accounts referencing such a celestial system must have existed at some point, detailing even its actual physical transformation. Were such found to complement the mathematics, the theory would be established beyond all doubt. The question is therefore, do records of such accounts exist that have survived down to the present age? The answer to this question is not as straight-forward as it may seem.

Myth: The 'Language' of the Ancients

From the study of many prominent civilisations from ancient times, researchers have well established the point that the 'world view' of the ancients was radically different to that which dominates this current age. Indeed, many of the so called 'creation stories' of the ancients that detail how the world was formed and ordered, including also the emergence of man, do not sit well at all alongside present day ideas such as the 'big bang' theory of creation, or the idea that man emerged as a chance event. Of astronomy in particular, those of the present age who study the celestial bodies of the heavens do so from a very different standpoint to those of the past. The modern scientist tends to look upon such objects as mere lumps of inert matter flying through empty space. Their courses are charted blindly, and their physical constituents determined in a most detached manner. In sharp contrast to this though, one finds that many ancient cultures believed in a continually active connection between events taking place in the celestial realm and events taking place upon the Earth, regulating human life itself even. As a result of this, the ancients constantly looked to the heavens for the signs and portents that would herald great changes upon the Earth.

With the world-views so radically different between those of this present age and those of the past, it should not come as a surprise to learn also that they differed greatly as to the very manner in which key concepts, ideas, or physical-historical facts, were actually communicated. In particular, when concerned with a description of such as astronomical phenomena, the terminology used by the ancients is not at all the same as that employed by such as modern astronomers. Indeed, through studying the writings of the ancients, scholars have found that when discussing such things as celestial bodies or elemental forces active in the universe, their work more often than not takes the form of mythical stories. This is of course in sharp contrast to the descriptive methods employed by present day scientists in their own analysis of the same sort of phenomena. Because of this, most modern researchers tend to be quite dismissive of the 'format' of the writings from ancient times, and thus the myths of the past are viewed as nothing but fanciful stories possessed of no real truth; at best, created purely for entertainment purposes. For the modern scientist only their own rigorous methods will do, and all those of the past are considered antiquated and patently unscientific. Such prejudices as these are however highly unproductive. The works of the past cannot truly be judged against the standards of the present. Instead, they must be considered from the viewpoint of the men who produced them; learned men at that. When the ancient mythical stories of the past are examined in this light, it may then be seen that they are just as valid a description of astronomical events as would be any dry academic paper on celestial mechanics produced by a university professor today. Truly it is a question of ages. The myths were essentially the scientific treaties of the past, and their authority was such that this was well known and accepted. It is most unfortunate then that over time such an understanding does appear to have been lost, as is evident from the fact that what once was obvious to the men of the past, has had to be re-discovered by modern day researchers of the present. Indeed, the actual recognition of a link between ancient myths and astronomy only emerged – or rather re-emerged – during the latter half of the 20th century. Prior to this time, scholars of the modern age tended to interpret mythical stories of the past in quite a different light.

In the early years of the 20th century, it was generally thought that the presence of mythical stories throughout the ages was due either to psychological or cultural factors. One prominent example of the former, was the view that they represented 'original revelations', actually associated in some manner with the human psyche; a viewpoint shared by certain followers of C. Jung, a noted psychologist of the time [1]. The primary emphasis of the theory was upon the unconscious element to the human personality coupled with an idea of universal symbolism. In contrast though to such an essentially psychological paradigm, others chose to examine ancient myths from the standpoint of anthropology and history [2]. Although such views did not appear to lack all merit, it was only later that researchers began to realise that the key to understanding the ancient myths of the past ultimately lay with the fact that they dwelt upon events taking place in the heavens. One of the best works of the recent period to actually note the connection between myth and astronomy was that produced by Giorgio de Santillana and Herta Von Dechend, called Hamlet's Mill. It was in this work that the authors explicitly stated that myth was indeed essentially cosmological [3]. Thus, whereas the psychological paradigm claimed that it was the innate psyche of all people each with a shared collective unconscious that was responsible for the creation of mythical stories similar in type the world over, the authors instead adopted the position that it was external cosmological events that were the basis of the mythical stories. Essentially, due to the fact that the same heavenly bodies can be viewed by all people upon the Earth regardless of their location, one would expect that all accounts of their changing configurations would at their core be identical. But critically however, each of the various cultures would use their own unique mythological symbolism and characters associated with their specific region as the means to describe the events in question.

In Hamlet's Mill the authors opened up their work by drawing attention to one myth in particular, as an example to demonstrate just how universal such stories can be. Their choice was that of Hamlet, by William Shakespeare. The story of Hamlet, briefly, is that of a king of some renown who is murdered by his brother, who then subsequently assumes his position of authority, taking as a wife even the widow of the man he has slain. On becoming aware of the crime his uncle has committed, which was concealed from all most carefully, Hamlet, the son of the murdered king proceeds to take revenge by ultimately exposing and killing the usurper. In doing this, Hamlet is cunning throughout. Not wishing to let his uncle know that he has become aware of his crime, he feigns madness up until the critical confrontation wherein he exacts his final act of revenge. This then is the essential theme of the play by Shakespeare, who himself lived in 16th century England. And yet the story itself, is not one original to Shakespeare. This very tale along with its primary elements appears to have been constructed by many other authors, even from quite diverse cultural backgrounds. One such author, Saxo Grammaticus (1150-1216 AD) wrote his own version of this story; located in books III and IV of his work entitled Gesta Danorum. It too began with two brothers, one of whom was killed by the other, who was then himself subsequently killed by his nephew as an act of revenge. The son of the murdered king in Saxo's version was named Amleth [4], which indeed bears a striking resemblance to the name Hamlet as chosen by Shakespeare. There is also a Finnish version of this tale known as the Kalevala, which was maintained through an oral tradition from very early times before eventually being committed to writing; the Hamlet character here having the name of Kullervo [5]. Yet further afield in the Middle East, in the region of Persia (modern day Iran) during the early 11th century, one of the most famous writers of the time, Firdausi, wrote the Shahmama, or Book of Kings, sections of which also dealt with the Hamlet myth. Translated into English the Book of Kings totals 9 volumes in all; detailing essentially the lives of various rulers from very ancient times right up until the time of writing. What is most intriguing though about this work is that it begins with a purely mythical narrative, only making an eventual transition towards historical reality as the work progresses. Indeed, the first 4 books of this work are completely within the realms of myth [6]. Of the rulers themselves thus noted within the work though, it is one Kai Krusrau whose life mirrors closely that of the Hamlet story. Indeed, there are great commonalities in particular to be found between Kai Krusrau and also the character of Amleth from Saxo Grammaticus' Gesta Danorum [7].

In view of the above, it is easily seen then that the Hamlet story appears throughout many regions of the world. However, it is important to realise that this particular story is certainly not a mere isolated case. From an extensive examination of a great multitude of different mythical stories worldwide, modern day researchers have determined that many are universal in nature. In each case, 'heroes' or characters culture-specific to a given region wherein the myth was known are found embedded into the central theme. Ultimately though, such stories were merely a means to describe the activity of celestial events or natural forces at work within the universe. Thus was the changing face of the heavens portrayed in terms of the actions of apparently intelligible beings having a certain 'poetic state of life', though oftentimes possessing strange fates, difficult to comprehend.

Gods & Heroes

When it comes to ancient myths, perhaps the most widely known, at least to Western civilisation, are those concerning the Greek gods. When Rome became dominant in the ancient world and spread its empire throughout much of Europe and North Africa, they adopted the gods of Greece into their society. Later, with the rise of Christianity though, such ancient gods ceased to be worshiped and were effectively abandoned over time.

In antiquity, many stories were recorded detailing the exploits of the Greek gods, their characters and even duties. They were conceived of as a pantheon of deities, immortal in their essential being. The greatest and most powerful of the gods, head of the entire pantheon and ruler over all, was Zeus. Other prominent gods included Poseidon, Hades, Here, Athena, Aphrodite and Ares, to name but a few; but there was also quite an array of lesser gods. Their place of residence was called Olympus, a lofty realm far removed from the world of men, though they did indeed interfere regularly in the affairs of mortals, being able seemingly to move with ease between their own realm and that of the Earth. Although they featured in stories compiled by many different authors, most would tend to agree that the greatest tales concerning the exploits of the Greek gods are those by Homer, as contained in *The Iliad* and *The Odyssey*. The former work is essentially an account of the closing stages of a great war that raged for ten years between the forces of the Greeks on the one side and the Trojans on the other, ultimately ending with victory for the Greeks who finally succeeded in their many attempts to capture the mighty Trojan city of Ilium. The latter work is somewhat of a sequel to the Iliad following the adventures of Odysseus, a prominent member of the Greek contingent on his return voyage to his homeland of Ithaca, carrying with him his share of the spoils from the Trojan campaign. The gods themselves play a major role in both stories, interacting with the human characters and interfering with their fates for their own ends. Indeed, many of the central characters of the stories are sons or daughters born as a result of the union between gods and mortals. Achilles, the greatest soldier that fought on the side of the Greeks is one such example; son of the goddess Thetis and the mortal Peleus. Such characters are strong favourites of their immortal parents who oftentimes 'look after them' and keep them alive during the major fighting, excepting when fate decrees the time at which they must die, as recounted particularly in the Iliad.

In general the gods of Olympus, described by a variety of ancient and most respected writers, not just Homer, are portrayed as somewhat supernatural beings, said to either actively control or personify elemental forces, including such as rain, thunder or sunlight; yet also such states as sleep, death, vengeance or love. There are even gods who preside over the souls of the dead in their afterlife realms. Yet, even when representative of such states of being or forces of nature, in the regular course of their existence, they oftentimes, as portrayed in tale, are able to interact intelligibly with humans. This is achieved through an ability to transform themselves outwardly into the form of mortal beings, even though remaining fundamentally different in their essential nature. The gods are not restricted either in their transformative abilities, for writers of antiquity often recount instances where gods have changed themselves into a wide variety of different animal forms. Indeed, at times they have even used their powers to transform various humans into certain types of animals, invariably to punish them for some misdeed. For the most part the actions of the gods with respect to mortals are quite arbitrary. They appear to favour certain of them on a whim over others, aiding them as they see fit, yet destroying those they choose to, also seemingly on a whim; a pattern of action that is extended even towards their dealings with one another. Indeed, the gods are in no way portrayed as being unified in their outlook at all; arguing and fighting frequently amongst themselves. Moreover, they oftentimes drag mortals into their disputes, exploiting them to fulfil their own selfish agendas. The war itself between the Greeks and Trojans described by Homer was in fact originally caused by certain prominent gods of the Olympian pantheon, who then took sides in the fighting, aiding their chosen side with their supernatural abilities as if it were a mere game to them and mortal beings simply their playthings.

Although the Iliad and the Odyssey are well known classic works it must be conceded that there is somewhat of a great mystery as to when they were written. No other documents contemporary to the works themselves make reference to them, and thus they exist in effect in a sort of vacuum [8]. Indeed, very little is actually known about Homer himself, with quite a number of different views existing as to his date and truth birthplace [9]. Some place him about the mid 8th century B.C. [10] yet at times modern day researchers have held to dates several centuries earlier than this. Also, what of the stories themselves? What of Ilium, and the long war between the Greeks and the Trojans? Is there historical truth to such a war? It was not until very recently that confirmation of the existence of the city of Ilium during ancient times emerged as a result of archaeological excavations. The site thought to be that of Ilium, also known as Troy, was uncovered through excavations at Hissarlik. Here it was revealed that one particular settlement at this location was highly walled and fortified, along with another that showed evidence of being destroyed by fire during the mid 13th century B.C [11]. Archaeological evidence does therefore provide some measure of support for the reality of the Greek-Trojan war, at least placing it within the bounds of possibility [12]. Indeed, if the war truly does have an actual historical basis, the manner in which it is recounted by Homer draws certain parallels with the Book of Kings by Firdausi. In the work by the 11th Century scholar from the Middle East, a transition was made from a time of myth through to true history, accompanied by real kings. In the case of the Iliad though both components appear to run in parallel; the actual war being a part of true history fought between real human forces, with the gods being the mythical element of the story. And yet, though the actual fighting amongst men and their actions towards each other can be conceived of with little trouble, even in poetic form, what is one to make of the stories concerning the gods? How did the actions of the gods themselves relate to an all too human war?

The solution would appear readily apparent once there is acceptance of the fact that the gods were indeed representative of celestial events unfolding in the heavens, as has been suggested by researchers. Indeed, with this understanding, one is able to decipher any number of what at first may seem baffling or even quite ludicrous tales concerning the Olympian gods as recounted by many writers, such as Homer.

The Affair of Ares & Aphrodite

Even with their seemingly higher state of being, from the accounts of the actions of the gods, they were still subject to certain criminal passions, just as mortals themselves are. Indeed, one of the most common acts of betrayal carried out by the gods was that of adultery, with many of the most prominent gods of Olympus including even Zeus himself having numerous illicit love affairs, even though bound by marriage. One story in particular that is of note, recounted by several authors from antiquity including Homer (in the Odyssey), concerned just such an affair between Ares, the god of war, and Aphrodite, the goddess of love. At the time Aphrodite was herself already married to another god, Hephaestus, when she embarked upon her affair with Ares. In all such mythical tales of illicit love, eventual exposure is practically guaranteed, and this indeed was certainly the case with Ares and Aphrodite, the result of a witness who then turned informer…

When Hephaestus was once absent from his palace, Ares and Aphrodite met in passionate embrace and made love to one another, even in the very marital bed of Hephaestus. However, the Sun himself, who sees all things before everyone else [13], *was witness to the event, and told Hephaestus of it. Upon learning of the unfaithfulness of his wife he immediately plotted revenge, his intention being to expose the pair, to publicly humiliate them. To achieve this, master craftsman that he was, Hephaestus went straight to his workshop wherein he began to forge an intricate network of chains that were woven together to form an elaborate net. The construction of the chains themselves was such that they could not be broken by force nor even undone, and yet so finely were they fashioned that they were invisible even to the gods* [14]. *When completed, Hephaestus took his chain netting and carefully arranged it about the legs and the rafters of his marital bed* [15]; *a trap for the unwary lovers. When set, he then made it known of his intention to visit the town of Lemnos upon the Earth* [16]. *Ares, taking careful note of this, seized the chance to once more be with Aphrodite in the absence of her husband. Meeting Aphrodite within her chambers, both goddess and god immediately succumbed to their desire to make love and thus lay down upon the bed. Almost instantly the chains fell upon them, capturing them in their embrace* [17]. *Unable to move, they had to remain in their prone position until Hephaestus returned. This in fact did not take long, for the sun, witness once more to their passions, intercepted Hephaestus, relating to him what had just transpired. Returning home immediately, he went straight to his bedchamber to observe the exposed lovers bound by his network of chains. Standing in the doorway of the room, he called out to all of the other gods* [18] *to come and witness the scene, to see with their own eyes the unfaithfulness of his wife. Many came to observe the captured pair, with some even praising the cunning of Hephaestus in finding them out, and all laughed loudly. However, out of modesty all of the goddesses stayed away* [19]. *Hephaestus himself stated that he would only release the pair were Zeus, the father of Aphrodite, to return all of the gifts that he had given him in order to win his daughter's hand in marriage* [20]. *It was only when the great god Poseidon, brother of Zeus, vowed that Ares himself would make restitution for his transgression that Hephaestus agreed to release them from his chains, for Poseidon personally guaranteed that if Ares absconded from his debt, then he himself would pay on his behalf* [21]. *With such a promise made by such an important god, Hephaestus could not refuse to release them. Once the chains binding Ares and Aphrodite were removed, both immediately fled the scene* [22].

Upon the surface, the above story is one that is easily conceivable in the mind in purely human terms. Yet when applied to the mythical gods of Olympus, it turns out that the whole event is not at all just composed of the actions of seemingly intelligible beings of a higher nature engaged in some sort of passion play. Even in antiquity there were scholars who fully understood the true meaning of this quite fantastic, even ludicrous tale. One such man, Lucian of Samosata, explained that the whole story is really to be understood as a conjunction of the planets Mars and Venus, involving also the Pleiades

[23]. It is indeed commonly recognised that the planets of the solar system (including moons) are each named after various Olympian gods. The planet Mars is representative of Ares, the god of war, whereas Aphrodite is linked to Venus. When carefully considered, one of the key elements of the story, providing a real hint as to its true interpretation, is the role of the sun. It is specifically stated that the sun was witness to the 'passionate embrace' of the god and goddess, seeing it before everyone else. This is perfectly understandable, as a conjunction itself is indeed the very alignment of a set of planets with the sun. Thus, from the perspective of the central solar body, looking outwards, the merging of two planets may well be taken as a sort of momentary embrace so to speak [24]. As for the Pleiades, quite a well-known constellation, this group of stars forms the backdrop to the whole scene; the intricate web of Hephaestus 'capturing' the planetary union as it occurs.

The above analysis well serves to illustrate the true nature of the stories of the gods as recounted by such writers as Homer. Within such tales the gods are not to be understood as supernatural characters at all, but instead as an active representation of changes that take place within the celestial realm itself. In effect, the gods and their actions are to be understood as signs or portents that may be observed in the night sky. Thus, it is quite conceivable that the actions of some of the gods, as recounted in the Iliad for example, were representative of certain events that took place in the heavens during the course of the actual fighting? This is certainly a distinct possibility. Moreover, if such an interpretation of ancient myth is correct, then there is much to link it to the realm of astrology, for the same sorts of principles apply. Astrology itself is ultimately the study of how the celestial bodies, including the stars and planets, affect the human condition. Astrologers look to the sky for the signs and portents that indicate the direction and flow of events unfolding upon the Earth, with relevance to either humanity as a whole or even one specific individual. In the myths of Homer then, as recounted in the Iliad, the very tide of the battle may have been thought to correlate with a whole series of changing events in the heavens, as conveyed by the poetic form of the work. The gods, portrayed as intelligible human-like beings with strange super-natural powers, were representative of nothing less than the celestial powers of the heavens, as were thought to dictate the very course of all human affairs on Earth.

It would seem quite clear then that there is indeed a distinct astronomical explanation lying behind the Greek myths. When properly decoded, they do appear to reveal the past actions of real observable bodies within the sky; a strong indication that myth itself ultimately does have at its core, physical truth. Of course, one could never claim with absolute certainty that all myths are representative of an underlying astronomical event.

However, it is those that are which are of greatest relevance to this present work. Most certainly, there are two further mythical stories from the past to consider that do stand out in particular as being highly significant in this regard; each possessing a most definite celestial feel. They are respectively, the tale of Phaethon, and that of the birth of the children of Geb and Nut. The essential theme of both of these myths is that of a general disruption to the orbits of the major celestial bodies within the solar system. An account of each is in order.

The Story of Phaethon

The main focus of the mythical tale of Phaethon is the attempt made by a mortal being to wield the powers of an immortal, and of how this leads to disaster upon a global scale through disruption of the normal course and ordering of the heavenly bodies. It is indeed a most intriguing tale that has been examined in detail by quite a variety of scholars over the years; the story itself being one that stretches far back into pre-history. Of the tale itself, perhaps the most elaborate account is that provided by Ovid, a poet from ancient Rome contemporary with Christ. Clear evidence does exist though to indicate that the story is in fact many hundreds of years older than Ovid's time. In drawing mainly from his version of the myth though, the story is related as follows:

Once, when in the company of Epaphus, the son of Io and Jupiter, Phaethon began boasting of his own parentage, bragging continuously about his father Phoebus, the Sun god, much to the annoyance of his companion. Reaching that point where he could take no more of such talk Epaphus spoke up to question the identity of Phaethon's parents, calling him a fool to believe his mother Clymene, who had told him that his father was the Sun god [25]. *Such a bold statement immediately silenced Phaethon in his boasting. Whereupon, after departing from his companion he sought out his mother to ask for proof of his noble birth, for doubt had now crept into his own mind over this issue. Upon meeting his mother, she swore a solemn oath to Phaethon with her arms stretched out to heaven, that the sun that guides the world and is seen by all is indeed his true father. However, should he still doubt this, she asked that he pay a visit to the Sun himself for confirmation* [26]; *and this he did.*

In his quest to find answers to the truth about his parentage, Phaethon went to the very palace of the Sun, a grand structure composed of high columns, bedecked with gold and silver. Upon entering, he found himself immediately in the presence of the Sun god, who was seated upon his throne and robed in purple. Standing to both his right and to his left were Day, Month and Year, including the Hours and Generations; all spaced out at equal intervals [27]. *Phaethon was well received by the Sun who confirmed that he was indeed his father, as Clymene his mother had said. However, should he still doubt this, he told Phaethon to ask for anything that he wished, and that he Phoebus, the Sun god, would grant it if it lay within his power. This promise he made with a most binding and sacred oath. In response, and without hesitation, Phaethon asked that he be allowed for one day to have control over his father's chariot and its steeds with their winged feet. Immediately upon hearing his reply the Sun realised what a terrible error he had made, for this indeed was the one thing that he would have denied Phaethon* [28]. *But alas, his promise could not be broken. Whatever Phaethon desired, Phoebus was compelled to grant. However, he did make one last plea to his son to choose something else, claiming that it was too dangerous an assignment, and that he had not the strength to carry it out being so young. Phoebus tried to explain to Phaethon that only he alone could manage his divine and fiery chariot. He emphasised the difficulty of the journey, of how the first stage was very steep as the chariot proceeded to the zenith of the sky* [29], *and that strong control over the steeds was required on the way down. Phoebus also told Phaethon to consider the spinning sky, carrying with it the stars of heaven. He told him of the great danger that existed from being swept away from the movement of the heavens* [30]. *Upon hearing all of this however, Phaethon was unmoved, and still refused to change his mind. Therefore, reluctantly, his father ordered his attendants to prepare the horses and the chariot, so as to be ready to go at the appointed time when Dawn awoke.*

When all the preparations had been made, Phoebus gave his son some last advice, telling him not to deviate from the clearly marked out path in the sky; its character being that of a broad curve. With this said he then transferred onto Phaethon's head his rays [31], *and giving him the reins of the horses, let him take to the sky. As Dawn emerged, the horses leaped into the air carrying both chariot and rider. Yet immediately they felt that their load was far too light, and the chariot was thrown about in a most unstable manner, resulting in a major deviation from its proper course. In consequence, the chilled stars in the Northern Plough grew hot,* [32] *being greatly disturbed from their regular condition; yet, onwards the chariot went. Reaching ever-greater heights, Phaethon became quite dizzy upon perceiving the Earth so far below him. He became terrified, dropping the reins even; the chariot was totally beyond his control, and moved close to the stars of heaven, following no set pathway. After reaching such an extreme altitude, it then turned once more in its course, and began hurtling downwards, this time moving dangerously close to the Earth, such that the moon itself was amazed by its position, even lower than its own* [33].

The Earth became ablaze with fire. Trees, plants and general vegetation, along with entire cities were consumed [34]. *The intense heat dried up rivers and forced oceans to contract, replacing many with large tracts of desert. Even cracks developed in the ground with shafts of daylight pouring through into Tartarus and the lower regions of the Earth* [35]. *With such widespread destruction occurring throughout the world, the Earth goddess herself made a plea to omnipotent Zeus for aid, for if the Earth, the seas and the heavens were all destroyed then this would mean a return to primeval chaos* [36]. *Upon hearing her plea, Zeus called upon all of the gods that they be witness to him that if he did nothing then the Earth itself would perish entirely. Then, removing himself to the highest part of heaven, he readied a thunderbolt, which he proceeded immediately to unleash against the chariot and its rider* [37]. *Upon impact Phaethon was thrown from the car, his life instantly taken from him; the chariot itself broken up by the force of the blast, its parts scattered widely. Freed from their yoke, the steeds leapt from the wreckage of the chariot, as the lifeless body of Phaethon in its turn hurtled downwards through the air, his glowing locks aflame, tracing out the very path of his fall. Landing eventually in a remote part of the world, in the river Eridanus, his body was buried by a group of Nymphs* [38].

Phaethon was mourned heavily by his father Phoebus, who was greatly angered at his death at the hands of Zeus, so much so that he refused to continue his fated task to provide light to the world because of what had happened to his son. It took much persuasion on the part of Zeus, including all of the other gods, to convince Phoebus to perform his regular duty. Yielding eventually to their pleas though, he did finally agree to re-commence his daily journey across the sky. Thus, rounding up his scattered horses [39] *he once more took to the air at the regular appointed times, so that the Earth could again receive its due measure of light.*

Upon examination it is easy to see that the story of Phaethon, though possessing an 'everyday' feel to it, contains astronomical associations that are undeniable. This indeed is something that the ancients themselves appear to have been all too aware of, being fully cognizant of its allegorical nature. The most decisive evidence in support of this is contained in one of the more prominent works of the great philosopher Plato (427-347 B.C.) entitled Timaeus & Critias. As with all of Plato's major works, the Timaeus & Critias takes the form of a dialogue, a conversation between several people, its format essentially being that of a script. Within this work, to aid him in his discussion of various philosophical and cosmological ideas, Plato has one of the characters of his dialogue relate the story of a true historical encounter that took place between Solon, the great lawgiver of Athens, who lived some two hundred years prior to Plato, and certain priests whom he met whilst travelling through Egypt. Speaking through one of his characters, Critias, Plato recounts that when Solon visited the Egyptian city of Sais, he held a discussion with certain priests there about events that took place during very early times [40]. In doing so he made mention of various stories, such as that concerning Phoroneus, the first man to have discovered the use of fire, stolen itself by Prometheus [41]. He spoke also of Deucalion and Pyrrha, both of whom survived the flood. As he related these events to the priests he even set his mind to try to calculate how far back in time they had occurred.

After Solon had spoken for some length of time upon such matters, it was then that an old priest present at the discussion made a very odd remark. He told Solon that the Greeks were as children, and that old Greeks did not even exist [42]. Upon hearing such a strange statement Solon asked for an explanation, whereupon the priest did indeed provide one. He told Solon that the knowledge of his own people was most ancient in comparison to that of the Greeks, for it was deeply rooted in traditions of old. The reason for this, said the priest, was that periodically the Earth is devastated by various natural forces, the greatest being fire and water, though many lesser ones exist also [43]. Most specifically, the priest openly gave the solution to the Greek myth concerning Phaethon, who lost control of his father's chariot destroying many parts of the Earth through fire. This story, the priest stated, was a mythical account of a certain astronomical truth: namely that at long intervals a shift occurs in the normal activities of the heavenly bodies, the result of which is extensive destruction over the whole Earth by means of fire. When such events as these occur only a mere remnant of mankind survives, composed for the most part of those who are uncultured and unlettered [44]. And thus essentially, the scattered remains of humanity must begin again like children, building anew their knowledge from the ground up.

The Myths of the Earth & Moon

From the above examples, it is quite evident then that certain ancient myths from antiquity do possess a definite astronomical basis, with the central characters actively representing various natural forces or known celestial bodies. With this understanding, and in light of the main objective of this present work, it would seem appropriate to consider then, if there are any mythical stories that relate specifically to both the Earth and the moon. Indeed, if the primary values of the Babylonian/Sumerian system of measure were established in a prior age in accordance with these particular celestial bodies, then such a physical truth could possibly be encoded in myth, as may even the very transformation of the system bodies themselves. Certainly, if a mythical account of the transformation of both bodies were to exist that was found to be in harmony with the proposed laws of proportion, as given previously, then their reality would seem assured. The question is then, are there any mythical stories from ancient times that would allow for such a determination? The answer, most emphatically, is strongly in the affirmative. The appropriate supporting myths are indeed found to exist, importantly, pre-dating the classical Greek period of Plato. They are to be found in the land of Egypt, many of them recorded during what scholars refer to as its dynastic period, from about 3100 BC through to 323 BC.

As with all such ancient civilisations of the past, Egypt too had its own set of mythological gods representative of the vast array of natural forces thought to be active in the world, including their own creation stories. Of all the myths of Egypt though, there is one particular story that would appear to possess special relevance to this present discussion. A story steeped in allegory, but which, when carefully analysed, would appear to be none other than an account of a series of physical-astronomical changes to both the Earth and moon, that are in exact accordance with the mathematical laws of transformation detailed previously. The myth itself concerns some of the most prominent members of the Egyptian pantheon of Gods, and is related here due to its obvious worth.

In a far off age during the time of creation, gods and goddesses of various types were still coming into being [45]. *As deities coupled with each other they gave birth to yet more gods, each of whom had one function or another. In the course of this process, the Earth god Geb, had an elicit affair with Nut, the sky goddess, who in turn subsequently became pregnant* [46]. *However, the Sun god Re, who ruled over the Earth, had heard of a prophecy that Nut's offspring would depose him from his throne* [47]. *To prevent this, Re cast a curse upon Nut so that she could not give birth to her children on any day of the year. At this time there were 360 days to one year* [48]. *However, Nut herself was not willing to give up on the prospect of child birth. Thus, she sought the aid of Thoth, the god of wisdom, who devised a means to out-manoeuvre Re. Thoth went to visit Khonsu, the lunar deity, whom he knew was quite partial to a game known as Senet. Thus, when Thoth challenged him to a match, Khonsu did not refuse. Thoth had in mind though a most unusual set of stakes for this match. He requested that they both play for the very light of the moon itself. Feeling confident, Khonsu agreed* [49]. *However, the game did not favour Khonsu at all, for Thoth thoroughly outplayed him. So much so that he won a substantial amount of the moon's light. The result of such a loss greatly weakened the Moon, forcing it to become smaller in the sky. It could no longer shine a full moon every night as was its custom. Instead it was able to reveal itself fully for only one night, after which it was forced to go into hiding in order to recuperate and build up its strength, allowing it to produce the next full moon* [50]. *As for Thoth, after winning the light of the Moon, he then transferred it over to the Earth. This had the effect of increasing its standard year of 360 days by 5 extra days. And because these were not a part of the normal year, but outside of it, Nut was able to give birth to her children upon these extra days despite the curse of Re. She gave birth to one child upon each of the additional days. They were Osiris, Isis, Seth, Nephthys, and 'Horus the Elder'* [51].

From a careful analysis of this tale, and with a complete awareness from the outset that it encodes a series of real physical-astronomical events, the myth is not as incomprehensible as it may seem. Essentially, the tale begins with the Earth possessing a full 360 days to one year. However, some sort of interaction between the Earth (Geb) and its sky (Nut) results in a change to the physical state of the planet. In order though that this change be allowed to progress, it would appear necessary that there be an increase in the Earth year. For this to occur though, certain changes are required also to the state of its orbiting moon. The careful examination of each of these key points allows for an understanding of the real physical processes that this myth would appear to be describing.

The starting point of the myth, as may be seen, is the intercourse of two deities, Geb and Nut. Their illicit affair is the catalyst for all that follows. Because of this, it is important to understand their relationship to one another, as revealed more generally in ancient Egyptian myth. It is of note that in the pantheon of Egyptian gods, Geb and Nut are depicted in ancient papyri as being separated by their father Shu, the god of air [52]. In certain images, Nut is portrayed as a woman arching over Geb, her male lover, who is himself lying prostrate on the ground, oftentimes with an erect penis. Shu, their father, is himself usually shown stood with his hands in the air holding up Nut, preventing her and Geb from uniting in intercourse [53]. For whatever reason though, Shu must have failed in this task at some point, for Nut does indeed become pregnant with Geb's children, as the myth relates. How can this be interpreted astronomically though? It is quite conceivable that Shu, representative of the air, is compressed, as Geb, the Earth, rises up to meet Nut, the sky, as she in turn descends upon Geb, in order that they may unite. This most definitely could be interpreted as a real physical process; one that produced a measurable increase in the Earth's size; perhaps even the formation of the atmosphere of the Earth if the appropriate air density of the planet was established at the same time. The very idea of such an increase is indeed strengthened, when one considers that the very condition of pregnancy in a woman is one of 'swelling up' i.e. becoming larger. Thus, interpreting this part of the myth as describing none other than a real physical increase in the size of the Earth would appear to have some validity. However, one must not forget Nut's 'children', namely the extra days added to the Earth year. The myth suggests that the physical increase in the Earth size has a causal relationship to the increased Earth year; both events essentially occurring as part of a single process. Such an association is in exact accordance with the mathematical law of proportion given earlier, specifying that a change in the Earth's physical equator is directly proportional to an increase in its tropical year. The form of this law indeed defines the exact manner in which the physical extension of the Earth, swelling up in 'pregnancy', is accompanied by an increase in its orbital period to compensate, exactly as the myth relates. In essence, as the Earth grew in size, it 'gave birth to children' in the form of additional solar days. Here then would appear to be ancient proof in the form of myth, supporting the validity of the first law of proportion put forth in this work. But what of the rest of the myth, and the other laws as stated? For this, one must look to the moon.

In the Egyptian myth there are two very specific changes to the moon that are spoken of. One of them would seem to hint at a shift in its orbit, whilst the other would seem to imply something altogether different. In dealing with the first of these, the myth states that the moon became smaller in the sky. From the view of the Earth itself, an increase in the distance at which the moon orbits its primary body would most definitely result in the moon becoming smaller in the sky, at least from the perspective of an observer upon the planet's surface, which would indeed seem a valid interpretation. However, when such an action is described as occurring alongside a simultaneous increase in the physical circumference of the Earth, clear support is to be had for the law of proportion linking a change to the semi-major axis of the moon, cubed, to a change in the physical Earth circumference, to the fourth power. Thus, taken as a whole, the myth would appear to describe a whole series of simultaneous transformations: as the moon extends its orbit the Earth is forced to extend its physical circumference, in addition to simultaneously increasing also its total orbital time i.e. its tropical year. The laws of proportion stated in this present work would seem then to rigorously capture the exact manner of these changes. With this, it would appear unquestionable that *the primary Babylonian units of measure were originally established to denote a series of real physical magnitudes specific to an 'ideal' Earth-moon system, as is noted in the Egyptian myth.* The initial state of these celestial bodies, as defined by the Babylonian values, would seem to be none other than that recognised as existing prior to 'Nut's pregnancy'; a time wherein the Earth itself possessed a 360-day year. It would seem then that the Egyptian myth is in point of fact an actual description of the real physical transformation of the Earth and moon from a state of initial harmony, as represented by the values of the sexagesimal system, to that existent in this present age. It is an account of an actual set of physical-orbital events that did once occur at some historical date in a past age, the very reality of which must confirm beyond all reasonable doubt that the stated laws of proportion as given are actively real, and do govern true physical change to the Earth-moon system as a whole, *even in an ongoing capacity in this present age*.

With the physical laws thus given, the extension of the moon orbit can be understood most exactingly. However, the secondary change to the state of the moon as detailed in the myth does not seem susceptible at all to a mathematical analysis in like manner. Indeed, the reason for this is that it would appear to be primarily qualitative as opposed to quantitative in nature. Specifically, it is that change to the moon which resulted in the celestial body 'losing its light' – yet one further consequence or apparent side-effect of being forced to extend its orbit. What then is to be made of this strange statement? Understanding its meaning is doubtless to be had from a consideration of the current moon phases, which would appear to have been established directly as a result of the loss of its light, as indeed the Egyptian myth intimates. In this regard one may note the fact well known to both lay people and professional astronomers alike, that the moon goes through various different phases whilst in orbit. As a result, over the course of a given month the moon is transformed from a full moon, through to a partially sighted moon, and then to a completely dark moon; from which the entire process is reversed until once more it becomes full again. According to the myth, prior to the loss of its light such phases did not exist, for the moon produced a full moon every night, and not just occasionally. But how could this be? Initially one may be tempted to think that a moon orbit of a radically different nature could perhaps account for such a difference. However, after much thought one must concede the inadequacy of such a proposal. The true solution to the mystery cannot lie with the orbital path of the body at all, but instead must lie with the fact that the phases of the moon actually occur because as a celestial body in space, it is a *reflector* of the sun's light towards the Earth. Thus, no matter what orbit it possesses, under such conditions it is physically impossible for it to produce a full moon every night. It lacks all ability to achieve this. In truth, there is only one way in which the moon could in the past have produced a full moon every night, and that is, the moon must have been a *source* of light itself. Much like the sun, though perhaps not with the same level of intensity, the moon must have been radiating light in the visible spectrum as it orbited the Earth. With this understanding it becomes perfectly clear what is meant by the moon 'losing its light' in the myth. Essentially, when it suffered its transformation and extended its orbit from the Earth, it also ceased to be a source of light, and became a reflector of light only.

Chapter 6

The Disruption of the Moon Orbit

The Transformation of the Month

With the mythical accounts clearly validating then the physical transformation of the Earth and moon in line with the specified laws of proportion, one is bound to suspect that perhaps there are yet more laws still to be identified that govern the transformation of additional features of these noted bodies, which at present remain undiscovered. With a continued focus upon the orbit of the moon, one may hope to draw out such further laws. For indeed, one must expect that the previously noted change to the mean distance separating the earth and moon, as found to accompany a tropical year change, would doubtless cause the moon orbital period to be transformed also. This being so, it would not be unreasonable to suspect that a proportional law, tied in to those already given, would be found to govern such a change. In exploring just this possibility, one must consider then just what the actual measure of a supposed 'ideal month' might have been, in contrast to its present value; the very starting point for such an investigation being a clear statement as to what constitutes the measure of the moon orbital period.

The Measure of the Moon Orbit

One of the most distinctive features of the moon orbit, as noted at the end of the last chapter, is that it is marked out by various phases. The moon constantly fluctuates between being completely full and being completely dark. Indeed, from historical times even unto the present, the measure of the length of a month has been the measure of a single phase-change cycle. In this regard, many cultures from the past including the Babylonians, Hebrews and Romans set the beginning of a new month when the moon was completely dark and just developing a slight crescent of light [1]; such a time being referred to as a 'new moon'. Thus, as the month progressed from this initial phase, the first half of the process saw a steady increase in the amount of sunlight reflected to the Earth culminating eventually in a full moon. The second half of the cycle was then a simple reversal of the first, with the moon gradually disappearing until it could be seen no more. In astronomy, this type of month, so defined as the time between successive new moons, is known as the synodic month. One must of course acknowledge that this is but one of several different types of month recognised by modern day astronomers. However, of them all, it is the synodic month that does appear to have been of greatest importance to the ancients.

Concerning the exact length of time actually required to complete a single synodic month, one must note that due essentially to a series of complex orbital factors, during the course of a year, the sightings of successive new moons are seen to vary; fluctuating between about 29 and 30 days. That being said, modern astronomers have been able to determine to a high level of accuracy an exacting measure for an average synodic month time interval whose value is in between the noted extremes. The currently established measure for the mean length of a synodic month is 29.530589 solar days [2].

The Synodic Month & Earth Year within the Egyptian Calendar

In the Egyptian calendar, both the tropical year and synodic month were tracked, each being considered highly important cycles of time; aspects of one unified system. For most of their history though, the Egyptians did not actually base the length of either of these time cycles upon actual observations as such. Instead they overlaid on to the true orbital periods of these bodies a fixed set of values to denote their measures. The Earth year was set at 360 days, and each month was fixed at precisely 30 days [3]. Marked out in the Egyptian calendar there were thus exactly 12 months perfectly integrated into each year. As a result of this, the calendar was not therefore in harmony with the true (observed) lengths of either month or year, and indeed, the Egyptians were forced in consequence to modify their calendar in order that they might correctly take into account the lengths of the actual cycles.

As they well knew, the true length of the Earth year was slightly over 5 days longer than their calendar ideal of 360 days, and thus they added on for practical necessity 5 extra days to compensate. The critical point to note though is that they considered the true 'foundation measure' of the year to be exactly 360 days, and set their calendar in accordance with it. The extra days, held to be strangely separate from the rest, almost anomalous, were not viewed as 'true' days. In consequence, not only were they considered to not form part of the Earth year as such, neither were they thought to form a part of the 12 months of the year [4]. Indeed, upon the issue of the synodic month itself as a cycle of time, it would seem that the views of the Egyptians were similar to those they had with regard to the observed length of the Earth year. Just as they considered the 365-day Earth year to be 'abnormal' in comparison to a 360-day year, any observed synodic month that did not match up with their set calendar ideal of 30 days was also thought to be somehow abnormal. Indeed, this was an idea that extended well beyond just Egypt. Other prominent cultures from the past shared practically identical views. The Babylonians are a good case in point. With the duration of the synodic month fluctuating between 29 and 30 days over the course of a year, they considered any observed month of 30 days to be 'full', and any that contained only 29 days to be 'deficient' [5].

A Law of Proportion for the Lunar Month

In view of the above, it is obvious that special importance was placed upon the Egyptian calendar values specifying a most harmonious Earth year - moon month configuration. A 360-day year combined with a 30-day month would seem to have been viewed by the ancients as a sort of ideal celestial template, and the fact that actual observations no longer corresponded to it was a point of some consternation. In light of this understanding, there is little doubt then that a value of exactly 30 days must have constituted the actual measure of an ideal synodic month. Consequently, uncovering a law of proportion to link the transformation of the Earth year to the transformation of the synodic month, must therefore involve the presence of this value. One is required thus to study the whole set of values for the ideal and current magnitudes of the orbital periods of both bodies:

	Presently Observed Values	Egyptian Calendar
Year	365.2421840 days	360 days
Month	29.530589 days	30 days

	Ratios of Change
Year	365.2421840 / 360 = 1.014561622461…
Month	29.530589 / 30 = 0.984352966666…

In view of the above, is it thus possible to demonstrate then a clear relationship between the ratios of change of the Earth and moon that is suggestive of the presence of another physical law of proportion? The answer would seem to be a resounding yes.

It so transpires that there is indeed a most exacting set of powers that do relate the moon synodic month ratio of change to that of the Earth tropical year. Interestingly enough, under the specified ideal configuration, it is of note that the moon possesses a total of precisely 12 months per year. The power relationship appearing to connect the transformation of the synodic month to that of the Earth tropical year has incorporated into it, this very number. Expressed as an equation, it is of the following form:

$$(\mathrm{TY} / ty)^{12} = 1 / (\mathrm{SM} / sm)^{11}$$

Where: TY = Tropical Year (Present)
SM = Synodic Month (Present)
ty = Tropical Year (past)
sm = Synodic Month (past)

By using known values for all of the above specified components with the exception of the synodic month ideal, it is possible to calculate the measure of this remaining value, and then compare it for accuracy against an exact value of 30 solar days. Using the above formula, this is achieved as follows:

1) Current Earth Tropical Year divided by Ideal Tropical Year:

$365.242184086309 / 360 = 1.014561622461$

2) Earth Year ratio to the 12th power:

$1.014561622461^{12} = 1.189436233796$

3) Reduction of answer to the 11th Root:

$$\sqrt[11]{1.189436233796} = 1.015895874668$$

4) Above answer multiplied by current Synodic Month:

$$1.015895874668 \times 29.530589 = \mathbf{30.000003541618} \text{ days}$$

The answer obtained is almost exactly 30, the degree of error being extremely small. In light of the previous laws put forth, it would seem highly likely then that a true relationship is revealed here by the specified powers. Consequently, this necessitates that the transformation of the synodic month over time is such that a value of exactly 30 days must have been present at some point in past history when the Earth itself had a value of 360 days per tropical year. Moreover, as it has already been shown that the transformation of the Earth year from an ideal of 360 days is linked to the transformation of the celestial equator of the moon orbit by a 4 to 3 power relation, the validity of the above law also implies that whilst possessed of a 30-day month, the celestial equator of the moon orbit was itself none other than 1296000 ideal geographical miles.

In view of these facts, one must conclude that the units of the Egyptian calendar were once a set of real physical measures. The values assigned to the Earth year and moon month were once representative of the true orbital periods for each of these bodies at some distant point in the past, and thus the units of measure they chose were ultimately based upon the reality of a once ideal Earth-moon system. The very strength of the mathematical evidence leaves little room for doubt then that the transformation of the orbital periods of the Earth and moon are indeed governed by the following physical law of proportion:

$$TY^{12} \propto \frac{1}{SM^{11}}$$

Where: TY & SM are ratios:

TY = Tropical Year (Present) / Tropical Year (Past)

SM = Synodic Month (Present) / Synodic Month (Past)

What one may note in particular from the form of this law, at least in contrast to the others given, is that it is indicative of an inverse transformation. Essentially, an increase in the length of the Earth year is accompanied by a decrease in the length of the lunar month, as determined by the specified power relationship.

By establishing the existence of this law, it is thus possible to derive from it a refined value for the moon synodic month using the ratio for the Earth year transformation, based ultimately upon the special ratio of increase discovered for the moon celestial equator, namely: 1100/1079. This may then be compared with the currently accepted value for the synodic month, as shown:

$$\frac{30}{\sqrt[11]{(365.2421840863 \;/\; 360)^{12}}} = 29.5305855137\ldots\text{days}$$

Compare with: 29.530589 days…

$$\begin{aligned}\text{Difference} &= (29.530589 - 29.5305855137) \times 86400 \\ &= 0.3012 \text{ seconds}\end{aligned}$$

It can be seen then that the above answer differs only very slightly from observations, so much so that it may easily be considered to be within an acceptable margin of error.

The Synodic Month & the Moon Celestial Equator

As a result of the above discovery, one is thus able ultimately to derive a law of proportion to link directly a change in the moon synodic month to a change in the value of its own celestial equator. Such can be achieved due to the fact that both of these features of the moon orbit have each separately been shown to relate to the Earth tropical year according to a distinctive set of powers.

Thus, it is possible to link them directly without reference as such to the tropical year at all. By combining therefore the law linking the Earth year and moon celestial equator with its 4 to 3 power relationship, and the law linking the Earth year to the synodic month with its 12 to 1/11 power relationship, a general law of proportion linking solely the moon celestial equator to its synodic month may be given:

$$CE^{9}_{(m)} \propto \frac{1}{SM^{11}}$$

Where: CE (m) & SM are ratios:

CE = Celestial Equator, moon (Present) / Celestial Equator, moon (Past)

SM = Synodic Month (Present) / Synodic Month (Past)

Chapter 7

Light Speed & the Ideal Earth Orbit

The Earth Celestial Equator Examined

With the analysis of the Babylonian numerical system and Egyptian calendar culminating in the discovery of a general law of proportion found to govern the simultaneous transformation of the orbital period of the moon (synodic month) and its celestial equator, it is quite natural for one to suppose also that a valid law of proportion of the same type exists for the Earth, governing changes to its own orbital period (tropical year) and celestial equator. In attempting to uncover the combination of powers inherent to such a law though, one is led by necessity to a most critically important discovery, concerning all of the laws of proportion put forth in this work; something that may not have been too apparent in the earlier stages of this investigation.

Static Laws of Proportion vs. Transformative Laws of Proportion

In considering just how the Earth orbital period may relate to the circumference of its celestial equator, one may immediately think that Kepler's Law should reveal the answer. For indeed, it has already been shown in a previous chapter that the planets of the solar system possess different semi-major axis values (the radius of their celestial equator), each dependent upon their precise orbital period, with Kepler's Law of Proportion appearing to govern their relations. It is easy to think then that the ratio between the current Earth tropical year and an ideal of 360 days, could be simply squared and then cube-rooted to produce a ratio specifying the corresponding measure of increase in the Earth's semi-major axis, that must have accompanied the change in its orbital period. By proceeding to then divide the current Earth semi-major axis value by this ratio one may hope then to reveal exactly what its mean distance from the sun would have been, when the Earth was in its ideal 360-day configuration. Although this would seem eminently reasonable, in truth, it would be a fatal mistake to think so; and herein is revealed the critical difference between Kepler's law of proportion, and the others put forth in this present work.

Essentially, Kepler's law is a 'static' law. In its given form it merely describes the character of a gravitational field surrounding a given object. Thus, practically employed, it can reveal the manner in which a gravitational field becomes weaker as one moves further away from the object responsible for generating the field. In the case of the sun, the orbiting planets taken together reveal the nature of the sun's extended gravitational field, each mathematically connected by Kepler's general inverse square law. Consequently, the law is 'instantaneously' correct in relating the planets in orbit of the sun. Yet, critically, it is implicitly assumed when making use of the law that the gravitational field of the primary body does not alter over time. Thus, a planet may in a sense be 'moved in one's own mind' merely as a conceptual exercise, using an inverse square relation to calculate what its mean distance from the sun would be, were it to suffer an increase or decrease in its orbital period by some or other distinct measure. In sum, the law demonstrates how the orbital period and semi-major axis of a celestial body must be altered with respect to one another as a result of an orbital change, when, and only when, *the change in question takes place within a gravitational field whose strength remains fixed throughout*.

In contrast to the above, the laws of proportion that have been stated in this present work differ not just with Kepler's law in that they possess different power combinations, but also, in that they are all *qualitatively* different to Kepler's law. In essence it would be wholly appropriate to consider them Transformative Laws of proportion, for they *do not represent or capture the actions of bodies that are merely manipulated in a static environment*. Instead, they describe the very unfolding of the universe itself as time passes. Thus, the 'instantaneously correct' relations of the type revealed by Kepler's general inverse square law are not present; or rather do not apply. Instead, the truth of the relations revealed by the laws of this present author, transcend physical associations inherent to a static system subject merely to artificial manipulation.

Correct understanding of this point is vital, though it may indeed be difficult to grasp at first. One should consider the example of a planet in a stable orbit about a star. If one chose to calculate in the *instantaneous present* what the orbital period of the planet would be were its semi-major axis twice its usual magnitude, then Kepler's law of proportion would be appropriate, as it would correctly model the gravitational field surrounding the primary body. However, suppose on the other hand, that the semi-major axis of the planet were doubled not through its manipulation in the 'immediate present', but instead by way of the *natural unfolding of the universe* itself. In such a case as this, planetary movement actually takes place as the universe as a whole is transformed; during which time, all existent celestial bodies are individually transformed also. It would not be unreasonable to think that the masses of celestial bodies would alter accordingly as this process unfolds, as so too would their gravitational fields. In this particular instance then, were a planet to extend its orbit by doubling its semi-major axis, such a change would occur alongside the simultaneous change in the strength of the gravitational field of the primary solar body about which it orbits. Consequently, Kepler's general inverse square law could not be employed to determine the planet's new orbital period. It would fail; providing a fraudulent answer. Instead, other laws based upon different power relations would apply; laws not of a static nature, but that capture something of the very transformation of the bodies in question as they move through an unfolding universe.

All of the laws of proportion put forth in this work, each relating the physical magnitudes of the current Earth-moon configuration to their ideals, are laws of exactly this type; being qualitatively different to that which was discovered by Kepler. Indeed, one may consider simply the law relating the synodic month to the celestial equator of the moon as proof of this. As one may recall, this particular law revealed that the moon, when possessed of a 30-day month, simultaneously had also a celestial equator of 1296000 IGM. This very law though could only be true if the gravitational field of the Earth was different at the time in question, to what it is at present. Such a fact must therefore necessitate that the actual extension of the moon celestial equator had to have occurred alongside the simultaneous transformation of the Earth's gravitational field at the very same time. As an example, this then clearly demonstrates that the movement of a celestial object within a gravitational field does not necessarily occur in accordance with a general inverse square law as a rule; not when the gravitational field itself is transformed at exactly the same time. In consequence, it would seem highly reasonable that the determination of a law of proportion linking a change in the Earth tropical year to its own celestial equator, is bound to be composed of power relations other than 2 to 3 as found in Kepler's law; for as the moon itself does not conform to such a general inverse square law, one must suspect that neither does the Earth.

In sum, the circumference of the Earth celestial equator at the time when it possessed a 360-day year cannot be derived using Kepler's law of proportion. A different law must be employed; one as yet to be revealed.

An Ideal Earth Celestial Equator

In order that a precise law of proportion may be discovered linking the transformation of both the Earth tropical year and celestial equator, one must of necessity know the ratio of change for each of these orbital features over the course of a set period. In this instance the period in question is defined as that between the Earth as it currently stands with its tropical year of 365.2421840 days, and that point in its past history when possessed of a 360-day year. The ratio of these two values as stated previously is some 1.014561622. What is sought then is a ratio of change for the Earth celestial equator that corresponds exactly with that for the Earth year transformation, when both are raised by a specific set of powers. With the value for the Earth year already stated, the primary focus here must be then upon the Earth celestial equator; examining its present value, and theorising as to its value when the Earth possessed a 360-day year. Consequently, a thorough account of the essential magnitudes of the current Earth orbit must be given, as was given in the study of the moon.

The Earth Orbital Ellipsoid

From current observations it is well established that the orbit of the Earth about the sun at the present time is just slightly elliptical. The mean distance of the planet's orbit, semi-major axis, has a value of some 81801193.269752711 ideal geographical miles. Moreover, as the Earth orbit is very close to circular, its eccentricity value is itself near to 0; its current value standing at about 0.016708617 [1]. Mathematically, the multiplication of these two values produces a value for the distance between the centre of the Earth orbital ellipse and either of its focal points; one of them being where the sun itself resides:

$$81801193.269752711 \times 0.016708617 = 1366784.808487275 \text{ IGM}$$

The semi-minor axis value is determined by the following formula, as discussed in a previous chapter:

$$\text{SMI} = \sqrt{81801193.269752711^2 - 1366784.808487275^2}$$

Semi Minor Axis = 81789773.930747140 IGM

Adding together both the semi-major and semi-minor axis and multiplying the result by 22/7 returns a value for the circumference of the 'instantaneous' ellipse associated with the Earth's actual orbit about the sun:

$$(81801193.269752711 + 81789773.930747140) \times (22/7) = 514143039.772999531 \text{ IGM}$$

The plane of the orbital ellipse is itself of course perpendicular to that of the Earth's circular celestial equator, which is the primary value of interest here, and calculated more simply by multiplying the Earth semi-major axis by 2 × 22/7:

$$81801193.269752711 \times 2 \times (22/7) = \mathbf{514178929.124159897} \text{ IGM}$$

With the value of the current Earth celestial equator determined then, attention must now turn towards the determination of its value when the Earth possessed a 360-day year, for only by knowing both, is it possible to derive its ratio of transformation, which in turn can then be examined to see if it is connected to the relevant Earth tropical year ratio, via a suitable series of powers.

On the surface, there would appear to be little to aid one in discovering the circumference of the ideal Earth celestial equator. In the case of the moon, the Babylonian units for angular measure proved decisive when used in conjunction with the Egyptian calendar for determining how its celestial equator related to its orbital period, expressed ultimately in the form of a valid law of proportion. Yet for the Earth, there is little aid from such sources. This being so, a certain ingenuity is required to determine just what the ideal Earth celestial equator once was. In essence, what is sought is some physical attribute associated with the Earth and its orbit that would appear in the present to be slightly inharmonious, but were it of another value, it would provide decisive insight into exactly what magnitude was possessed by the Earth celestial equator under ideal conditions.

After very careful consideration, it would appear that there is indeed a physical magnitude strongly related to the Earth whose transformation does provide the critical breakthrough: the speed of light.

Light Speed & the Earth

The primary light source existent within the solar system is without doubt the sun. The light from the sun reaches all of the planets, continuing on even into deep space. Modern scientists have become quite adept in measuring the speed of light here on Earth, determining its rate of propagation even under vacuum conditions, mirroring to the best of their abilities the environment of space. As a result, they have been able to calculate quite accurately the transit time required for the light of the sun to reach the Earth, covering the distance of its mean point of approach i.e. its semi-major axis.

As noted previously, the speed of light has been determined to be about 186282.397 statute miles per second. When converted into ideal geographical miles (186282.397 × 0.88) its value becomes 163928.509 IGM per second. Using this speed in conjunction with the known distance for the Earth's semi-major axis, the time required to cover this distance may be calculated as follows:

$$81801193.269 / 163928.509 = 499.00529 \text{ seconds}$$

Noting that the transit time is almost exactly 500 seconds, one may suspect then that when the Earth possessed its past ideal orbital configuration, that 500 seconds was indeed exactly the measure of time required for light to cover the mean distance between the Earth and the sun, but, as with so many other things (as relate to the Earth) it now deviates slightly from such an exact figure. And yet, if 500 seconds is the correct answer, such a time would still need to be multiplied by an existent speed of light in order that the ideal semi-major axis could be determined; from which the celestial equator could then be derived by multiplying the given answer by 2 × 22/7. However, should such a proposed ideal transit time actually be multiplied by the current speed of light? When considered, the current value, expressed even in ideal geographical miles, seems hardly remarkable; the sequence of numbers denoting the speed none too impressive. Thus, one is led to suspect that not only is the transit time of light now different to what it once was, the same must be said also of the speed of light itself. Essentially, what is proposed here is that when the Earth possessed its ideal orbital configuration in harmony with the moon, the speed of light was of a different value to that which it presently holds, whether propagating either through free space in the general vicinity of the solar system, or even over the surface of the Earth, were it to be measured passing through a vacuum.

The very idea that light speed was somehow different in the past when compared to the present is a bold claim indeed. Is there any evidence to back it up? In general, in the realm of physics, light is thought by many scientists to be somewhat of a constant of nature; fixed in value. Such a belief emerged primarily in the modern era as a result of the development of the Special Theory of Relativity in 1905 by Albert Einstein; a theory upheld by many scientists even today. It is axiomatically assumed in relativity theory that the speed of light is independent of the speed of any object that emits it. Also, all observers are thought to measure light passing them by at the same fixed speed regardless of their own speed of movement [2]. Such are some of the strange effects imputed to light under this theory. Whilst it is not necessary to further expand on these particular points in any greater detail, there are certain aspects of light propagation of great relevance to this present discussion that must indeed be elaborated upon. For instance, it should be noted that light propagates through different mediums at different speeds. In general, the denser the medium is, the slower light travels through it. For example, when measured passing through water, the speed of light is slightly slower than when passing through air [3]. Indeed, when light is therefore measured moving through a vacuum, theoretically a region containing the least amount of matter possible, the speed recorded is considered to be the fastest speed at which it can physically travel. The mathematical symbol in physics used to denote light travelling at this particular speed is C. The most accurate (as currently determined) value for the speed of light propagating through a vacuum is 163928.509 IGM per second, as previously noted.

As light itself must cover the distance between the Earth and the sun in the vacuum of space, it is this particular speed of light i.e. the supposed maximum, which must be of primary importance here. Just how exactly though can it be proven, that during an earlier stage in the unfolding of the universe, the speed of light through space, at least in the general vicinity of the solar system, was of a fundamentally different value to that of this current age? The answer it would appear, is to be found when one considers the relationship between light speed, and gravity. An examination of some of the relevant research in this area is most definitely in order.

Light Speed Measurements & Gravitational Acceleration

During the early part of the 20th century, technological innovations had advanced far enough such that it became possible for researchers to measure the speed of light to a level of accuracy that they had not previously been able to achieve. One of the great pioneers of such research was Professor Albert

Michelson. In 1926, he conducted an experiment to measure the speed of light between two mountains 22 miles apart. Placing a mirror upon each mountain peak, light flashes were observed passing between them. The result yielded a value for the speed of light of 186284 miles per second [4]. However, the experiments went further still. The researchers, including Michelson, Dr Francis Pease and Fred Pierson next turned their attention to measuring the speed of light passing through a vacuum. This was done by measuring its passage through a tube a mile in length and three inches in diameter [5]. The results of this experiment however demonstrated a marked difference in the speed of light. To make certain that this was not in some way related to the tube itself, U.S. Coast & Geodetic Survey controllers were enlisted. Measuring the tube several times though, they revealed no errors [6]. What the researchers had essentially found was that the speed of light was continuously changing over time. On some days when measured, it travelled up to 12 miles per second faster than on other days [7]. Such differences though were not random. Clear cycles were evident. Its speed seemed to change in accordance with the seasons. Moreover, in addition to this even, a strange two week cycle of change in the speed of light was also detected, though its source remained a mystery. In light of these findings, scientists decided to average out all of the readings to establish a mean value for the speed of light. The value in question was determined to be 186271 miles per second [8]. It was only in the decades that followed, that some sort of solution eventually presented itself to account for the continuously changing behaviour of the speed of light. One particular individual, who showed the way forward, was the American Physicist, Dr. Saxl, who conducted a number of experiments using torsion pendulums. As a result of his work, researchers were able to determine that during a New Moon phase when the moon is completely dark to an Earth-based observer, gravitational acceleration on the Earth suffers a slight increase, and that this in turn influences the speed of light [9]. This would appear to indicate then that gravity acceleration and light speed possess a close affinity, such that a change in one results in a change in the other. The truth of this indeed has definite implications when one considers how both gravity and light may be related to one another with respect to the Earth itself as a physical body.

From the study of the Earth over the recent historical period, it has been well established that acceleration due to gravity is known to vary naturally over the whole of the Earth's surface with latitude alone. This is due in large part to the fact that the Earth is an ellipsoid and not a perfect sphere. Consequently, the gravitational pull experienced at the equator is significantly different to that at either pole. As a matter of fact, as one progresses throughout a full 90 degrees from the equator to either the North or South Pole, its value progressively increases. Presently, it has been established that the acceleration due to gravity at the equator of the Earth has a value of about 9.78039 metres per second squared (32.0878937 feet per second squared). In contrast, upon the Earth's poles, its value is significantly greater; about 9.83217 metres per second squared (32.2577755 feet per second squared) [10]. These are the two extremes upon the surface of the Earth. Between them, gravitational acceleration values covering all other latitudes vary, with each possessing its own unique value somewhere within this defined range. Consequently, if gravity is tied in to the speed of light, then just as the former possesses a fundamentally different value upon any given latitude, so too must the latter. Of course, ordered fluctuations from orbital bodies such as the moon would be embedded into any readings, but the mean values derived for light speed should clearly be transformed in accordance with the changes in gravitational acceleration felt at different points upon the Earth as one moved through different latitudes. Has anyone to date been able to demonstrate this though, that the speed of light does indeed vary according to latitude upon the surface of the Earth, and that such variations correspond with simultaneous changes in gravitational acceleration? Indeed they have. In point of fact, the researcher in question who has been able to demonstrate such a link is none other than Bruce Cathie, whose discoveries have been mentioned several times throughout this work.

In one of his books, *The Harmonic Conquest of Space*, Cathie relates how he discovered a means by which he could derive a value for the speed of light at any given latitude upon the Earth's surface, purely by knowing the gravitational acceleration value present at the latitude in question. However, the connection only revealed itself to him though when light speed was expressed in nautical miles per grid second, and, when gravity acceleration was itself expressed in terms of 60 nautical miles per grid second squared [11]. Without going too deeply into the technicalities of his procedures, what he demonstrated was that the magnitudes of both light speed and gravity acceleration were inversely related to one another, such that a change in the acceleration value of gravity resulted directly in a change to the speed of light. As the value of one increased, the value of the other decreased in direct proportion, and vice versa [12]. Moreover, through his studies, Cathie appears to have refined the known gravity values at both the equator and the poles; this being in addition to determining the light speed values associated with each acceleration value itself. By using his method to convert from one to the other, he determined that gravity acceleration at the equator is about 32.08789707… feet per second squared, and that the accompanying light speed value existent at this latitude upon the Earth's surface was 143791.3643… nm/g sec. In contrast, at the poles, gravity acceleration was refined to a value of about 32.2578977… feet per second squared. The corresponding light speed value here was shown to be 143033.5767... nm/g sec [13]. Now, as may be noted from these light speed values, both of them, especially the latter, significantly differ from the speed of light passing through a vacuum, at about 143888.868 nautical miles per grid second. This being said, the differences are marginal and not too great. Indeed, in his book Cathie does demonstrate how they may be linked together in a harmonic sense. Thus, he does provide some sort of connection between the light speed values derived from gravity and the actual speed of light (C) measured though a vacuum. For a full account of the technical details of Cathie's light-gravity transformations one should of course consult his books [14]. It is enough to state here though in view of his work, that on principle, a valid physical connection between gravity acceleration and light speed does appear evident. And from this, it must follow, that were gravity acceleration in the vicinity of the Earth fundamentally different in a prior age, then the speed of light also would have been different. Combining this with the previously established finding that a given change to the moon orbit *in a transformative manner* is not governed by a general inverse square law, it is quite clear that when the Earth was in its ideal state, its gravitational field must have been significantly different to that of this current age. And consequently, the speed of light in the vicinity of the planet at such a time must also by necessity have been different.

An Ideal Speed of Light at the Equator of the Earth

The question that must now be posed therefore is just what exactly was the speed of light under an ideal Earth configuration? The key to solving this, it would seem, appears to be had from a more in-depth examination of the thoughts of Cathie upon what he regards as the theoretical maximum for light speed. Already it has been stated that the fastest known measure for the speed of light (using Cathie's preferred units of measure) is about 143888.868 nautical miles per grid second, as measured in a vacuum. However, as noted in a previous chapter, Cathie does not appear to regard this as its true maximum. Instead, he proposes that the true maximum speed of light is 144000 nm/g sec; a value that he associates directly with the surface of the Earth at the equator [15]. When one considers this idea most carefully though, especially in light of the findings of this present work, one is bound to suspect that the number 144000, should not actually be associated with the Earth in its current configuration at all; but instead with it's ideal. In exploring this possibility, one is thus led to develop a counter argument against Cathie's theory that the speed of light maximum associated with the equator of the Earth (at this present time) truly is 144000 nautical miles per grid second.

There are several points indeed that one should consider with regard to this.

Firstly, one may recall the 3-4-5 extended triangle examined earlier, with its side lengths of 360, 288, and 216. The very analysis of the triangle clearly indicated its intimate association to the ideal Earth configuration, not that of its current state. The longest side of 360 units was shown to be linked to the ideal Earth tropical year, which indeed is transformed in direct proportion to the Earth's equatorial circumference; itself linked to the value of the triangle's shortest side, 216. Yet further to this though, there is the remaining side length of 288. When halved, one can clearly see that its value is 144, a harmonic of 144000 using a simple base 10 transformation (144 × 10 × 10 × 10). One is bound to suspect then that light speed, expressed by the number 144000, must be associated not with the current Earth at all, but the ideal. And, that any such association that it may have, would not be with the pole to pole circumference of such an Earth, but rather with its equatorial circumference, whose value has been proven to be 21600 ideal geographical miles, each of which are 6000 feet in length. In addition to such points, one must also concede that the nautical mile itself as a unit of distance is a derivative of the size of the current Earth, using the elliptical circumference of the planet as its base. However, with the size of the ideal Earth upon the equatorial plane shown to have been less than that of the present age, one must suspect that its pole to pole circumference would also have been smaller [16]. As a result, the nautical mile would not and could not even exist as a unit length of 6076 feet, based upon or derived from the physical size of the ideal Earth in any way.

When all of the above points are carefully considered, one is led to think almost naturally, that the speed of light upon the equatorial plane of the Earth in its ideal state under vacuum conditions, must have been, using at least Cathie's preferred units of time, grid seconds, none other than 144000 ideal geographical miles / grid second (162000 IGM / standard second); a value significantly lower than the current speed of light measured in a vacuum. Indeed, for comparison, employing the same units of measure, the current speed of light is 145714.230 IGM / grid second. Quite clearly then, if the proposed ideal of 144000 IGM / grid second is correct, then the maximum speed of light for this current age is greater (faster) in absolute terms - at least in the general vicinity of the solar system under vacuum conditions - than in the age when the Earth and moon were each in their ideal states. The truth of this certainly would seem to fit most harmoniously with all the evidence examined. Even so, this is not to suggest that the associations that Cathie revealed linking gravity acceleration to the speed of light over the Earth's surface are wrong, nor to suggest that the harmonic linkages that he makes between the light speed values he derives from gravity, and the speed of light measured in vacuum, 143888.868 nm/g sec, which is slightly larger, are incorrect. However, it is hereby suggested that there is no physical link at all between the value 143888.868 nm/g sec, and a slightly higher value of 144000 nm/g sec; the reason being, that whilst the former is valid as a real physical magnitude – one that can be experimentally measured – the latter value has no physical validity at all. Thus, the idea of a theoretical light speed maximum of 144000 nautical miles / grid sec associated with the equator of the (present) Earth would seem simply to be without foundation. Instead, the true significance of the number 144000, as representative of the measure of the speed of light, is realised only when associated with units of *ideal geographical miles* (6000 ft), grid seconds, and critically, with the Earth in its ideal state.

The Transformation of Light Speed: Ideal to Present

If the transformation of the moon orbit, as specified previously, did indeed coincide with the transformation of the gravitational field of the Earth, one would expect that as there is an intimate link between gravity acceleration and light speed, then a lawful transformation of the latter at the surface of the Earth must also have occurred at the same time. By way of a transformational link uncovered between the speed of light and the equatorial circumference of the Earth, this can actually be proven quite decisively:

1) Current Earth physical equator divided by ideal:

$$21914.531045178562 / 21600 = 1.014561622461970$$

2) Above answer to the 9th power:

$$1.014561622461970^9 = 1.138953184475610$$

3) Reduced to the 11th root:

$$\sqrt[11]{1.138953184475610} = 1.011898372868578$$

4) Current speed of light in vacuum (ideal geographical miles per second) divided by this ratio:

$$163928.509 / 1.011898372868578 = 162000.961$$

5) Converted into ideal geographical miles per grid second:

$$162000.961 \times (8 / 9) = 144000.854$$

As may be seen from this analysis, by employing *exactly the same combination of powers i.e. 9 and 11, found previously to govern the very extension of the moon orbit,* one is able to uncover a most exacting relationship between the transformation of the speed of light at the equator of the Earth, and also an increase in the actual size of the planet. From this it is clearly seen that the transformation of the Earth's gravitational field, as evidenced by the extension of the moon orbit, is most precisely linked to an increase in the size of the Earth, which in turn is also associated with the transformation of the speed of light itself, employing exactly the same combination of powers. As a result of this, one is thus able to propose the existence of a further law of proportion:

$$LS^{11} \propto PE^{9}_{(e)}$$

Where: LS & PE (e) are ratios:

LS = Light Speed (present) / Light Speed (past)

PE (e) = Physical Equator, Earth (present) / Physical Equator, Earth (past)

NB: Speed of light is that taken through vacuum

Further to this, one may also link the speed of light to the Earth tropical year using exactly the same powers, for a change to the Earth's physical equator is directly proportional to its tropical year. Therefore the following law is also valid:

$$TY^9 \propto LS^{11}$$

Where: TY & LS are ratios:

TY = Tropical Year (present) / Tropical Year (past)

LS = Light Speed (present) / Light Speed (past)

The existence of these laws of proportion would appear to strongly confirm then that the speed of light present at the equator of the ideal Earth was exactly 144000 IGM per grid second, or 162000 IGM per standard second. Moreover, using the above established ratio of increase for the Earth tropical year, it is now possible to derive a most accurate value for the speed of light for the current age, using the expressed law:

$$\text{Lightspeed (present)} = \sqrt[11]{(365.2421840 \ / \ 360)^9} \times 144000$$

= **145713.36569** IGM / grid second, or

= **163927.53640** IGM / second

The Transformation of the Earth Celestial Equator

With the above analysis complete, in returning to the issue of the determination of the ideal Earth celestial equator, one is now able to propose a value for the key measure of the ideal Earth semi-major axis, critical to the overall calculation. By combining an ideal light speed of 162000 IGM/sec with the previously proposed ideal light transit time of exactly 500 seconds between the Earth and Sun along the full length of semi-major axis separating them, the length of this latter component is determined as follows:

Ideal Earth semi-major axis = 162000 × 500 = 81000000 IGM

The actual circumference of the ideal Earth orbital celestial equator can be determined by simply multiplying this distance by 2 × 22/7, which gives the following:

= 509142857.142857142857142857... IGM
509142857 & 1 / 7 as an exact fraction

With this answer given, it is here then, that this present discussion must return to the central issue of this chapter; namely the determination of a valid physical law of proportion governing the simultaneous transformation of the Earth celestial equator and Earth tropical year. What is therefore sought is a suitable combination of powers to relate the critical ratios of increase of the two relevant variables:

Tropical Year ratio of increase
= 365.2421840 / 360 = 1.014561622461970…

Earth Celestial Equator ratio of increase
= 514178929.12415 / 509142857.14285 = 1.009891274935199214…

It is here that one must speculate to a degree.

Although it is by no means certain, there is indeed a power combination that would seem to link the above ratios of increase; the strength of which does suggest the active presence of a physical law of proportion. Formally expressed, it is as follows:

$$CE^{47}_{(e)} \propto TY^{32}$$

Where: CE (e) & TY are ratios:

CE (e) = Celestial Equator, Earth (present) / Celestial Equator, Earth (past)

TY = Tropical Year (present) / Tropical Year (past)

Although the expressed powers do seem somewhat high, this does not imply the falsity of the law by any means. Indeed, one may note of the above law that it differs only very slightly from Kepler's static general inverse square law. For best comparison, one may present Kepler's law alongside the above in a slightly modified form using a higher series of powers, such that one still preserves the expressed mathematical relationship. Thus can the simple powers of 3 and 2 employed by Kepler each be multiplied by 2×2×2×2, capturing the inverse square law with a combination of powers that are much closer to those employed by the expressed transformative law:

$$CE_{(e)}^{48} \propto TY^{32}$$

Upon comparison, the actual difference between the two laws rests upon merely a slightly reduced CE transformation with respect to that of the tropical year. It is precisely this though that allows for the ratio of increase of the Earth CE to match that of its TY increase with a margin of error much smaller than were Kepler's law itself employed to relate the two ratios of change. Actively employed, the newly stated transformative law in equation form returns a value for the current Earth celestial equator as follows:

Current Earth Celestial Equator =

$$= \sqrt[47]{(TY_{(p)} \,/\, TY_{(i)})^{32}} \times \text{Ideal Earth CE}$$

$$= \sqrt[47]{(365.2421840 \,/\, 360)^{32}} \times 509142857\frac{1}{7}$$

= **514178995.31251** Ideal geographical miles

Compared with the value for the Earth celestial equator given previously, the difference is only very slight, thus supporting the stated law of proportion:

514178995.31251 – 514178929.12415 = 66.1883 IGM

Chapter 8

The Ideal Earth Form: A Perfect Sphere

Temperature & Physical Transformation

So far the laws of proportion that have been put forth allow only for a descriptive understanding of the transformation of the Earth and moon, detailing only the relative changes of various spatially extended magnitudes associated with their respective forms and orbits, albeit as wavelength components linked to the fundamental frequencies of the noted bodies. What one lacks though is actual insight as to their true physical state, as naturally formed entities. The reason, is that the frequency measures found to be associated with them say little in and of themselves concerning the physical condition of the noted bodies, being in effect somewhat of an abstraction. What is thus needed is a way to link the discovered frequencies to something in nature, allowing for the development of a far greater awareness of their inherent properties.

That something is temperature.

Careful study would seem to indicate that the frequency values associated with the noted bodies, particularly the Earth, appear to be strongly related to certain naturally occurring intervals embedded within various well established temperature scales. Such a fact is of great importance, when one considers that the physical phenomenon that is temperature is fundamentally tied in to the actual transformation of the very elements existent in nature. Numerous studies of the basic atomic elements over the years have readily confirmed that each possess a unique set of temperature thresholds, which when breached, affect a fundamental change in their physical state. If one were thus able to relate such changes at the atomic level, to the values and ratios inherent to the numerical progression of the sexagesimal system, particularly relative to the Earth, then it should be possible to reveal something of the actual physical state of the planet in an elemental sense, as opposed to knowledge simply of its form. In order that this idea may be fully explored, a broad understanding of certain well known and widely used temperature scales employed in the sciences today is required.

Temperature Scales and their Basis

In the physical sciences, the most commonly used temperature scales are the Centigrade and the Kelvin scales. They actually use the same fundamental unit, known coincidently, as the degree. This unit is indicative of an identical change in temperature for both systems. The main difference between the two scales is simply that they are each set according to a different starting point. In the case of the former, the Centigrade scale is established in accordance with an important phase change in the state of water; a value of 0 degrees set against the point wherein water is in a transition state hovering between being a liquid and being frozen. Water is of course one of the most abundant natural substances upon the planet, and has been identified as a molecule two parts Hydrogen and one part Oxygen (H20). The degree unit itself, as a temperature measure, is actually derived from a combination of two phase-change states of water. As is quite well known, heating up water in the form of a liquid will eventually result in it boiling and turning into a vapour. It is the naturally occurring transition points of water hovering between being frozen/liquid and liquid/vapour that are used to establish the degree unit. These two distinct phase-change states of water are divided up into 100 equal units of measure, each individual unit being one single degree. However, though the centigrade scale is fixed at 0 degrees with respect to a change in the state of water; physically there are far lower temperatures that can be obtained by certain substances. Thus, in describing the measure of certain temperatures using the centigrade scale, the numbers can go into minus figures. Progressing in such a manner, scientists have been able to establish that in nature, the lowest known temperature that can be achieved, at least theoretically, is approximately –273.16 degrees centigrade. This is known as Absolute Zero, a point at which it is theorised there is a complete absence of heat [1]. Such thinking was established from studies of gases wherein it has been found that when a fixed volume of gas is cooled, there is a reduction in both its temperature and pressure. At zero pressure, the temperature of a gas would be that of absolute zero [2], and a 'perfect' gas would at this point possess zero volume [3]. It is here also, upon the point of absolute zero, that the Kelvin temperature scale is established; set at 0 degrees [4]. Thus, –273.16 degrees centigrade is equivalent to 0 degrees Kelvin.

An Ideal Absolute Zero

Upon seeing the value in degrees ascribed to the physical state of absolute zero using the centigrade scale, one is bound in light of all the previous elements of this work to be somewhat curious about it. Indeed, if one were to take the reciprocal of this value and examine it carefully, one may very well suspect that perhaps this value, like so many others that have been studied, is ever so slightly out of harmony:

$$1 / -273.16 = -0.00366085$$

Were one to multiply the value for absolute zero given in degrees centigrade by the ratio of change for the Earth tropical year - that which transformed it from an ideal of 360 to its currently value - one can see that the reciprocal is close to an exact value of –0.0036, a simple harmonic of 360, the very frequency of the ideal Earth:

$$-273.16 \times 1.014561622 = -277.13765$$
$$1 / -277.13765 = -0.003608315$$

From this, one cannot help but wonder if the fundamental measure of zero temperature, relative to the frozen/liquid transition state of water, has somehow been transformed over the course of time in direct proportion to the Earth tropical year. Could there at one time have been exactly 277.77777... degrees (277 & 7/9) separating the condition of absolute zero from the frozen/liquid state of water, as opposed to the current 273.16 degrees? In light of previous analyses this is not an unreasonable position to hold at all. Working then from this premise, and with reference to the Earth tropical year, one may derive a new value for the current standard of absolute zero:

$$-277.7777777\ldots / (1.014561622\ldots) = -273.7909375 \text{ degrees}$$

Considering such a fundamental change, one should not expect it to be restricted to the locality of just the solar system, for indeed, temperature related physical transition states among the elements are thought to hold constant throughout the entire universe when set against absolute zero. This is suggestive of the fact that the changes to the Earth and moon may well have occurred during a major transformation of the wider universe itself. Such an argument is not without support when one considers the previous study of light. The very fact that light appears to have undergone a general increase to achieve its present value is something that one would assume would apply to the universe as a whole and not just the immediate solar system. And if this is true of light speed, then why not the physical properties of all the elements?

An Ideal Highest Melting Point in Nature

It would seem undeniable from the above, that the temperature separation expressed in degrees between absolute zero and certain key phase-changes in the state of water are of great significance; a fact that quite naturally leads one to consider if there are other similar elemental phase-changes to be found in nature which may also be relevant to this present work. With recognition of absolute zero as an important fundamental marker in nature, one would assume that it should bear some harmonious relationship to yet higher levels of temperature. Such a view would not appear to be in error, for indeed the actual reciprocal of the noted ideal temperature separation between absolute zero and the frozen/liquid state of water, almost naturally leads one on to discover the next element, the like of which almost intuitively, would seem to have a most profound association to the ideal Earth state.

Of all the elements studied over the past few centuries, and all the tests to determine their innumerable physical properties, including temperature phase-change threshold points, there is one in particular that is of special significance: carbon. Being one of the key elements that make up the human body carbon is a critical ingredient of life itself. Of great importance to this present work though is the fact that carbon possesses the highest melting point of all of the elements in nature, whose value is about 3550 degrees centigrade; this being the melting point of carbon in the form of diamond [5] – a mineral of pure carbon, and the hardest natural substance known to exist [6]. In noting the temperature of its melting point, one cannot help but see how close it is to an exact value of 3600, leading one to consider that the element may have undergone some sort of physical shift over time, much like the temperature separation between absolute zero and the frozen/liquid state of water. Indeed, suspecting this, were one to multiply the current melting point of carbon by the ratio of decrease for the noted temperature separation in question (identical to that governing the increase in the Earth tropical year), the answer is very close to a precise value of 3600:

$$3550 \times 1.014561622... = 3601.69375 \text{ degrees centigrade}$$

The answer here must of course lead one to consider that there is a principled relationship between the current melting point of carbon and what must be an ideal melting point for this element, this latter being physically existent in some past age. And once more, the age in question could be none other than that when the Earth possessed a 360-day year. Thus, the very transformation of the carbon atom, along with the other noted temperature shifts, must have occurred simultaneously in line with all of the other physical and orbital changes discussed, each shown to be governed by a unique set of laws (of proportion). Holding to the view then that the true and exacting melting point of carbon in its ideal state was once precisely 3600 degrees centigrade, and also that the ratio of its transformation is identical to that shown to relate to the Earth year change, then a more refined value for the current melting point of carbon may be determined:

$$3600 / 1.014561622... = 3548.33055 \text{ degrees centigrade}$$

With the idea put forth then that carbon as an atomic element once possessed an ideal melting point of 3600 degrees, one must ask how such a physical fact as this may be related to the Earth itself under ideal conditions, where it possessed a harmonic of this measure in the form of its orbital period/ frequency, i.e. its 360 days per year. In the former case, one is speaking of a microscopic entity; in the latter case, a macroscopic entity upon a celestial scale. Is there a valid physical connection between the two?

Carbon – Its Form and Properties

The idea which would seem to present itself in light of the above is that the Earth, as a formed body, may in some way be a 'giant carbon atom'; the celestial equivalent so to speak of the microscopic entity. Of course, at the present time given both an increased Earth year and a decreased melting point for the carbon element on the atomic scale, if the Earth were indeed a macroscopic carbon body then one would not be inclined to think it of exacting 'purity' in its current state, as it too has undergone its own celestial changes. And yet, in the past, when the planet did once possess exactly 360 days per year, at which time it is proposed that the atomic melting point of carbon was itself precisely 3600 degrees C, is it not conceivable that the Earth truly could have been at such a moment in time, the real celestial equivalent of precisely this very element, existing in pure form as a large macroscopic entity? There is the suspicion that perhaps this is so...the suspicion. Yet is there any evidence that could be presented to prove the truth of such an idea? Quite remarkably, it would seem so. An examination of the work of one Walter Russell, a man who studied the elements of nature most deeply during the early part of the 20th century, would appear to reveal strong support for such a contention. It is highly appropriate therefore to introduce this man, and to relate those works of his that bear strongly upon this present discussion.

Although there are many who have studied the fundamental elements of the natural world over the centuries, Walter Russell stands out well beyond the crowd, due essentially to the almost unique brilliance of his discoveries. Indeed, during his lifetime Russell was regarded by many as a sort of universal genius, who excelled in practically every field of endeavour that he turned his attention to. Concerning his work in the field of the natural sciences, he was nothing short of divinely inspired in developing his ideas as to both the structure and function of the universe as a complete entity. Yet moreover, he was also mindful to speak most crucially of man's place within its bounds. Taken as a whole, the scope of Russell's writings is quite extensive, and it is difficult to give here a complete rendition of all of the key concepts that he developed relating to the physical world. However, those concepts and ideas that are most relevant to this present work shall be relayed. Indeed, upon this point one may rightly ask just what it is that makes the work of Walter Russell so important to this present investigation, in comparison to say, the work of others who have also studied most extensively the natural elements. The answer quite simply is that whereas most scientists who study the elements tend to focus merely upon their chemical aspects taken almost in isolation, Russell provides crucial insight into their geometrical formation. It is an understanding of the geometry of the elements combined with that of their physical properties that makes Russell's work of critical importance to this present discussion. Indeed, concerning most specifically the carbon element introduced above, Russell has much to say of its geometry. Before proceeding to such specifics though, a more general overview of some of Russell's ideas concerning the universe shall be given.

Overall, the universe, as composed of matter, was held by Russell to be one of motion; indeed, that matter itself is *motion* [7]. However, Russell sets the condition of motion, and thus of matter, against the condition of absolute rest or stillness. All motion is said to originate from a condition of stillness, and ultimately to return to such a condition; a process repeated eternally, without end. Indeed, a cycle of alternation between these two conditions forms the very activity of the universe according to Russell, with matter and motion coming into temporary existence to form the universe, only to withdraw from existence into the realm of stillness. Moreover, the very condition of absolute rest is held to be a realm invisible to the senses; one that Russell characterises as being an omnipresent universal vacuum [8]. It is a realm that exists in a state of complete equilibrium; of zero tension. Indeed, the very division of this realm to produce tension is what gives rise to the condition of motion and to the emergence and formation of matter. Matter itself may be seen then as something of an abnormal condition; one that requires effort to establish, but that returns to a state of rest without any extra effort required at all. Thus, every particle of matter be it large or small, atomic or celestial, is in a continual state of alternation between a condition of motion and rest; a cycle that is repeated eternally at many levels, such that microscopic pulsations of matter may enter and withdraw from existence having a life span on the order of the merest fraction of a second. Yet by contrast, the measure of the existence of great celestial objects such as suns may exceed millions or even billions of years [9]. All objects regardless of apparent size thus constitute the 'stuff' of the universe, and are visible to the senses. Outwardly, all matter is the thought-form of an idea contained within the invisible realm of stillness that is given temporary expression in the realm of motion. Thus, all of the elements of the periodic table are each an expression of a particular universal idea, formed by being projected from the condition of universal stillness to become imaged motion that one labels matter.

In further considering the differences between the two fundamentally opposed states of existence, rest and motion, Russell states also, that there is an associated geometry characteristic of each realm, the former being that of zero curvature, the latter, that of curvature. Indeed, Russell holds to the view that the zero curvature geometry of the realm of rest is cubic in its essential form. Cubes appear in his system to be the invisible templates from which curved forms are projected into existence. The cube itself he specifically states is composed of invisible magnetic white light [10]. Of crucial importance though to this present work is his notion that there is unity between both the cube and the sphere. According to Russell, both forms are indeed one. Cubes are in point of fact frozen spheres, and by contrast, the sphere is an incandescent cube [11]. Thus, in the realm of matter and of motion and curvature, the most extreme form of expression in opposition to the idea of a cube is that of the sphere. Indeed, it is the express view of Russell that the true sphere produced in nature is the maturity point of an imaged form [12]. It is precisely this very point that allows for an understanding of the geometry of the carbon atom and how it fits in with Russell's system overall; and indeed, why it is this particular element that has the highest melting point of all the elements. Essentially, the carbon atom is exactly the element Russell identifies in nature as actually manifesting in the form of a perfect sphere; one that reaches the highest point of maturity attaining also the condition of being a perfectly balanced element: a compressed cube in the form of a hot sphere.

The Geometry of the Ideal Earth Form

From Russell's work one can clearly see the significance of carbon; as the most mature element in nature, and that possessed of the form of a perfect sphere. Indeed, as a fact of nature one may well regard this as something quite profound, even remarkable. However, what is even more so is the fact that the ancients themselves would also seem in some way to have been cognisant of this. That is not to say of course that they knew of the existence of the carbon element like modern-day scientists do. However, from their writings there are certain subtle indications that they were aware of a link between the value given to denote the ideal melting point of carbon, and the condition of perfect circularity. Indeed, in the Sumerian language, the value 3600, depicted as a large circle called a SAR, was viewed as a "princely" or "Royal" number [13], whose actual stated meaning was, "a perfect circle" or "a completed cycle" [14]. Given this fact, it is not outside the realm of possibility that as the number 3600 is a simple harmonic of 360, that both values may have been held by the Sumerians to have been indicative of exactly the same physical state of being; one whose characteristic form was that of a sphere. The very truth of this would of course offer a significant degree of support for the idea that the physical Earth itself when possessed of a 360-day year was of exactly this form; being manifest at the maturity point of its own existence; the celestial equivalent of the carbon atom, so to speak. Such indeed would tie in well with the ancient belief in celestial harmony. However, one would of course require proof. Evidence of a most decisive nature would need to be produced to firmly establish that the form of the past ideal Earth was indeed that of a perfect sphere. Quite remarkably, such evidence does appear to exist, and once again, it is in the form of a law of proportion, linked most exactingly to those already stated.

A Once Spherical Earth

Previously it was stated that in its current ellipsoid form the centre to pole radius of the Earth, or semi-minor axis of the planet, is significantly shorter than its equatorial radius, or semi major axis, by approximately 12 IGM. If the Earth were once a perfect sphere, both would of course have to be identical. Now, as the equatorial radius of the Earth under a 360-day tropical year has been shown to have been 3436 & 4/11 IGM (21600 / (2 ×(22/7))), if the planet were perfectly spherical at exactly this time, then its centre to pole radius must have been of this value also. One may conclude then that the transformation of the Earth polar radius from an ideal of 3436 & 4/11 IGM to its current value could not possibly have occurred in direct proportion to the Earth year change, for if it had, then it would be of the same value as the equatorial radius, which evidently it is not. Consequently, the transformation of the polar radius must therefore be of a more complex nature than the equatorial radius. Careful analysis would appear to confirm this fact. Indeed, close scrutiny of the relevant values of the current Earth compared with those of the suggested ideal do appear to reveal quite decisively a most exacting relationship between the two noted axis components, the like of which is indicative of the presence of yet a further law of proportion. The analysis begins as follows:

Current equatorial radius = 3436 & 4/11 × 1.014561622…
= 3486.402666… IGM

With this established, one must next determine the length of the semi-minor axis component of the current physical Earth form i.e. its centre to pole radius. This is best achieved by calculation, through making use of what astronomers refer to as the measure of 'flattening' of the Earth. Essentially, this is another means of simply expressing the eccentricity of the Earth when viewed 'side-on' as a two-dimensional elliptical object. In most reference books the actual measure of flattening (f) of the Earth is given as a fraction, its value being: $f = 1 / 298.257$ [15]; the flattening measure itself derived from a very precise relationship between the equatorial radius of the planet and its centre to pole radius i.e. the semi-major and semi-minor axis components of the planet's physical form:

$$f = (\text{SMA} - \text{SMI}) / \text{SMA}$$

Where: SMA = Semi-Major Axis SMI = Semi-Minor Axis

Transposition makes SMI the subject of the formula:

$$f = (\text{SMA} - \text{SMI}) / \text{SMA}$$
$$f \times \text{SMA} = \text{SMA} - \text{SMI}$$
$$(f \times \text{SMA}) + \text{SMI} = \text{SMA}$$
$$\text{SMI} = \text{SMA} - (f \times \text{SMA})$$

SMI calculated as follows:

SMI = 3486.402666 – ((1 / 298.257) × 3486.402666)
Semi-Minor Axis = 3474.713409…IGM

From the above one may thus table all of the relevant values:

	Ideal Earth	Current Earth
Semi-major axis	3436 & 4/11	3486.402666
Semi-minor axis	3436 & 4/11	3474.713409

All values in Ideal Geographical Miles

From a careful study of the noted values an almost incredible association presents itself, that the transformation ratios of each component appear to be linked by a set of powers identical to those already found to govern the relative transformation of the moon celestial equator and the Earth physical equator: namely a 3 and 4 power combination. In the form of an equation, the relationship is detailed as follows:

$$(\text{SMA} / \mathit{SMA})^3 = (\text{SMI} / \mathit{SMI})^4$$

Where:

SMA = Semi Major Axis (present)
SMI = Semi Minor Axis (present)
SMA = Semi Major Axis (past)
SMI = Semi Minor Axis (past)

Using this equation, it is possible to derive a measure for the current centre to pole radius of the Earth that very closely matches that derived using the expressed flattening measure formula:

$$\left(3486.402666 \,/\, 3436\tfrac{4}{11}\right)^3 = \left(\text{SMI} \,/\, 3436\tfrac{4}{11}\right)^4$$

$$\text{SMI} = \sqrt[4]{\left(3486.402666 \,/\, 3436\tfrac{4}{11}\right)^3} \times 3436\tfrac{4}{11}$$

Semi Minor Axis = 3473.825009…IGM

When compared to the value given previously, the difference is less than 1 ideal geographical mile:

3474.713409 – 3473.825009 = 0.8884 IGM

As a result of such a close match between these two values one may posit the existence of yet a further law of proportion; one that would appear to actively describe the complete physical transformation of the planet, dictating thus the manner in which the Earth suffers expansion (or contraction?) upon the plane of its equator, precisely relative to the length of its axis:

$$\text{SMA}^3_{(p)} \propto \text{SMI}^4_{(p)}$$

Where: SMA (p) & SMI (p) are physical ratios:

SMA (p) = Semi Major Axis (present) / Semi Major Axis (past)

SMI (p) = Semi Minor Axis (present) / Semi Minor Axis (past)

The reality of this law would appear to confirm then that the Earth at that point in its history when possessed of a 360-day orbit, was also perfectly spherical in its form, as would be required were the Earth to manifest as a 'macroscopic carbon atom'. The subsequent transformation though of the tropical year led to the planet losing its exacting spherical state, becoming the deformed ellipsoid that it presently is. What is of course interesting to note in this is that the very reality of this law suggests a most profound physical dynamic between the extension of the orbit of the moon and the outward form of the Earth. Indeed, the proportional increase of the moon semi-major axis with respect to the physical equatorial radius of the Earth would appear the very inverse of the relationship between this latter noted component and the physical polar radius of the planet. The 3 to 4 power relationship is reversed respecting purely the physical axis values of the Earth, as compared with the physical and celestial axis values of the Earth and moon.

A Refined Nautical Mile

In view of the completed analysis of the physical transformation of the Earth form, one is able to end this section by returning to a point first raised in the previous chapter, concerning the actual validity of the nautical mile unit as it pertains to both the current Earth form, and to the suggested maximum speed of light of 144000 nautical miles per grid second. What allows for a deeper consideration of these matters at this time is the fact that the newly stated proportional law as given above, now enables one to conduct a principled and exacting calculation of the full measure of the elliptical circumference of the Earth, and thus derive what may be a definitive measure for the nautical mile unit as it relates to the Earth.

In the previous chapter it was noted that the researcher Bruce Cathie placed great emphasis upon the nautical mile as being an important unit of measure linked to both the speed of light and gravity acceleration. However, as mentioned at the beginning of this work, the actual length of the nautical mile itself is not one upon which there appears to be principled consensus. One may cite the fact that the international nautical mile (6076.11549 ft) that was referenced earlier has certainly not been adopted by all countries. Moreover, Cathie himself does not seem to employ it either in his work, appearing to make use of an exact figure of 6076 ft for many of his own transformations [16]. Indeed, the truth of the matter is that depending upon the exact elliptical circumference of the Earth that one chooses to accept, one may derive a variety of different values for the nautical mile, all of which would hover close to 6076 ft. That said, in light of the proportional law now found to govern the relative transformation of both the polar radius and the equatorial radius of the Earth, it is now possible for one to derive a measure for the nautical mile unit, which may trump all others in terms of its validity and truthfulness. Using the Earth radius values generated by the above stated law, a measure for the elliptical circumference of the Earth may be had, that further allows one to derive a value for the nautical mile unit itself:

Current Equatorial Radius = 3486.40266627…IGM
Current Polar Radius = 3473.82500901…IGM

Therefore, Elliptical Earth circumference =

$$(3486.40266627 + 3473.82500901) \times (22/7) = 21875.001265\ldots \text{IGM}$$

And, 1 Nautical Mile =

$$(21875.001265 \times 6000) / 21600 = 6076.38924032\ldots \text{ feet}$$

Upon first glance, one must confess that the newly derived value is somewhat unremarkable. What then, is to be made of it? Is it a more significant, more definitive measure, than such as the internationally recognised nautical mile, or a value of precisely 6076 ft? On this point it is interesting to contrast the nautical mile so generated with the ideal geographical mile of 6000 ft. Concerning this latter unit it has been demonstrated quite conclusively that not only does it possess physical validity with respect to the ideal Earth, but also, it appears to retain its physical validity even in the current age; the fact that the present day semi major axis of the moon expressed in IGM contains the very ratio of its increase being strongly suggestive of this. By contrast, the nautical mile so determined above is derived solely from the elliptical circumference of the present Earth. Under the ideal spherical Earth such a measure does not exist separately from the IGM, as both are identical. As a result of this, one is bound to suspect initially that the nautical mile based upon the current Earth has no association whatsoever with the ideal Earth or with the primary values or close derivatives of the sexagesimal system at all. However, this is in fact not true. Indeed, it would appear that there is a profound harmony to be had between the nautical mile of 6076.38924032 feet and the Babylonian values. Such is revealed by simply multiplying the nautical mile value by 8 and then dividing it by the number 7 to uncover a very intriguing association to the ideal light speed harmonic of 144:

6076.38924032 × (8 / 7) = 6944.44484608…

1 / 6944.44484608 = 0.00014399999…

The linkage is not exactly precise, admittedly. But such perfection can indeed be achieved were the elliptical circumference of the Earth taken to be exactly 21875 IGM. With this slight correction, one may derive a nautical mile in perfect harmony to the 144 value:

1/ 0.000144 = 6944.4444444…

6944.4444444… × (7 / 8) =

6076.3888888… (6076 & 7 / 18 as a perfect fraction)

In view of the above analysis one is able to suggest then that the value of precisely 6076 & 7/18 ft is more truthfully representative of exactly 1 nautical mile with respect to the Earth, than any other value, which is to imply that it possesses a physical validity that the others lack. Moreover, as this value would appear to be tied in to the ideal measures linked to the ideal Earth, whilst also maintaining its validity with respect to the current Earth, as is the case with the Ideal Geographical Mile, it would seem most appropriate to so designate the unit value of 6076 & 7/18 ft as the Ideal Nautical Mile or INM.

Chapter 9

The Lost Harmony of the Sun & Moon

Extended Moon Orbital Transformations

Up to this point much has already been said of the transformation of the orbit of the moon, concerning both the increase to its semi-major axis and the decrease of the synodic month. However, there is yet a further aspect of the moon orbit also worthy of note: the inclination of its orbital plane, and how it intersects that of the Earth's own orbital plane about the sun. Indeed, this is a matter of most critical importance when concerned with timing the passage of solar eclipses, wherein a momentary conjunction occurs between the sun, Earth and moon, with the latter of these bodies briefly passing precisely in between the other two. Such exacting events as these are of course to be clearly distinguished from what may be termed merely apparent conjunctions, as with the timing of regular successive synodic months themselves, which rest simply upon a 'plan view' conjunction of the relevant bodies. Indeed, it is precisely due to the fact that the orbital plane of the moon is slightly offset from that of the Earth's own about the sun; the angle of intersection between them being about 5.145396 degrees [1], that not all synodic months when complete at apparent conjunction actually result in an eclipse. Rather, the complex manner of the moon orbit itself dictates that they occur anywhere from between 4 times per calendar year, which is a minimum, through to 7 times, which is more infrequent, occurring only every 31 years [2]. Of the events themselves though, when they do indeed occur, from a visual standpoint, at the very moment of an eclipse, one is able to note a most intriguing observation; that the size of the lunar disc is such that it covers the disc of the sun with hardly any excess at all. It is an almost perfect fit; a result of the fact that the moon, though being much smaller than the sun, just happens to be almost exactly the right distance from the Earth to match its apparent size in the sky at the moment of crossing.

In carefully considering the noted relations, as with so many other aspects of these celestial objects, one is bound to think it a distinct possibility that solar eclipses may once have been of a truly exacting harmony; one that no longer exists in this present age, but is merely closely approximated. Were this so, then there may exist a precise set of physical laws that would account for such a loss. In exploring this, it would seem appropriate initially to examine the possibility that successive solar eclipses in a past age were based upon a time cycle far different from that current to this present age. Indeed, one may very well conceive of the possibility that perhaps they were in fact once tied in exactly to the synodic month cycle. This would imply of course that the orbital planes of both the Earth and moon would have to have been exactly aligned, and one would thus need to demonstrate the precise manner in which such an alignment was lost. An evaluation of the orbital planes of each body to test such an idea does not appear to go unrewarded. Indeed, the relevant orbital values would seem to be highly suggestive:

Angle of inclination: Moon orbital plane to Earth orbital plane:

= 5.145396 degrees

360 (degrees) – 5.145396 = 354.854604 degrees
360 / 354.854604 = **1.014500011**

Compare with Earth tropical year ratio of increase = **1.014561622**

Accepting the slight margin of error, the ratios of change are almost identical; enough to suggest that a valid law of proportion is at work in actively governing the transformation of this component of the moon orbit. One may conclude then that at that time when the Earth was possessed of a 360-day year, there was no angular inclination existent at all between the plane of the Earth's orbit about the sun and the plane of the moon's orbit about the Earth. They shared exactly the same plane, and thus consequently, there was perfect synchronicity between the completion of each 30-day Synodic month, and the occurrence of a solar eclipse.

The law that would evidently seem responsible then for transforming the frequency of successive solar eclipses from a 30-day interval to one consisting of a much more varied pattern, is as follows:

$$TY \propto \frac{1}{OI_{(m)}}$$

Where: TY & OI (m) are ratios:

TY = Tropical Year (present) / Tropical Year (past)

OI (m) = Orbital Inclination, moon (present) / Orbital Inclination, moon (past)

With a valid physical law set forth to cover the transformation of the inclination of the moon's orbital plane, one is bound to consider if certain other additional celestial features of the Earth, moon and sun, linked to an idealised solar eclipse event, may themselves also have been transformed by physical laws to a state of slight disharmony from a once exacting ideal. Indeed, there is the point that was raised earlier of the close though not exact match between the size of the moon as viewed from the Earth, and the apparent size of the sun disc at the moment of an eclipse. Could it be that from the perspective of the Earth there was once an exacting apparent circumference of these bodies at the moment of such an event? The thought that this might once have been so must lead one to consider that perhaps both the sun and moon, like the Earth itself, have also suffered a transformation of their actual physical size. Such an idea is indeed not lacking support, even from the ancients, for once more what today are mostly dismissed as mythical stories, reveal decisively that the relevant bodies in question do indeed appear to have undergone some sort of real volumetric change; one that is finely synchronised to all of the other transformations detailed thus far.

Ancient Harmony between the Sun and Moon

Although the sexagesimal system is most readily identified with the Sumerians and Babylonians, the numerical system they employed is not so dissimilar to many that were in use by a variety of different cultures during ancient times. As noted previously, the Egyptians themselves employed some of the exact same numerical values as their neighbours. However, they were not alone in this. Another most prominent civilisation from the past, a contemporary of ancient Babylon, also shared some of the same mathematics and measures. Far to the east of the great empire of Babylon lived the ancient Hindus, who populated much of the Indian subcontinent and its surrounding areas. These were a people also with a most refined culture for this period, and like all others they too possessed their own versions of the timeless universal myths that seem to permeate every age.

Concerning the mathematics of the Hindus one may readily note that just as with the Babylonians and Sumerians, they too placed special meaning upon certain numerical values. Greatly relevant to this present investigation are the various meanings they attached to one number in particular, namely 108. This was a number most auspicious to the Hindus. On a practical level it is the number of beads used in their rosaries [3]. However, it is also tied strongly into Indian cosmology as a whole. Indeed, the ancient Indians were of the view that the number 108 was the exact number of times the diameter of the sun fitted into the distance between the sun and the Earth. Moreover, there was even a definite spiritual aspect to this belief, for they also held to the view that the outer cosmos in some way reflected the make-up of the inner human cosmos [4]. Thus was the number 108 viewed in some manner as the distance between God and the human devote in a spiritual sense, in addition to simply linking the sun diameter to the Earth-sun distance [5]. This said; it is of course the celestial relations of the number 108 that are of concern for the moment. Thus, in setting aside the spiritual aspects of the value and focusing upon the physical, is it truly possible to derive a value of 108 or close to it from the existing value for the size of the sun and also the earth orbital distance? Remarkably, this does indeed prove to be the case:

Current diameter of the sun = 1392000 kilometres [6]

1392000 × 0.6213711922 × 0.88 = 761154.855 IGM

Current Earth celestial equator = 514178995.31251 IGM

Therefore, current Earth semi-major axis value =

514178995.31251 / (2 × (22/7)) = 81801203.79971 IGM

And: 81801203.79971 / 761154.855 = **107.4698574966**

It may be seen then from the calculations that the ancient Hindu belief that the sun diameter fits into the distance between its own centre and that of the

Earth does have support from actual observations. However, one can take this analysis further. Firstly, although the number 108 is given as the ratio between the Earth-sun distance and the diameter of the sun, would it not be more appropriate to divide the full diameter or major axis of the Earth orbit by the sun's own physical diameter, rather than calculate the ratio between a radius and a diameter? It would certainly seem so. It would follow from this then that the precise value the Hindus would recognise as being correct to their system would be twice that of 108, namely 216. Now, it is easily noted that this new value is most definitely apart of the sexagesimal system of the Babylonians and Sumerians. However, all the numerical values of this system have been shown previously not to be descriptive of the current Earth-system configuration at all, but rather that of a harmonious ideal. Consequently, it would be highly appropriate to integrate the exacting value of 216 with what has previously been determined to be the full major axis of the ideal Earth celestial equator, 162000000 IGM. If the Hindus were exactly right that their number (hereby doubled) was indeed once the precise ratio (answer) between the relevant diameters, then the possibility exists to derive a value for the ideal physical diameter of the sun itself, allowing one to compare it against the currently known value. Simple calculation reveals the following:

Ideal physical diameter of sun = 162000000 / 216 = 750000 IGM

Before proceeding directly to a comparison between the ideal sun diameter thus calculated and its present diameter, it would be well to note something of the previous analysis concerning the change to the Earth's own physical form. Earlier it was shown that the current geophysical equatorial radius of the Earth as given in standard textbooks is about 3487.609361 IGM (6378.14 kilometres), but that the precise value obtained by multiplying the ideal of 3436 & 4/11 IGM by the exact ratio of change for the Earth year, produces 3486.402666 IGM. The latter value though is representative of the Earth radius in terms of a precise frequency measure; the total number of ideal geographical mile wavelength units contained within its expressed radius. It is therefore a measure associated with the very energy configuration of the planet, and must differ from a more crude or rather static geophysical measure of the planet's gross material radius. Given this, it would seem most appropriate then before comparing the ideal sun diameter to that of its present diameter, to reduce the latter by the same ratio as that existent between the current Earth physical radius and the current Earth 'energy waveform' radius. Therefore:

3487.609361 / 3486.402666 = 1.000346115

761154.855 / 1.000346115 =

760891.4993 IGM, current energy waveform diameter of sun

Thus, the ratio of sun transformation = 760891.4993 / 750000 = **1.014522**

Compare once more with the Earth tropical year increase = **1.014561622**

Accepting the slight margin of error between the two figures, one may conclude that the physical transformation of the sun is directly proportional to that of the Earth over time. Thus, a new law of proportion is revealed connecting both the physical Earth-form and the physical sun-form:

$$PE_{(s)} \propto PE_{(e)}$$

Where: PE (s) & PE (e) are ratios:

PE (s) = Physical Equator, sun (present) / Physical Equator, sun (past)

PE (e) = Physical Equator, Earth (present) / Physical Equator, Earth (past)

Moreover, given the above law one may seek to combine it to one previously stated detailing the transformation of the speed of light. Already it has been shown that there is a definite link between the increase in the speed of light and the increase in the length of the Earth tropical year. But, as the sun itself is the most obvious emitter of visible light in the solar system, it would seem then most appropriate to state the relationship between the transformation of the sun's own physical equator and that of the speed of light. Thus, the following law is also valid:

$$PE_{(s)}^{9} \propto LS^{11}$$

Where: PE (s) & LS are ratios:

PE (s) = Physical Equator, sun (present) / Physical Equator, sun (past)

LS = Light Speed (present) / Light Speed (past)

Given the above, one may now see upon careful examination the means by which a value for the ideal moon diameter can be obtained. The solution, hinted at previously, would appear to rest upon the almost exact match between the apparent size of the moon and the disc of the sun at the moment of a solar eclipse. This may be comprehended most clearly by examining the angular diameter of both the sun and moon as set against the centre point of the Earth. Basic trigonometry reveals how close the match is between these bodies:

Angular Diameter of Present Sun:

$$\text{Angle (Tangent)} = \frac{\text{Radius of Sun}}{\text{Sun Centre to Earth Centre Distance}}$$

= (760891.4993 / 2) / 81801203.79971
= 0.004650857592
= 0.266472589 degrees (radius)
= 0.266472589 × 2
Angle = **0.532945179** degrees (diameter)

Angular Diameter of Present Moon:

$$\text{Angle (Tangent)} = \frac{\text{Radius of Moon [7]}}{\text{Moon Centre to Earth Centre Distance}}$$

$= 950.3499561 / 210194.6245$
$= 0.004521285729$
$= 0.259048825$ degrees (radius)
$= 0.259048825 \times 2$
Angle $=$ **0.51809765** degrees (diameter)

As may be seen from the values given, the angular diameter of both the sun and the moon are very close, being just slightly above half of one degree. This is due essentially to the fact that the moon, though being a small celestial body, is relatively close to the Earth, whereas the sun is an extremely large body that is much farther away from the Earth. The physical dimensions and distances of the respective bodies though are such that the angles are a close match. Given this fact however, one must strongly suspect that a valid physical measure for the ideal moon diameter could be derived by working with the assumption that under an ideal celestial configuration the angles would have been an exact match. In this case then, the ratio between the ideal moon radius and ideal Earth-moon distance would be identical to that of the ideal sun radius and ideal Earth-sun distance. With this relationship assumed, and by working with known values, as previously determined, the ideal moon radius may be calculated as follows:

Ideal Moon Semi-Major Axis = 1296000 / (2 × (22/7)) = 206181 & 9/11 IGM

Ideal Sun Radius = 750000 / 2 = 375000 IGM

Ideal Earth Semi-Major Axis = 81000000 IGM

Therefore:

$$\frac{375000}{81000000} = \frac{\text{Ideal Moon Radius}}{206181\,{}^{9}\!/_{11}}$$

$$\frac{375000 \times 206181\,{}^{9}\!/_{11}}{81000000} = \text{Ideal Moon Radius}$$

Ideal Moon Radius = 954.54545454…or 954 & 6/11 IGM

Diameter = 954 & 6/11 × 2 = 1909.09090909… or 1909 & 1/11 IGM

Circumference = 1909 & 1/11 × (22/7) = **6000** IGM

For comparison, the present moon circumference =

950.3499561 × 2 × (22/7) = **5973.6282954** IGM

There is no question that an ideal moon circumference of precisely 6000 IGM is a highly significant measure. Indeed, if one were to divide the ideal Earth equatorial circumference by that of the moon, the obtained answer is a harmonic of the very measure of the ideal Earth tropical year: 21600 / 6000 = 3.6. Moreover, one can see also that the result implies that there has been a physical reduction in the size of the moon over time; the exact opposite of the transformation of the Earth. Indeed, the laws of proportion found to govern the Earth-form indicate that an increased Earth tropical year is accompanied by an increase in its physical size. For the moon however it would appear to be the reverse. The decrease of the synodic month would seem to be accompanied by a reduction in the outward physical size of the moon.

In order to fully complete one's understanding of the transformation of the moon, it would of course be desirable to discover the precise physical law responsible for governing the ongoing transformation of the body itself. What combination of powers dictated its reduction in size? In truth, one can only at this point make a best guess. There is no real hint or indication as to exactly what the power values associated with such a law may be. However, if one were hard pressed to give an answer, a law based upon the following power combination would not seem too far off the mark:

$$SM^{9} \propto PE_{(m)}^{32}$$

Where: SM & PE (m) are ratios:

SM = Synodic Month (Present) / Synodic Month (past)

PE (m) = Physical Equator, moon (present) / Physical Equator, moon (past)

Accepting the truth of this law allows at least for the determination of a somewhat speculative value for the equatorial circumference of the current moon:

As previously established:

Present synodic month = 29.53058551 days
Ideal synodic month = 30 days

Therefore:

$29.53058551 / 30 = 0.98435285$

$0.98435285^9 = 0.867675288$

$\sqrt[32]{0.867675289} = 0.995574268$

$0.995574268 \times 6000 =$ **5973.445611** IGM

Compare with **5973.6282954** IGM, as given above

Chapter 10

A Once Harmonious Solar System

With a full set of laws now given covering the physical and orbital transformations of the Earth, sun and moon, the precise manner in which each of these celestial bodies have undergone developmental changes over the course of time is well established. So too is the very unity of the system itself; evident from the fact that the laws are all perfectly integrated, such that the slightest change to any one aspect of any one particular body, results in a precise series of changes affecting every other feature of the system. One can see then that the laws are obviously a part of a single unified set that actively govern the simultaneous transformation of each of the three noted bodies. Moreover, the back transformation of the current physical magnitudes associated with the Earth, sun and moon, in accordance with the laws of proportion, decisively confirms that they are indeed returned to a most harmonious configuration in line with the measures of the ancients. Even with all that has so far been achieved though, one can still go further to ask if the past harmony of the Earth, moon and sun, was itself experienced by the other planets within the solar system. Could they too have been integrated into the key frequency measures associated with these three celestial bodies?

The Correct Use of Measure

If one does suspect then that in some forgotten age, when the Earth, moon and sun were manifest so harmoniously, the other planets of the solar system also existed in a similar 'perfected' state; one is bound to inquire if it is possible to actually derive what would have been their key orbital-frequency values also. Through careful study of the current values of the other solar bodies, it would seem that this is indeed very much achievable, provided of course certain key provisions are met at the outset.

From the study thus far conducted of the Earth and moon, it is quite clear that the fundamental unit of time, the solar day unit, associated most especially with the former body, has remained constant throughout the physical transformation of the Earth itself. And indeed, that the solar day unit, as a pure frequency, is also most intimately associated with the spatially extended distance unit so designated the ideal geographical mile. Indeed, the analysis so far presented appears to imply that both such measures, the solar day and IGM, possess an almost profound physical – even universal? – validity. With particular reference to the IGM, it would have been impossible to have determined the true ideal semi-major axis components of the Earth and moon using any other unit. The use of Kilometres for example would never have yielded the true spatial associations that would have led to the truth. Consequently therefore any systematic study of the other planets of the solar system to determine their ideal orbital magnitudes could not make use of any arbitrary units of distance measure, such as Kilometres represent, or any arbitrary time based units. For any chance of success, one must insist upon the combined use of both solar days (24 hours) to express the orbital periods of the planets, and ideal geographical miles (6000 ft) to express the relevant spatial dimensions of their orbits.

The Fundamental Standard upon which Change is Based

In any study of the possible orbital changes that the planets may have been subject to, it is vital that a central standard of some sort be established, such that all potential transformations can have some point of reference. There would seem to be none better than the established ratio of transformation for the increase in the actual physical size of the Sun, itself the primary central body of the entire solar system. Indeed, as has been demonstrated previously, the ratio of increase of the physical equator of the sun is identical to that of the physical increase of the Earth including also it's orbital period, namely 1.0145616224619...One must clearly suspect then that the change to the sun was compensated for by the Earth, and thus both bodies accommodated one another in their transformations. With such a close tie-in evident then between a change to the primary central body and an orbiting planet, one may wonder if the same can be demonstrated respecting the remaining bodies. Such indeed will be the character of the investigation to follow.

With a standard now set, the next issue to consider is how one may best guide oneself from the outset, in determining the ideal orbital periods and semi-major axis (SMA) values of the known planets. On the face of it, the entire endeavour seems almost monumental; to determine a set of physical magnitudes for each and every planet. What could such ideal planetary values possibly be? How could one begin to determine them? In respect of this, the answer must lie with an initial study of what can only be described as 'values of interest'. As one may recall, the ancient Sumerians and Babylonians used a base-60 system. Through use of a primary value of 1, by multiplying successively by 6 and 10, they were able to generate a whole series of values, the like of which have been clearly demonstrated in this present work to be representative of certain key physical magnitudes linked to the Earth, sun and moon. It would seem most reasonable to suspect then that the exact same set of values in such a series, including their close derivates, may very well constitute the key frequency measures of the other bodies of the solar system.

With the significance of the number 6 established, it would thus seem only fitting that one examine how this particular value of the sexagesimal numerical series may be modified to produce other values of interest; a none too outrageous start being to successively multiply and divide the number 6 by the numbers 2 and 3, both of which indeed can easily be used to produce the primary value itself through simple multiplication i.e. 2 × 3 = 6. The following table reveals the numbers that may be generated in this way:

Base 6	**Base Numbers × 2 Progression >**				
6	12	24	48	96	192
36	72	144	288	576	1152
216	432	864	1728	3456	6912
1296	2592	5184	10368	20736	41472
7776	15552	31104	62208	124416	248832

Base 6	**Base Numbers / 2 Progression >**				
6	3	1.5	0.75	0.375	0.1875
36	18	9	4.5	2.25	1.125
216	108	54	27	13.5	6.75
1296	648	324	162	81	40.5
7776	3888	1944	972	486	243

Base 6	**Base Numbers × 3 Progression >**				
6	18	54	162	486	1458
36	108	324	972	2916	8748
216	648	1944	5832	17496	52488
1296	3888	11664	34992	104976	314928
7776	23328	69984	209952	629856	1889568

Base 6	Base Numbers / 3 Progression >				
6	2	0.6666667	0.2222222	0.0740741	0.0246914
36	12	4	1.3333333	0.4444444	0.1481481
216	72	24	8	2.6666667	0.8888889
1296	432	144	48	16	5.3333333
7776	2592	864	288	96	32

Of the above tabled values, one can easily see a variety of numbers previously associated with the ideal Earth-moon-sun system. In each such instance, the fundamental number values are modified through use of the number 10, enabling one to produce a 'proper' physical magnitude. For example, the primary sexagesimal value of 1296 may simply be multiplied by 10 a total of 3 times to give a value of 1296000 – the noted measure of the ideal moon celestial equator (in IGM). Also, starting out with the number 6, and multiplying by 3 a total of 3 times, one produces 162. If this is itself then multiplied by 10 a total of 6 times, 162000000 is produced, revealed previously to be the measure of the ideal orbital major axis of the Earth (in IGM). As a result of such connections, the choice to use the numbers 2, 3 and also 10 to modify the primary values of the sexagesimal set is, in a most profound way, physically valid; clearly evident from the fact that the values so generated from these numbers are decisively linked to the real physical magnitudes of known celestial bodies in their ideal state.

The Ideal Measures of the Planets

In light of the above, the character and direction of the investigation into the ideal orbital periods and SMA values of the planets is now set; to proceed by examination of the currently established values of the planets expressed in terms of solar days and ideal geographical miles, to see if there exists a close affinity between them and any of the given primary values of the sexagesimal base-60 set, including close derivatives. The affinity that is sought is that whereby one is able to demonstrate a clear link between the ratio of change between a suspected ideal and a currently established planetary magnitude, and the ratio between the already established ideal physical equator of the sun and its current physical equator, *via some specific set of powers*. As with the analysis detailed in previous chapters, what one is attempting to uncover are a set of precise mathematical laws of proportion with an associated combination of powers. Indeed, the laws as derived must be decisive, as must also the selected ideal planetary values; a most subtle judgement call to be sure. The reader is of course to be the judge of all that follows [1].

The Venus Connection

Venus: Current orbital values:

Orbital Period: 224.69543369 Solar Days
Semi-Major Axis: 59169182.520 IGM

In beginning the study of the other planets, it would seem perhaps fitting that one start with one of the Earth's closest neighbours, the planet Venus, which is oftentimes referred to as a 'sister' planet to the Earth. Indeed, the most striking point of similarity between the two planets is that they are extremely close to one another in size. The diameter of the Earth is some 6972.8 IGM compared to that of Venus, which is only slightly less at 6618.32 IGM. One may suspect then that perhaps there is a close affinity between these planets with regard to their transformation from an ideal state. Working upon such a notion one is able to uncover very quickly a most interesting relationship.

Studying the current value of the semi-major axis of Venus one cannot help but notice how close it is to an exact value of 60000000 IGM. Indeed, one may be greatly intrigued to find that by dividing 60000000 by the current SMA the result is 1.014041388. Such a ratio is very close to that of 1.014561622, the ratio of the physical expansion of the sun. However tantalising this may be however, were one to accept an ideal standard of 60000000 IGM, and assume an inverse relationship between the expansion of the sun and the orbit of Venus, one is forced to accept quite a high margin of error:

60000000 / 1.014561622 = 59138842.5 IGM

59169182.52 – 59138842.5 = 30340.0 IGM difference

As one can see, a considerable level of error does exist here; an error that is some 4 and a half times greater than the very diameter of Venus itself. As a result, one may have little confidence in the reality of such a relationship. However, with a slight alteration in one's thinking, a most remarkable connection can indeed be made between the simple expansion of the sun and the orbit of Venus. This is revealed by assuming that instead of the SMA of Venus being reduced by the ratio of 1.014561622 from a higher level ideal, that it has in fact been *increased* by this ratio from a lower ideal standard. Therefore, by dividing the current SMA of Venus by the sun ratio, one should reveal the ideal Venus SMA:

59169182.52 / 1.014561622 = 58319949.43 IGM

Checking the table given above, one may note that by starting out with the sexagesimal value 216, and multiplying by 3 a total of three times, one achieves the value 5832. If this is further multiplied by 10 four times more, this returns a value of 58320000, which differs from the above derived Venus SMA by only a little over 50 IGM:

58320000 – 58319949.43 = 50.565 IGM

Accepting this margin of error one may in principle state then that the ideal SMA of Venus was precisely 58320000 IGM, and that this ideal was held at the exact same time as when the Earth itself held a SMA of 81000000 IGM, as demonstrated in a previous chapter. Moreover, with respect to both of these ideal values one can see a very close connection between them and the number representative of the ideal Earth year:

58320000 / 81000000 = 0.72

0.72 / 2 = 0.36, a harmonic of 360

With these relations established, one is able to propose a definite law of proportion linking the increase in the SMA of Venus to that of the physical expansion of the Sun:

$$PE_{(S)} \propto V_{(SMA)}$$

Where: PE (S) & V (SMA) are ratios:

PE (S) = Physical Equator, Sun (present) / Physical Equator, Sun (past)

V (SMA) = Semi-major axis, Venus (present) / Semi-major axis, Venus (past)

Using the above law one is able to generate a value for the current SMA of Venus of 59169233.82 IGM.

The reality of the proposed law offers yet further insight into the workings of the solar system, and of how closely related some of the planets are. The idea that there is some sort of compensation effect at work with respect to the planets is strongly reinforced. For as this new law indicates, just as the Earth increased its orbital period in direct proportion to the sun expansion, Venus chose to respond by increasing its distance from the sun, also in direct proportion to the sun expansion. But what though of the orbital period of Venus? How might it have been transformed?

Of the current orbital period of Venus it can be seen that a complete circuit about the sun is just slightly less than an exact figure of 225 days; standing presently at about 224.69543369 days. Checking the values of interest in the table above, it can be seen that the division of 36 by 2 a total of 4 times returns a value of 2.25, which is a clear harmonic of 225. This indeed is very suggestive, and does lead one on to suspect that the true ideal orbital period of Venus when possessed of a SMA of 58320000 IGM was exactly 225 days. Is there however a decisive combination of powers that can relate the sun expansion ratio to such a subtle change? Indeed there is, and it is revealed as follows:

1.014561622… to the 3rd power = 1.044324078…

1.044324077… reduced to the 32nd root = 1.001356226…

225 / 1.001356226 = 224.6952622 days

Level of Discrepancy (LD):

(224.69543369 – 224.6952622) × 86400 = 14.8 seconds

Using then the power combination of 3 and 32, a remarkably accurate transformation of the orbit of Venus is to be had with an error of only 14.8 seconds from that established from observations. Such would seem well within an acceptable margin. Furthermore, the powers used are those that have already been used elsewhere. The power of 3 for example is to be found in many of the laws of proportion already established, and 32 has itself been shown to be linked to nothing less than the transformation of the Earth's own orbit (the Tropical Year – SMA law of proportion, as given in a previous chapter). That being said, the idea of a general law being based upon these powers may still seem somewhat speculative at this point. What is needed therefore is a further orbital connection to Venus of some sort; one that decisively establishes the above relationship beyond all doubt. To uncover this, a brief detour is necessary at this point, towards one of the most intriguing objects within the entire solar system: the Asteroid Belt.

Lying between Mars and Jupiter, the asteroid belt is a large debris field of a most remarkable nature; a field of scattered rocks that range in size from mere dust particles to large bodies many miles across. Indeed, that the asteroids themselves throughout the region are so very widely dispersed, and with most being practically impossible to detect with the naked eye, the belt itself was only discovered in the modern era of astronomy. In terms of the general boundary of the field, using the Earth SMA value as a standard unit of 1, the region of the asteroid belt is approximately between 2.1 and 3.5 unit distances from the sun (as a note, Mars is 1.523 and Jupiter 5.202). Of the belt itself, it is estimated that there are several hundred thousand asteroids within it, if not millions, though only a small few are known to be of a very large size. As of the present time, no more than 26 have been detected that exceed a diameter of more than 124 miles, or 200 kilometres [2]. Furthermore, the total volume of all of the asteroids of the belt is less than one tenth of the total volume of the Earth's own moon.

In modern astronomy there are two main theories to account for the asteroid belt. One is that it consists of the fragments of a planet that was never able to form when the solar system itself was in the early stages of its development. The other is that the sum total of the mass of the asteroid belt did at one time constitute a full though very small planet in orbit between Mars and Jupiter, but that for some unknown reason, it became unstable and exploded. The scattered fragments of rock that make up the asteroid belt are thus held to be the remains of such a planet. In light of this present work, it is the latter idea that would seem to be of greatest interest.

With the main aim of this section being to determine the potential orbits of the planets of the solar system in their ideal state, the thought that the sum total of the matter of the asteroid belt did once constitute a definite planetary form with a distinct orbital period and semi-major axis, does lend itself to further study. In view of this, taking the position then that the matter of the belt did indeed at one time form a unique planetary body, how might one determine its orbital characteristics? One idea would be to take the dispersed field and work out a mean value as to its overall distance from the sun, and from this determine its orbital period through a general inverse square law. However, it would be difficult to see just how one could obtain accurate orbital values here, were one inclined to explore this avenue. That being said, there is indeed an alternative method, which would appear to overcome this specific issue.

As was noted above, though the vast majority of asteroids within the belt are very small, there are a few that have been detected that are very large. Indeed, the initial discovery of the asteroid belt occurred in 1801 with the sighting of the largest of the asteroids contained within it. It was of course at the time the easiest to detect. The name assigned to the first observed asteroid, which holds even today, is Ceres. By some measure the largest of all the asteroids within the belt, Ceres accounts for approximately 1/3rd of all of the matter of the entire belt. Also, rather than being simply an irregular deformed rock, Ceres is spherical in its shape. As a result it has the distinction of being referred to as a Minor Planet. The actual diameter of Ceres is approximately 509.951 IGM [3]. And, like with all the asteroids of the belt it has its own unique orbital period.

With the very fact of the existence of such a body as Ceres, it would seem far more reasonable to analyse the orbital parameters of this distinct body as opposed to speculating loosely about the entire debris field. Indeed, if the suspicion is correct that the remains of the belt did once form a single body, then perhaps Ceres, being the largest remnant, is the most truthful standard by which to set the deviation of such a body. One may speculate then that at that time when the solar system was in a state of harmony, the asteroid belt did not exist as such. Instead, a distinct planet of a given mass orbited the sun at an ideal standard between Mars and Jupiter, and that the resulting loss of harmony led to instability at this particular distance from the sun, such that the planet exploded, leaving its largest remnant, Ceres, to depart to a new orbital distance and coalesce into a smaller planetary body in its own right, much removed from the general fragments of rock that compose the field overall. Working with this particular idea, one may then study the current orbit of Ceres in much the same way that Venus itself was studied.

Ceres

Current orbital values [4]:

Orbital Period: 1680.09156 Solar Days
Semi-Major Axis: 226245511 IGM

Examining first the orbital period, and considering what possible ideal may once have sustained the complete planetary form existent between Mars and Jupiter, initially one may suspect a value of perhaps 1620 days, derived by multiplying the value 1296 by 10 and then dividing the answer by 8 (2 × 2 × 2). However, it so transpires that a value of exactly 1440 days, the speed of light harmonic first suggested by Cathie, is vastly more intriguing. Choosing this value, one is able to determine that *the exact same set of powers* appear to link it to the ratio of the sun expansion that were found to transform the orbit of Venus:

Sun expansion ratio: 1.014561622…

1.014561622… to the 32nd power = 1.588216747…

1.588216747… reduced to the 3rd root = 1.166728814…

1440 × 1.166728814… = 1680.089493…days

LD → (1680.09156 – 1680.089493) × 86400 = 178.5 sec

As can be seen from the above, the transformation of the orbital period of Ceres makes use of exactly the same powers as with Venus, except that Venus itself suffers a slight decrease to its orbital period from a precise 225 day orbit, whereas Ceres suffers a major increase of some 240 days or so. The powers for such a connection are quite exceptional to yield such a highly accurate association for both Venus and Ceres. As a result, they decisively establish each other. Thus, one may hold now with confidence the Venus orbital period law, and add to it the law that would now appear to govern the transformation of the orbital period of Ceres:

$$PE^{3}_{(S)} \propto \frac{1}{V^{32}_{(OP)}}$$

$$PE^{32}_{(S)} \propto C^{3}_{(OP)}$$

Where: $PE_{(S)}$, $C_{(OP)}$ & $V_{(OP)}$ are ratios:

$PE_{(S)}$ = Physical Equator, Sun (present) / Physical Equator, Sun (past)

$C_{(OP)}$ = Orbital period, Ceres (present) / Orbital period, Ceres (past)

$V_{(OP)}$ = Orbital period, Venus (present) / Orbital period, Venus (past)

Yet more evidence is to be had then from the above, that the planets do appear to have complemented one another most harmoniously in their transformations as the very structure of the solar system itself was altered.

The remaining aspect of the orbit of Ceres is its semi-major axis. Concerning this, one cannot fail to notice that when expressed in ideal geographical miles the current value is a close harmonic of 225. Indeed, with the orbital period connection existent between Venus and Ceres, an ideal SMA of 225000000 IGM for Ceres is obviously the main contender. A systematic study of the power combinations that would appear to back transform Ceres to this SMA reveals a close match with a low margin of error. The powers themselves admittedly are unremarkable, yet they do not appear to be outrageously so:

1.014561622… to the 8th power = 1.122606216…

1.122606216… reduced to the 21st root = 1.005522476…

225000000 × 1.005522476…= 226242557.28 IGM

LD → 226245511 – 226242557.28 = 2953.7 IGM

The law of proportion that results from the above is expressed as follows:

$$PE^{8}_{(S)} \propto C^{21}_{(SMA)}$$

Where: PE (S) & C (SMA) are ratios:

PE (S) = Physical Equator, Sun (present) / Physical Equator, Sun (past)

C (SMA) = Semi-major axis, Ceres (present) / Semi-major axis, Ceres (past)

As one can see, the law implies that the expansion of the sun caused a general increase in the distance between the sun and Ceres.

Mercury

Current orbital values:

Orbital Period: 87.96843536 Solar Days
Semi-Major Axis: 31665071.629 IGM

Leaving behind Ceres and Venus, what then of Mercury, the nearest planet to the sun? What ideal orbital values might this body once have possessed? The first step towards isolating them would appear to be had from a most distinctive relationship between both the current semi-major axis and the current orbital period of the planet. The ratio between these values proves to be highly suggestive:

31665071.629 / 87.96843536 = 359959.4729

Immediately upon viewing one is bound to suspect that in its ideal configuration, the ratio between the SMA of Mercury and its orbital period would have been exactly 360000. Accepting this, whatever ideal value one was inclined to choose for either the orbital period or SMA, the other would be determined. So, what could guide any decision upon them? One particularly simple relationship between the orbital periods of both the Earth and Ceres may well hold the key. As established, the ideal orbital period of Ceres has been found to have been exactly 1440 days. If one divides this by the number 4, the answer is 360 days – the ideal Earth year. If one extends this process one stage further however and divides the Earth year by 4, the result is 90 days. Such a value is very close to the current value for the orbital period of Mercury, and thus at this stage can be tentatively accepted as being the ideal. This of course results in the acceptance also of the ideal Mercury SMA as per the reality of the above suggested ratio: 90 × 360000 = 32400000 IGM.

Is there however yet further evidence to be had to reinforce these chosen values for Mercury? Indeed, it would seem that the current values held by another planet in the solar system do in fact help to establish more forcefully the selected values for the orbit of Mercury. The planet in question is that which is at the farthest end of the solar system: Pluto.

Pluto

Current orbital values:

Orbital Period: 90589.5972 Solar Days
Semi-Major Axis: 3234798228 IGM

There is little doubt upon viewing the current orbital values of Pluto that they are strikingly similar in a harmonic sense to the suggested ideal values for Mercury. The given period is only slightly over an exact 90000 days whereas the SMA is only slightly under an exact value of 3240000000 IGM. In the case of the former, this is 1000 times that of the suggested Mercury ideal. With respect to the latter, the suggested SMA of Pluto is exactly 100 times that of Mercury.

With both Mercury and Pluto being amongst the smallest planets in the solar system, a connection between them does feel appropriate. Is there however any decisive relationship between them respecting the potential powers that may transform the orbits of these bodies to the suggested ideals? In this case, admittedly, the answer would appear to be no. A close study of the possible combinations of powers found to link the sun equator expansion ratio to that of the orbital values of Mercury and Pluto, does not appear to give rise to a set of powers that reinforce one another, such as was evident in the case of Ceres and Venus. That being said, taken in isolation, a series of powers for each of the planets are to be found, producing a set of current orbital values that do differ only slightly from observed values. Indeed, in the case of Pluto, one must emphasise quite forcefully that by assuming the suggested ideals of 90000 days and 3240000000 IGM, two sets of powers are to be found that transform Pluto to its current orbit with an absolutely exceptional level of accuracy, equal to if not greater than that of Venus. This is all the more remarkable considering that the powers used are relatively low. This is important, for truth be told, by far and away the main criticism to be levelled at any suggested set of powers is that they be too high. Indeed, it is very much the case that by using a high enough combination of powers, one can 'prove' any suggested ideal value with an exceptional level of accuracy. Thus, one must always be wary of high power combinations. The lower and more accurate they are, the more likely they are to be correct, though there are of course no guarantees. That said, the existence of certain decisive connections between several planets can reinforce a proposed set and give it more weight. This sometimes is all that can be achieved.

Concerning then both Mercury and Pluto, one may offer the following sets of powers that would appear to closely link the current orbital values of these planets to the suggested ideals. From them one may derive four new laws of proportion to be accepted, albeit in this case somewhat tentatively:

Mercury

Orbital Period:

1.014561622… to the 30th power = 1.542953758…

1.542953758… reduced to the 19th root = 1.023088754…

90 / 1.023088754 = 87.96890748…days

LD → (87.96890748 – 87.96843536) × 86400 = 40.79 sec

Semi-Major Axis:

1.014561622… to the 27th power = 1.477466422…

1.477466422… reduced to the 17th root = 1.023226136…

32400000 / 1.023226136 = 31664554.74 IGM

LD → 31665071.629 – 31664554.74 = 516.88 IGM

From the above, the following laws of proportion may be derived:

$$PE_{(S)}^{30} \propto \frac{1}{M_{(OP)}^{19}}$$

$$PE_{(S)}^{27} \propto \frac{1}{M_{(SMA)}^{17}}$$

Where: PE (S), M (OP) & M (SMA) are ratios:

PE (S) = Physical Equator, Sun (present) / Physical Equator, Sun (past)

M (OP) = Orbital period, Mercury (present) / Orbital period, Mercury (past)

M (SMA) = Semi-major axis, Mercury (present) / Semi-major axis, Mercury (past)

Pluto

Orbital Period:

1.014561622… to the 14th power = 1.224328685…

1.224328685… reduced to the 31st root = 1.006550155…

90000 × 1.006550155… = 90589.51397…days

LD → (90589.5972 – 90589.51397) × 86400 = 7190.91 sec

Semi-Major Axis:

1.014561622… reduced to the 9th root = 1.001607581…

3240000000 / 1.001607581… = 3234799794.38 IGM

LD → 3234799794.38 – 3234798228 = 1566.38 IGM

From the above, the following laws of proportion for Pluto may be derived:

$$PE^{14}_{(S)} \propto P^{31}_{(OP)}$$

$$PE_{(S)} \propto \frac{1}{P^{9}_{(SMA)}}$$

Where: PE (S), P (OP) & P (SMA) are ratios:

PE (S) = Physical Equator, Sun (present) / Physical Equator, Sun (past)

P (OP) = Orbital period, Pluto (present) / Orbital period, Pluto (past)

P (SMA) = Semi-major axis, Pluto (present) / Semi-major axis, Pluto (past)

Mars

Current orbital values:

Orbital Period: 686.9297110 Solar Days
Semi-Major Axis: 124638662.50 IGM

Before proceeding to the large planets just beyond the asteroid belt, the so called gas giants, there is one remaining planet to consider of the inner part of the solar system; the fourth planet from the sun: Mars. Concerning the time element of its orbit first, unlike the other planets so far considered, the study of Mars to determine its possible ideal orbital period would seem to lead one to the conclusion that the ideal Mars orbit did not consist of a whole number of solar days. Indeed, a very careful study of the planet reveals two possible ideal periods closely related to the primary sexagesimal series that rank above all others by some measure. One of them is 666.6666… days (666 + 2 / 3) a harmonic of the base number 6 divided by 3, twice: 6 / (3 × 3) = 0.66666… (2 / 3). The other is 691.2 days, a harmonic of 216 × 2 a total of 5 times i.e. 216 × 2 × 2 × 2 × 2 × 2, which equals 6912. By far and away both of these ideal values appear to be vastly more suggestive than any other alternatives. However, when set against one another, it would seem that the latter has the advantage. Of the value 666 + 2 / 3 days, were one to compare the ratio it has with respect to the current orbital period of Mars, and seek to link it to the sun expansion ratio, a favourable power combination of 29 and 14 can be seen to produce a value to match the current Mars orbit with an error of about 240 seconds. In this instance, the powers are reasonably low, as too is the time discrepancy; proportionally so, even better than that obtained from the acceptance of the law for Mercury, as given above. However, if one chooses an ideal of 691.2 days, one can cut the time discrepancy by more than half, and do so by using an extremely low combination of powers. The strength of this leads far more to the acceptance of this value for the ideal orbital period. This can be seen as follows:

Orbital Period:

1.014561622… to the 3rd power = 1.044324077…

1.044324077… reduced to the 7th root = 1.006214927…

691.2 / 1.006214927… = 686.9307751…days

LD → (686.9307751 – 686.929711) × 86400 = 91.94 sec

With a set of powers far lower than many presented already for the other planets, a high level of confidence can be had with this combination; enough to accept a proportional law based upon the stated powers:

$$PE^{3}_{(S)} \propto \frac{1}{Ma^{7}_{(OP)}}$$

Where: PE (S) & Ma (OP) are ratios:
PE (S) = Physical Equator, Sun (present) / Physical Equator, Sun (past)
Ma (OP) = Orbital period, Mars (present) / Orbital period, Mars (past)

Turning next to the semi-major axis of Mars, once more there is to be had an ideal value that again yields high accuracy and makes use of a very low combination of powers. Due to this, from among several possible alternatives it stands out significantly from the rest. The value itself makes use of a simple harmonic of 125. This is demonstrated as follows:

Semi-Major Axis:

1.014561622… reduced to the 5th root = 1.002895507…

125000000 / 1.002895507… = 124639106.47 IGM

LD → 124639106.47 – 124638662.50 = 443.97 IGM

This relationship indicates then that the SMA of Mars was precisely 125000000 IGM at such time when the planet was possessed of an orbital period of exactly 691.2 days. The law of proportion detailing the transformation of the Mars SMA is as follows:

$$PE_{(S)} \propto \frac{1}{Ma^{5}_{(SMA)}}$$

Where: PE (S) & Ma (SMA) are ratios:

PE (S) = Physical Equator, Sun (present) /
Physical Equator, Sun (past)

Ma (SMA) = Semi-major axis, Mars (present) /
Semi-major axis, Mars (past)

Jupiter & Saturn

With the smaller planets of the solar system having thus been dealt with, attention now must focus upon the larger bodies. The first set of what are generally referred to as the gas giants are Jupiter and Saturn. Both of these bodies are to be found in orbit about the sun just beyond Ceres and the asteroid belt. Their primary values are as follows:

Jupiter: Orbital Period: 4330.595763 Solar Days
Semi-Major Axis: 425578718.806 IGM

Saturn: Orbital Period: 10746.940442 Solar Days
Semi-Major Axis: 781602216.1604 IGM

Concerning oneself first with the orbital periods, one cannot help but quickly perceive how close they are to certain values easily derived from the primary sexagesimal set, namely 4320 and 10800 (days) respectively: 216 × 2 = 432 and 216 / 2 = 108. Studying the current values in light of such a proposed ideal set, one is able to rapidly reveal a most simple and elegant relationship between the relative transformation of the orbital periods of both bodies. This is detailed as follows:

4330.595763 / 4320 = 1.002452722…

1.002452722… squared = 1.004911461…

1.004911461…× 10746.940442 = 10799.72362…days

From this association one is able to suggest a principled relationship between Jupiter and Saturn governing the relative transformation of the orbital periods of both bodies. Expressed as a distinct law of proportion, this may be stated as follows:

$$J^2_{(OP)} \propto \frac{1}{S_{(OP)}}$$

Where: $J_{(OP)}$ & $S_{(OP)}$ are ratios:
$J_{(OP)}$ = Orbital period, Jupiter (present) / Orbital period, Jupiter (past)
$S_{(OP)}$ = Orbital period, Saturn (present) / Orbital period, Saturn (past)

In acceptance of the stated law, one must further inquire then as to how individually the respective transformations of both Jupiter and Saturn may be linked to the primary sun expansion ratio. In doing so, the result leads to a most remarkable connection; an association to the very law of proportion previously established linking the Earth's own orbital period to its semi-major axis, the very character of which strongly reinforces the truth of the above proposed law. Indeed, from careful study, the suggested powers that would appear by far to best associate the sun expansion ratio to the noted orbital changes above are as follows:

Jupiter:

Orbital Period:

1.014561622… to the 8th power = 1.122606216…

1.122606216… reduced to the 47th root = 1.002463731…

4320 × 1.002463731… = 4330.643319…days

LD → (4330.643319 – 4330.595763) × 86400

= 4108.8 seconds, or 68.48 minutes

Saturn:

Orbital Period:

1.014561622… to the 16th power = 1.260244717…

1.260244717… reduced to the 47th root = 1.004933532…

10800 / 1.004933532… = 10746.97942…days

LD → (10746.97942 – 10746.940442) × 86400

= 3368.27 seconds, or 56.137 minutes

Using the above combinations of powers, 8 and 47 for Jupiter, and 16 and 47 for Saturn; with the proposed ideal orbital periods of 4320 and 10800 days for each noted planet, one is thus able to derive a set of current values for these bodies that differ only very slightly from observations. Furthermore, in the case of both planets, one can clearly perceive a connection to the powers found to govern the relative transformation of the Earth semi-major axis and orbital period. For again one can see a power value of 47, as is present in the Earth law itself. But in addition, one is able to see a further link involving the power value of 32, being the remaining value of the Earth law. For indeed, 32 divided by 2 gives 16, which divided by 2 again gives 8; a numeric progression linking most harmoniously to the noted remaining powers of Jupiter and Saturn, as given.

With an acceptance of the above mathematical associations, a full set of transformational laws for the orbital periods of both Jupiter and Saturn may be given as follows:

$$PE_{(S)}^{8} \propto J_{(OP)}^{47}$$

$$PE_{(S)}^{16} \propto \frac{1}{S_{(OP)}^{47}}$$

Where: PE (S), J (OP) & S (OP) are ratios:

PE (S) = Physical Equator, Sun (present) / Physical Equator, Sun (past)

J (OP) = Orbital period, Jupiter (present) / Orbital period, Jupiter (past)

S (OP) = Orbital period, Saturn (present) / Orbital period, Saturn (past)

Upon the issue of the semi-major axis values of the above planets, one may find yet additional associations to those already presented. To begin with

Saturn, by far the strongest contender for the value of ideal SMA is 781250000 IGM, which indeed has a most striking association to the numerical values of the '×2' series noted above. In starting out from a value of 1, the sequence progresses as follows: 2, 4, 8, 16, 32, 64, 128… Upon reaching the value of 128, one may note that the reciprocal of this number is of the same sequence as the suggested value for the SMA of Saturn: 1 / 128 = 0.0078125. But in addition to this, is the fact that by accepting a value of 781250000 IGM for the ideal Saturn SMA, a transformational power of extreme accuracy presents itself, also a part of the noted '×2' progression, leading to a proportional law that one would be hard pressed to think invalid:

Semi-Major Axis:

1.014561622… reduced to the 32nd root = 1.000451871…

781250000 × 1.000451871… = 781603024.56 IGM

LD → 781603024.56 – 781602216.1604 = 808.40 IGM

From the above, the general law of proportion is as follows:

$$PE_{(S)} \propto S^{32}_{(SMA)}$$

Where: PE (S) & S (SMA) are ratios:

PE (S) = Physical Equator, Sun (present) / Physical Equator, Sun (past)

S (SMA) = Semi-Major axis, Saturn (present) / Semi-Major axis, Saturn (past)

In the case of the ideal semi-major axis of Jupiter, though there are a number of possible alternative answers, to the mind of this present author, the best value would appear to rest upon the following noted associations:

1.014561622… to the 24th power = 1.414758554…

1.414758554… reduced to the 7th root = 1.050814475…

405000000 × 1.050814475… = 425579862.72 IGM

LD → 425579862.72 – 425578718.806 = 1143.92 IGM

Here then one can see a suggested ideal of 405000000 IGM for the semi-major axis of Jupiter; a harmonic of the primary base 60 value of 1296 divided by 2 × 2 × 2 × 2 × 2 i.e. 32, which gives the answer 40.5. Also of note, a value of 405000000 is exactly equal to 5 times the Earth's own ideal SMA. As a result of such connections, the value does present itself as a strong potential candidate, though one must concede that the associations are not as powerful as those uncovered with respect to Saturn. Nevertheless, based upon such connections, one may tentatively put forward the implied law of proportion for the transformation of the SMA of Jupiter:

$$PE^{24}_{(S)} \propto J^{7}_{(SMA)}$$

Where: PE (S), & J (SMA) are ratios:

PE (S) = Physical Equator, Sun (present) /
Physical Equator, Sun (past)

J (SMA) = Semi-major axis, Jupiter (present) /
Semi-major axis, Jupiter (past)

Uranus & Neptune

Beyond both Jupiter and Saturn lie a second set of giant planets: Uranus and Neptune. Compared with the first set they are not nearly so large, in that Uranus and Neptune are both less than half the size of Saturn and Jupiter. That being said though, there is a very close match between the planets themselves; the diameter of Uranus being about 27777 IGM, only just slightly greater than that of Neptune, at some 26574 IGM [5]. Their primary orbital values are as follows:

Uranus: Orbital Period: 30588.740341 Solar Days
Semi-Major Axis: 1572090231 IGM

Neptune: Orbital Period: 59799.900443 Solar Days
Semi-Major Axis: 2463063084 IGM

Beginning with a consideration of the orbital periods, what values tend to present themselves as possible ideals? Initially one is bound to suspect a combination of 30000 and 60000 solar days. However, in truth, a careful study reveals several other possibilities for both planets. Indeed, there is a good selection of ideal values for each of these planets that can be linked to the sun expansion ratio with a fair amount of accuracy and a reasonably low set of powers. As a result, it is difficult to choose an orbital period value for either planet based purely upon these considerations. However, extending the scope of study of both planets to consider more subtle orbital elements does reduce the field of possibilities to a very specific set of values, most likely to be true in the face of all others.

In addition to just the orbital periods of the planets, one may also point towards the existence of additional time cycles built in to their movements; most notably as a result of periodic conjunctions. An examination of the ideal orbits of Jupiter and Saturn serve as a good case in point, the stated values as previously given for their ideal orbital periods being 4320 and 10800 days respectively. Given these values, if both planets were initially aligned in conjunction with the sun and allowed to progress through space, there would come a point in time when Jupiter, due to its faster orbit, would rapidly outpace Saturn such that it would eventually overtake and lap the planet. At such a moment in time, at the point of overtaking, once more the two planets would be in conjunction with the sun. In order to determine the time taken for such periodic conjunctions between any two bodies, known as synodic conjunctions, a precise formula is used:

$$(P1 \times P2) / (P1 - P2) = \text{Time between successive conjunctions.}$$

Where: P1 = orbital period of one planet
P2 = orbital period of a second planet

Using the above formula, in the case of Jupiter and Saturn, the time between successive conjunctions is calculated as follows:

$$(10800 \times 4320) / (10800 - 4320) = 7200 \text{ days.}$$

The formula given one must understand is for periodic conjunctions of any two planets wherein they successively conjunct 'at one side of the sun'. Essentially, the alignment itself involves one of the two planets being in the middle, with the other planet and the sun at either side. However, technically, one should realise that the time value derived from this equation can be halved. The reason is that in point of fact the first conjunction to occur between the three bodies involves the Sun being in the middle of the set. Thus, in the example of Jupiter and Saturn, from a starting point of both planets being aligned to the sun with the sun itself at one end of the arrangement, after 3600 days, the faster movement of Jupiter has taken it to the other side of the sun, such that all three bodies initially conjunct with the sun in the middle. It is only after a further 3600 days that the original arrangement is established once more.

One further point to note about synodic conjunctions that involve the above stated formula is that the planets in question do not with each successive conjunction, always return to their original starting points. Again, using the ideal orbits of Jupiter and Saturn by way of example, one may note that from an initial alignment, with the next conjunction 7200 days later, neither planet has returned to its original starting point. Specifically, after 7200 days, Jupiter will have completed 1 & 2/3 of 1 orbit, and Saturn 2/3 of 1 orbit, placing both planets at an angle of 240 degrees from a zero starting angle. In order to actually determine just when exactly they will re-conjunct at their initial starting positions, one must know the lowest common multiple (LCM) of the two planetary orbital periods. This value, the LCM, is defined as the smallest number that a set of other numbers will divide into evenly with no remainder. In the case of 10800 and 4320, the LCM is 21600 days, for after exactly 21600 days have passed, Saturn will have completed precisely 2 orbits about the sun from an initial starting point, and Jupiter will have completed a total of 5 orbits. Thus, although both planets do conjunct successively (in a synodic manner) every 7200 days, they only conjunct at their original starting points every 21600 days, which is the LCM of the two orbital periods. Indeed, one may note of Venus and Ceres, whose ideal orbital periods have been given as 225 and 1440 days respectively, that their LCM is 7200 days (their synodic conjunction period being 266 & 2/3 days).

From the above it can be seen that several planets appear to be strongly associated with the number 7200; both Ceres and Venus, including Jupiter and Saturn. The possibility thus exists that perhaps this number will also be found to relate to the second set of gas giants, Uranus and Neptune. Indeed, a careful study of the possible ideal orbital periods for these two planets reveals a most intriguing association. It turns out that by choosing a value of 32000 days for Uranus, and one of 57600 days for Neptune, one can derive a higher harmonic of a 7200 day synodic period, as found to be embedded in the orbits of Jupiter and Saturn. This is revealed as follows:

$$(57600 \times 32000) / (57600 - 32000) = 72000 \text{ days.}$$

Thus, every 72000 days, a synodic conjunction between Uranus and Neptune would occur, were they of the stated ideal orbital periods. And moreover, careful study reveals that such a conjunction would occur every 90 degrees from an initial starting angle. Thus, the orbit would be 'split-up' into quarters. This indeed is a most interesting finding when one compares it to the conjunctions of the first set of gas giants. In the case of Jupiter and Saturn, successive conjunctions divide their orbits into thirds. Also, on the question of the LCM of Uranus and Neptune, the numbers 32000 and 57600 produce a value of 288000 days. Half of this value is 144000 days, which is itself exactly 100 times the orbit of Ceres.

As a result of such relations, the proposed ideal values for Uranus and Neptune are indeed highly suggestive. Can a set of powers be found though to link them via their current values to the sun expansion ratio? One may provide the following in answer:

Uranus

Orbital Period:

1.014561622… to the 78th power = 3.088286392…

3.088286392… reduced to the 25th root = 1.046137337…

32000 / 1.046137337… = 30588.71799…days

(30588.740341 – 30588.71799) × 86400

= 1930.84 seconds, or 32.18 minutes

Neptune

Orbital Period:

1.014561622… to the 70th power = 2.750997051…

2.750997051… reduced to the 27th power = 1.038191364…

57600 × 1.038191364… = 59799.82257…days

(59799.900443 – 59799.82257) × 86400

= 6727.7 seconds, or 112.1 minutes

From the above, a set of laws can be derived as follows:

$$PE^{78}_{(S)} \propto \frac{1}{U^{25}_{(OP)}}$$

$$PE^{70}_{(S)} \propto N^{27}_{(OP)}$$

Where: $PE_{(S)}$, $U_{(OP)}$ & $N_{(OP)}$ are ratios:

$PE_{(S)}$ = Physical Equator, Sun (present) / Physical Equator, Sun (past)

$U_{(OP)}$ = Orbital period, Uranus (present) / Orbital period, Uranus (past)

$N_{(OP)}$ = Orbital period, Neptune (present) / Orbital period, Neptune (past)

With the above established, what then of the semi-major axis values of the planets? In the case of Uranus, the most obvious value to suggest itself would be twice that of the ideal SMA of Saturn, being 1562500000 IGM, only slightly under 1572090231 IGM, its present value. For Neptune, 2531250000 IGM would seem most favourable; achieved by simply multiplying the ideal semi-major axis of Pluto (3240000000 IGM) by 0.78125, a basic harmonic of the ideal SMA of Saturn. Based upon such values, the following may thus be had:

Uranus

Semi-major axis:

1.014561622… to the 11th power = 1.172364701…

1.172364701… reduced to the 26th root = 1.006135004…

1562500000 × 1.006135004… = 1572085945.14 IGM

1572090231 – 1572085945.14 = 4285.85 IGM

Neptune

Semi-major axis:

1.014561622… to the 17th power = 1.278595925…

1.278595925… reduced to the 9th power = 1.027683200…

2531250000 / 1.027683200… = 2463064490.97 IGM

2463064490.97 – 2463063084 = 1406.97 IGM

The laws of proportion derived from the above are as follows:

$$PE^{11}_{(S)} \propto U^{26}_{(SMA)}$$

$$PE^{17}_{(S)} \propto \frac{1}{N^{9}_{(SMA)}}$$

Where: PE (S), U (SMA) & N (SMA) are ratios:

PE (S) = Physical Equator, Sun (present) / Physical Equator, Sun (past)

U (SMA) = Semi-major axis, Uranus (present) / Semi-major axis, Uranus (past)

N (SMA) = Semi-major axis, Neptune (present) / Semi-major axis, Neptune (past)

Summary Analysis of the Proposed Ideal Solar System

From a careful study of the above proposed laws, one may readily apprehend that of the stated relations, some are obviously far better and more forceful than others. And thus consequently, some laws doubtless will be subject to future revision. That being said, one must concede that certain connections established are so decisive that they would appear to confirm beyond all doubt that physical laws of proportion per se are operative within the solar system at large. Indeed, upon this very point, the strongest laws to note are those shown to link the sun expansion ratio to the orbital periods of the Earth, Venus, Ceres, Jupiter and Saturn, including also the semi-major axis values of Venus, the Earth, and Saturn. Of these one may consider the Venus-Ceres orbital period associations in particular to be utterly decisive. And thus, in all, one can claim exceptional support for the view that when the Earth itself was possessed of a 360-day year in some remote age, it was in striking harmony with the other planets of the solar system at the same time. The later transformation of the system though altered everything.

With the laws of proportion now revealed, one is able to draw certain definite conclusions about the way in which the solar system as a whole operates, and how the planets suffer orbital shifts. The first point to note from the discovered relations, especially those that link up the primary characteristics of the Earth, Venus and Ceres, is that it is clearly evident that a general inverse square law does not on principle link the orbital periods of the planets to their semi-major axis values. What can be seen instead is that the planets appear to be transformed by a distinct set of laws, as are unique to each celestial body. A general inverse square law merely *approximates* their relations, and is thus an incomplete law. Indeed, the implication here, is that there are a set of 'invisible bands' or regions of extension from the sun, wherein specific laws with their unique power combinations dictate the manner in which subtle changes to a planet's orbital period affect changes to its SMA. The precise manner of the transformation of a planet is thus determined by the laws responsible for it; which itself is determined by how far the planet resides from the sun, and thus which band it operates within. Of course, where exactly the boundaries are between each of the different bands, as yet remains to be uncovered. However, the very issue of the boundaries may well have a bearing upon the important matter of the formation of the asteroid belt.

Of the noted proportional changes to have affected the orbits of the planets, it may be seen that the increase to the orbital period of Ceres, the largest remnant of the exploded planet, is far greater than that experienced by any other body. Now, as one may recall from earlier, it was demonstrated respecting the Earth (including the moon also) that changes to the orbit of a celestial body are linked via proportional laws to the size of the body also. One may theorise then that quite possibly the shift in the orbit of Ceres – in its unexploded state – caused such an extreme proportional change via the laws governing its physical size, that the planet suffered nothing less than catastrophic failure. Pushed beyond a critical threshold, it could no longer physically maintain itself, the result being that Ceres was literally ripped apart [6].

Chapter 11

Precession of the Equinoxes

Precession: The Orthodox Theory

So far in this work the longest celestial time cycles that have been examined are the orbital periods of the primary bodies of the solar system. However, it is well established that there are quite a number of other astronomical cycles that operate over far greater lengths of time. One of the most widely known, as is held indeed to be directly associated with the Earth, is the precession of the equinoxes, which is generally considered to be a subtle movement of the orientation of the axis of the Earth with respect to the background stars, tracing out a continuous circular path in the sky over roughly a 25770 year period.

Diagram 1 (overleaf): In addition to the more obvious planetary motions, including the daily rotation of the Earth on its axis and the yearly orbit about the sun, the Earth is also thought to be acting much like a spinning top, with the axis of the planet itself changing its orientation in space over time by slowly tracing out a circular path in the sky. This has the effect of 'drawing out' two inverted cone shapes, as pictured. The total time required to complete one full cycle is just over 25770 years. As a result of such a motion the Pole or North Star does not maintain its position directly overhead of the Earth continually, but will periodically move away from this spot, only to move back into alignment every 25770 years.

Earth Orbital Elements:

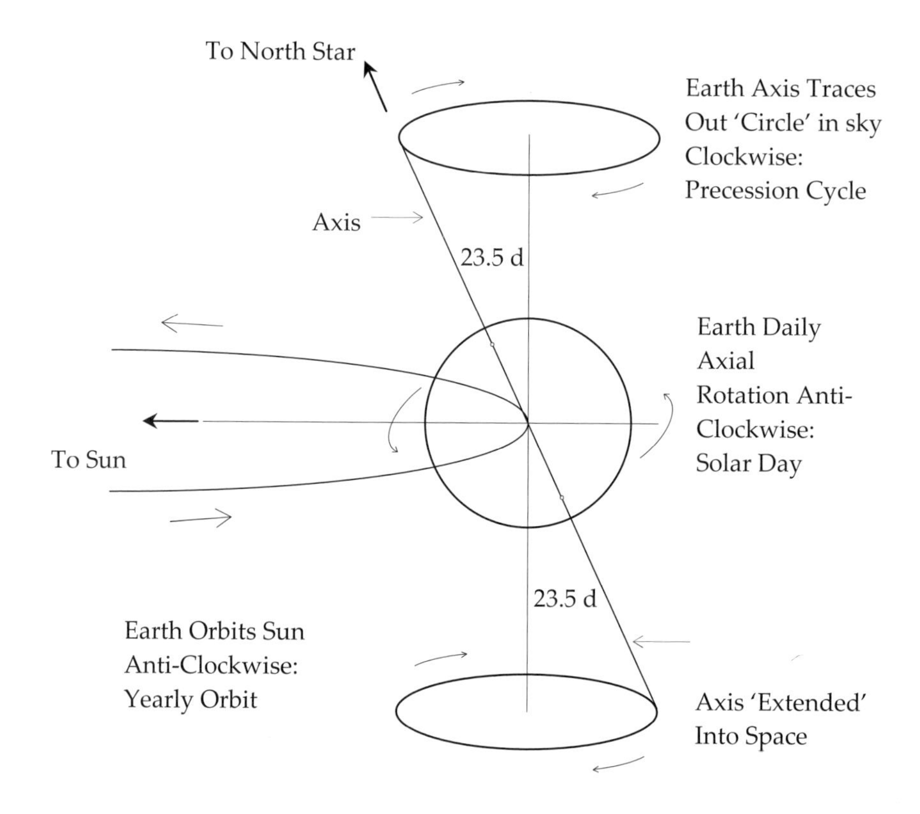

The idea that the Earth is actually engaged in this type of motion rests upon certain anomalies associated with the observation of the background stars. Essentially, though the Earth appears to make one complete orbit about the sun, from vernal equinox to vernal equinox for example, the background stars do not after the same measure of time realign themselves back relative to the sun. Instead, a further 20 minutes or so of extra time is required in order for them to return to the exact same starting position in the sky. As a result of this fact astronomers have been led to derive two entirely different values for the apparent length of the yearly Earth orbit; one based upon the seasonal markers associated with the sun, and the other based upon the background stars. The time period of the former marks out the time between successive vernal equinox points for example, and has a duration of approximately 365.242184 days, which is the length of the Earth Tropical Year as previously detailed. By contrast, the time required for the background stars to return to the same position in the sky is observed to be about 365.256363 days [1], a time period known as the sidereal year. Indeed, it was the desire to account for this very discrepancy that led to the development of the theory of precession itself. A view of the Earth orbit from above helps to explain this in more detail:

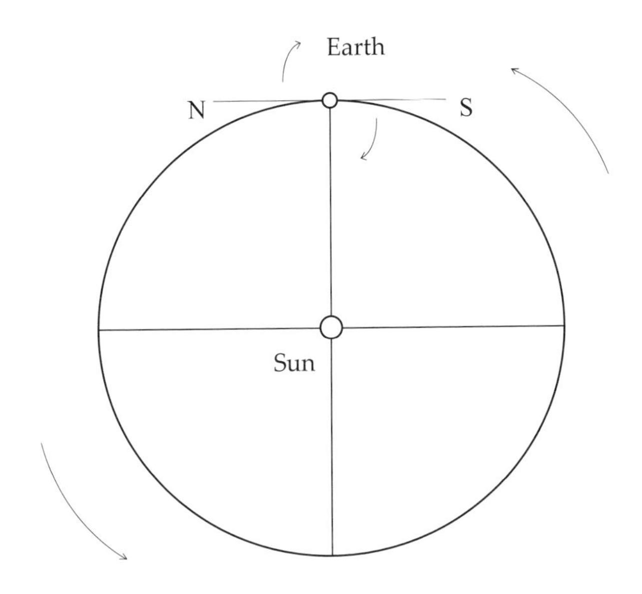

Diagram 2a (not to scale): The Earth is shown at the precise point of the vernal equinox. On the surface of the planet at a point on the equator, the sun is directly overhead casting no shadow. Looking down upon the entire orbit, were one to extend the axis of the Earth into space as a straight line, this line would be at 90 degrees to the line connecting the centre of the Earth to the sun, albeit from the perspective of a visual illusion. From the overhead view the Earth orbits the sun anti-clockwise. In addition to this though, the Earth's axis is thought to be simultaneously moving slowly in a clockwise direction.

Diagram 2b (below): As a result of the two combined motions, the next successive vernal equinox will occur slightly earlier than the time required to actually move a full 360 degrees about the sun:

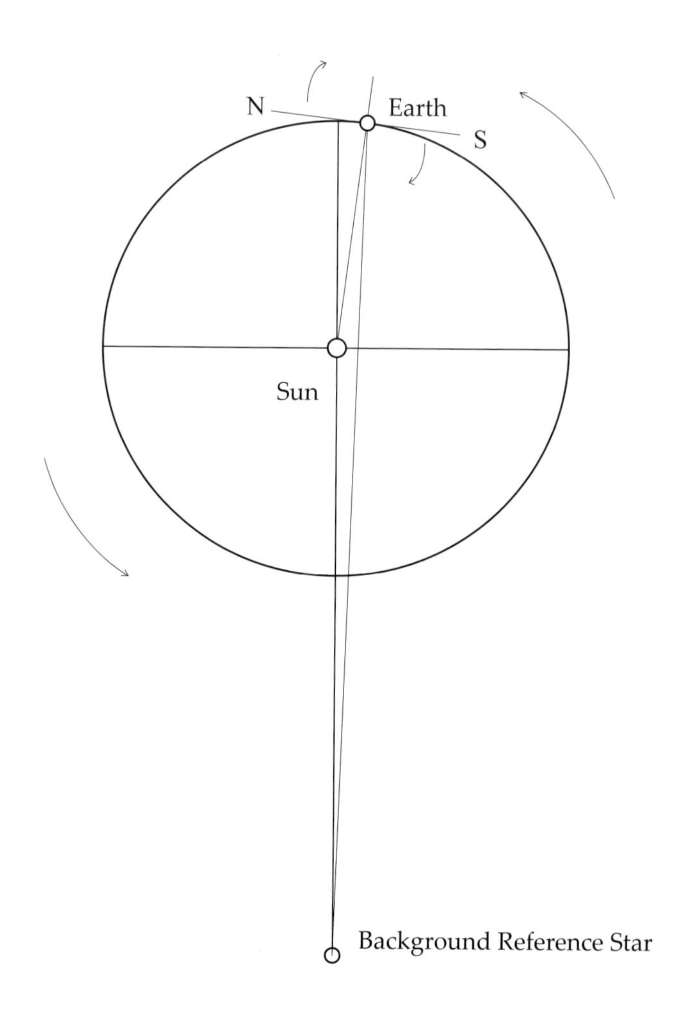

Due to the shortfall, whereas the sun will have returned to the next vernal equinox point after 365.242184 days, the background stars relative to the sun, will not have returned to their original positions. In order for this to occur, extra time is needed, as will allow for the Earth to return to that point in its orbit as will attain once more the starting conjunction between it, the sun, and the background stars. The time required for this to occur will be some 365.256363 days.

As one may surmise from the above description, precessional theory requires that one consider the sidereal year of 365.256363 days to be the time taken for the Earth to complete one full 360 degree angular sweep about the sun, whereas the completion of the tropical year involves a shortfall; being completed about 20 minutes earlier ((365.256363 – 365.242184) × 24 × 60 = 20.4177 minutes of time), equivalent to an angular sweep of about 50.29 seconds of arc (less than 360 degrees). And indeed, with such a shortfall continually repeated every orbit, the point at which the planet achieves each successive vernal equinox shifts in a clockwise direction; a fact that gives rise to the very name of the phenomenon itself: precession of the equinoxes.

An Alternative Theory of Precession

On the face of it precessional theory does seem plausible. The idea that the Earth is engaged in continually changing its orientation as it moves through space would appear to account quite well for the apparent difference between the tropical and sidereal orbital years. However, as a theory it is certainly far from proven, and though one must concede that it is the most dominant explanation for the noted time difference between the two types of Earth year, a strong alternative theory does exist to rival it; one that is also very capable of accounting for the observations. This opposing theory, mooted by a number of individuals and research groups, holds that the sun, about which not just the Earth but all of the other planets orbit, is itself engaged in orbital motion about yet another celestial mass; a nearby star [2]. As a result of this, the time required to complete one full precessional cycle of 25770 years is actually the time required for the sun to orbit once about this object. Thus, the suggestion that the Earth is itself engaged uniquely in an extra movement whereby it continuously changes its orientation in space is entirely discarded. Essentially, such a motion under this alternate theory is considered purely fictitious and held not to occur at all. The discrepancy is thus accounted for due to the orbital characteristics of the Earth and sun interacting with an alternate object, which one may model as follows (NB: In the diagrams to follow only the Earth is considered, the other planets not being required for proof of principle) [3]:

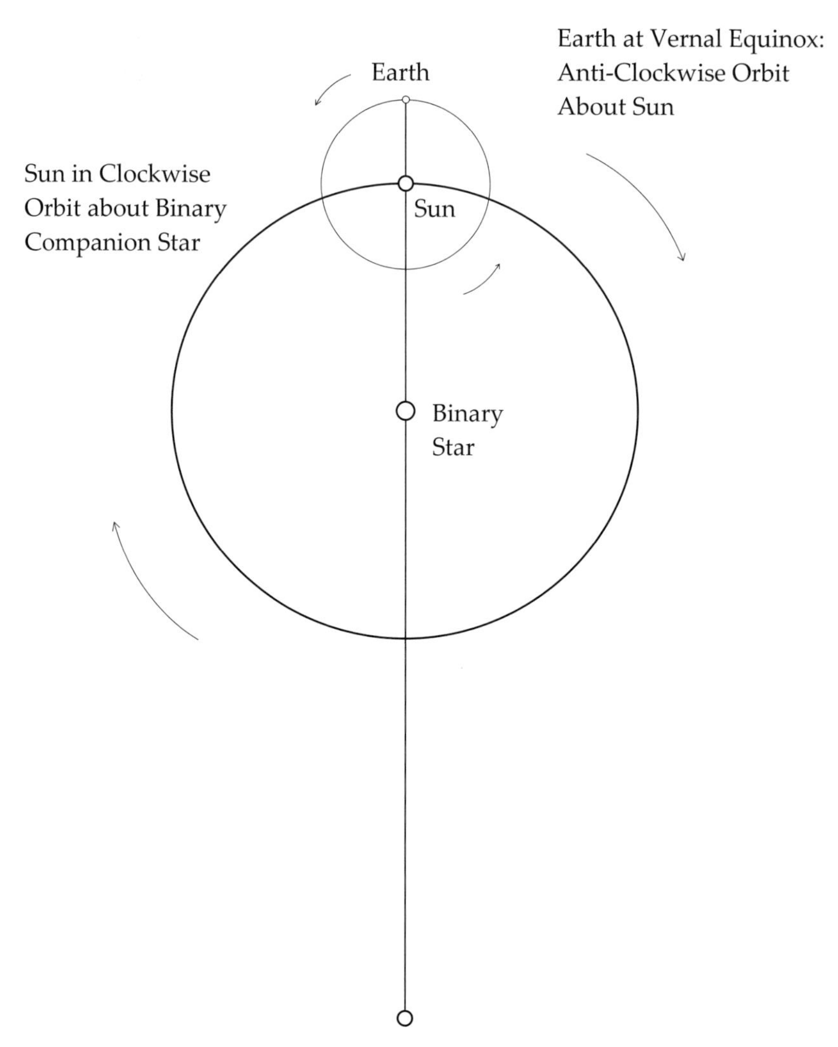

Diagram 3a (Not to scale):

In the arrangement as is detailed above (Diagram 3a), the Earth is aligned to the sun and its binary companion star, and also, a 'far away background star' taken as a reference object from the edge of the visible celestial field.

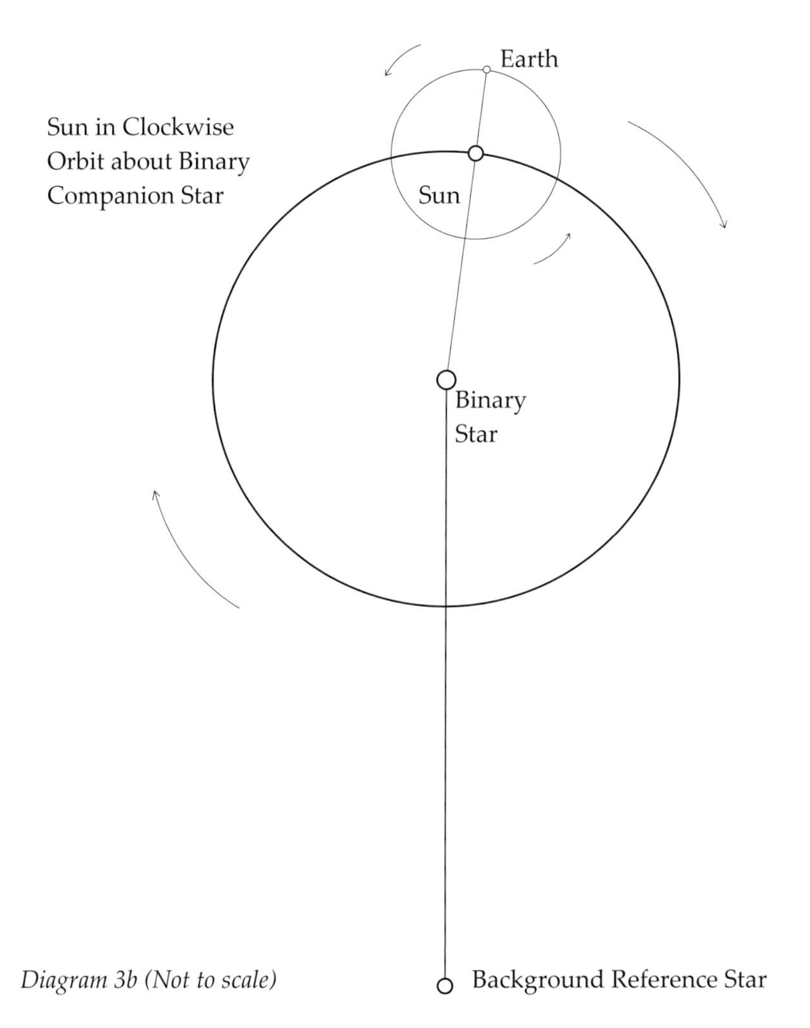

Diagram 3b (Not to scale)

From the initial conjunction, the Earth proceeds on its yearly orbit about the sun travelling anti-clockwise (Diagram 3b) whilst at the same time the sun orbits its binary companion star in a clockwise orbit. In doing so, there will come a point when the Earth will achieve its next successive vernal equinox, at which point it will also re-align with the sun and its binary companion. The time interval between these two events will be equal to the tropical year of 365.242184 days. However, the background reference star will not be a part of this new conjunction.

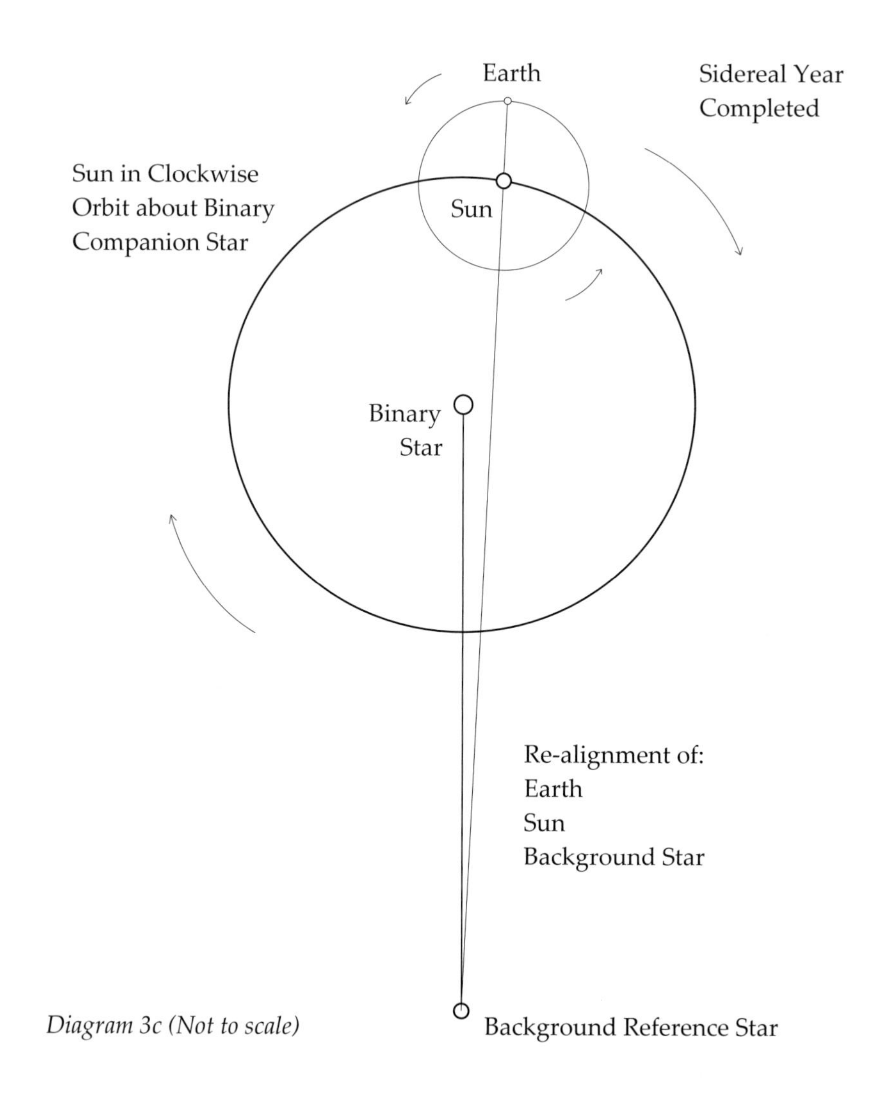

Diagram 3c (Not to scale)

Diagram 3c: With the Earth orbit about the sun being significantly faster than that of the sun about its binary companion star, about 20 minutes or so of extra time is required for the Earth to re-align once more with the sun and the background reference star – this time to the exclusion of the sun's binary companion. The time interval thus separating the initial vernal equinox and this latest conjunction is equal to the sidereal year of 365.256363 days.

Diagrams: 3a, 3b & 3c above not to scale and not to true proportions. Angles exaggerated for clarity of explanation.

The above relations as detailed indicate that the tropical year is thus defined by a successive conjunction between the Earth, sun, and the sun's binary companion star. Whereas the sidereal year is defined by successive conjunctions between the Earth, sun, and any given 'far away' background star. Essentially therefore, under this system, the relationship between the tropical Earth year and the sidereal year is exactly identical to that existent between the moon synodic month and the moon tropical month. Thus, just as the time discrepancy between the synodic month and the tropical month is due to the curvature of the Earth's movement through space i.e. the fact that it is in orbit about another object; the time discrepancy between the tropical year and the sidereal year is due to the very same thing: the fact that the sun (and thus the whole solar system) is itself engaged in a curved motion though space, in this instance about another local body, which is in fact its binary companion.

The Alternative Theory of Precession & this Present Work

In certain forums much has been written comparing each of the two theories of precession, with evidence evaluated in respect of them both. For those interested in the fine details of such analyses, such sources should be consulted [4]. For the purposes of this present work though it is enough to simply summarise the main point over which they appear to disagree, which is that of the timing of past lunar and planetary conjunctions involving the Earth. Essentially, some of the best evidence held to support the alternative theory of precession is that concerning recorded instances of eclipses, and also of the direct transit of the planet Venus between the Earth and the sun [5]. In the case of both types of events, those who uphold the alternative theory state after analysis, that their precise historical timings cannot be reconciled with the view that the Earth completes a full 360 degree orbit about the sun over the course of a sidereal year of 365.256363 days. Rather, the timings of such events, particularly Venus transits, only match up to an astronomical model assuming the completion of a 360 degree orbit about the sun over the course of a tropical year of 365.242184 days. Indeed, the central point of this argument is that conjunctions per se cannot in any way be dependent upon the changing orientation of the Earth axis through space, which is the proposed movement said to account for precession according to orthodox theory. Of course, those who do hold to the standard theory that such a motion is indeed a part of the Earth orbit, claim that the very same data covering past eclipses and Venus transits reveals no discrepancy at all, and that it actually supports the established theory, as opposed to falsifying it. The result of this is complete deadlock between the two camps. They both look at exactly the same observational data and claim that it supports their particular theory [6].

As with all matters concerning the truth of a state of affairs, one is always required to make ones own judgement in light of the evidence as examined by ones own mind. In this, the present author is of the view that the alternative theory of precession is sound, and that the central contention that the Earth completes a 360 degree orbital sweep about the sun in a tropical year is correct. By contrast, the opposing idea that such a sweep is completed in a sidereal year, as claimed under the standard theory, is thus false. Such a view held by this current author is due not just on account of the arguments developed by others for the alternative theory, but also, and not so surprisingly, the fact that the findings of this present work do themselves support the alternative precessional theory. The basic reasoning as given is that the current value of the precessional cycle is closely related to the basic sexagesimal values of the Babylonian system, strongly hinting at a once existent ideal cycle, and also of a law of proportion having been directly responsible for transforming it to its present value.

Now, under the alternative theory of precession, the actual time period as given for the completion of one cycle of 25770 years would be the measure of a distinct orbital period, being that of the sun about a binary companion. It is easy to understand then that as an orbital period, a law of proportion would apply to it. However, under the standard orthodox theory, with precession held to be a subtle movement of the axial orientation of the Earth, one would not be inclined to think the presence of a proportional law. A review of certain records surviving from the ancient libraries of Mesopotamia, in conjunction with a detailed examination of the time value of the precessional cycle itself, would indeed appear to confirm a once ideal standard and a governing celestial law, both of which by their character lend further support to the alternative theory of precession.

The Transformation of Precession

It of course goes without saying that were precession to have had an ideal standard at some time in the past, then it would have been at that very same time as when the Earth itself was possessed of a 360 day year. This being established, just what then would the value of an ideal precessional cycle have been? At least one scholar would appear to have uncovered the answer.

From the study of quite literally thousands of mathematical tables that survived from libraries located in the ruins of the ancient Mesopotamian cities of Nippur, Sippar and Nineveh, one H. V. Hilprecht concluded that the numerical value 12960000 – already shown previously to be a part of the sexagesimal series – was a value strongly associated with the precessional cycle. Essentially, this value appeared to be representative of exactly 500 precessional cycles [7]. Such an understanding is indeed most intriguing because of the way in which this number is tied not only to the ideal moon, as discussed earlier, but also certain astrological ideas. Indeed, within the domain of astrology the sky is usually split up into 12 regions, each being associated with a particular constellation. Thus do astrologers speak of the 12 houses of the zodiac, mathematically idealised as each being a 30 degree sweep along the equator of the celestial sphere. Moreover, to astrologers it is precession itself that is actually held to be the 'marker' denoting which zodiacal house dominates a given age. Over a complete cycle, the sun, rising at the vernal equinox each year, successively 'moves through' each of the 12 houses. Such ideas indeed appear to possess a certain natural linkage to the moon, as the number 1296000, containing one less zero than that identified by Hilprecht, has itself already been shown to be the measure of the ideal moon celestial equator. In addition to which, the time when this was so was when there were precisely 30 days to one synodic month, when a full 12 months were contained within a single 360-day Earth year. One can see then that under this arrangement the sky is split up most naturally into 12 sections by the orbit of the moon; a fact that may even have served as the original basis for the very establishment of the 12 Houses of the Zodiac.

With the stated number 12960000 thus identified with a full 500 cycles of precession, the simple division of 12960000 by 500 must produce a value associated with a single precessional cycle. The answer obtained from this calculation is 25920. Upon viewing this number one is of course bound to suspect that it is a measure of Earth years as would complete one full precessional cycle. Critically however, as the number 12960000 is linked to the ideal Earth-moon configuration, one cannot fail to recognise that the number 25920 must itself denote the measure of a number of ideal Earth years, each being 360 days in length. With such established, one is thus able to give an exacting value for the length of a complete ideal precessional orbital period expressed in solar days, so as to be able to compare it with the currently known value. Both are derived as follows:

Ideal Precessional Orbital Period:

25920 × 360 = **9331200** solar days

Current Precessional Orbital Period:

General precession in longitude (as an angle) per Julian Century = 5029.0966 seconds of arc [8]

5029.0966 / 100 = 50.290966 seconds of arc, or 0.013969712 degrees, per Julian Year

0.013969712 / (365.25 / 365.2421840863) = 0.013969413 degrees per Tropical Year

Therefore, one full precessional period in current years = 360 / 0.013969413 = 25770.58737 Tropical years

25770.58737 × 365.2421840863 = **9412505.617** solar days

With a value obtained for the suspected ideal precessional orbital period, one is able to set it against the currently established value and thus obtain a ratio of transformation:

9412505.618 / 9331200 = **1.008713307…**

As is evident from the ratio, the length of the precessional cycle has increased. Is there any way though to establish that this increase is linked to that of the Earth's own orbital period? Indeed there is. It would appear that there is a very simple and accurate relationship between the precessional ratio of increase and that of the Earth tropical year; one that suggests yet again the existence of a physical law of proportion governing both changes simultaneously. Starting with the established value for the Earth year increase one may obtain a value very close to the above noted precessional ratio, by way of the following manipulations:

$$365.242184 / 360 = 1.014561622$$

$$1.014561622^3 = 1.044324077$$

$$\sqrt[5]{1.044324077} = \mathbf{1.008711699...}$$

Using this ratio, a new value may then be calculated for the length of the precessional orbital period existent in the current age:

$1.008711699 \times 9331200 = 9412490.614$ solar days

(25770.54629…in current Earth years)

Compared with the presently established value the difference is only a little over 15 days:

$9412505.618 - 9412490.614 = 15.00306$ days

In a cycle lasting just over 9.4 million days, such a discrepancy is extremely low. One may thus be confident that a valid law of proportion with the powers as specified does actually link the transformation of the Earth year and precessional orbital period:

$$TY^3 \propto P^5$$

Where: TY & P are ratios:

TY = Tropical Year (present) / Tropical Year (past)

P = Precessional Period (present) / Precessional Period (past)

Chapter 12

Global Destruction in Ancient Times

Universal Harmony in which past Age?

With an ideal celestial configuration thus established for the primary solar bodies, one is of course bound to ask quite naturally just when exactly such a system was truly manifest within the universe. Upon this point, careful consideration of the evidence presented so far, would appear to suggest that it is most likely to have been in an age far removed from the present day. Indeed, an estimate of several million if not billion years into the past would seem eminently more reasonable than a few thousand or even several tens of thousands of years. Evidence in support of this comes first and foremost from the writings of the ancient Egyptians themselves, who as already noted, held that the specified changes to the Earth-moon system – as were responsible for destroying its harmony – occurred in a remote age when 'gods and goddesses' were still coming into being [1]. Such indeed is highly suggestive of the fact that the changes themselves took place when the universe as a whole, or at the very least the immediate solar system, was still in the early stages of its physical development. Further to this though, by far the greatest support for the idea that the transformation of the solar system occurred in a very remote age, results from a study of the actual impact of the noted celestial changes, with particular reference to the physical transformation of the Earth itself. Under such an examination, the unified nature of the laws of proportion cannot be understated.

An increase of about 5 & ¼ days to the Earth's orbital period, accompanied by a slight extension of the distance between the planet and the sun may seem only to be a minor set of changes. But the accompanying increase to the physical size of the Earth, that must match any corresponding change to its orbital period, would undoubtedly result in widespread destruction over the entire planet without exception, were it to occur rapidly over a very short space of time. Could anyone living upon the Earth even survive such an extreme process of physical upheaval, even a remnant? Moreover, consider also how the moon itself is transformed. Once again, a slight orbital shift may seem only to be of minor significance to the Earth, but the fact that this body prior to such a shift was a visible source of light could be of immense significance. The implications for life itself upon the Earth in the face of this may be extremely profound. What affect would a moon constantly radiating light have upon humans alive upon the Earth at such a time? Could they have even existed under such conditions? It is not an idle question.

When one considers carefully the physical impact of the unified changes to the Earth and the moon, it is very difficult to conceive of the idea that they took place in very recent times. Rather, the evidence appears to favour far more the contention that the ideal celestial configuration was manifest at a point in history many, many years into the past. Indeed, the possibility that human life in its current form did not exist upon the Earth at the time of the changes cannot be ruled out also. In view of this, any suggestion that people living upon the Earth just prior to antiquity could possibly have lived through such orbital-physical changes is highly unlikely. One may thus discount the theory that some civilisation from the past could possibly have survived the transformation of the ideal Earth-moon system, as established its present configuration. Such indeed has an important bearing upon another issue of great relevance to this work; confirmation of the fact that knowledge of an advanced nature was once held by those living prior to recorded history.

Lost Knowledge of the Ancients

From a purely technical standpoint, the ability to determine from observations a good set of values for the duration of the tropical year and synodic month does not necessarily imply that those able to achieve such a feat must be a part of an advanced civilisation per se. Such may be accomplished with only a basic understanding of the heavens and without use of any highly sophisticated equipment. That the ancients could thus have obtained such measures and to a significant level of accuracy does not therefore seem too remarkable on its own. However, what does seem remarkable is that they appear to have been fully cognisant of the fact that the observed values existent in their own time, comparable even to those of this present age, were profoundly different from those of a still far more remote age, even to them, and that without ever physically observing them, they knew precisely the exact values that once constituted the key frequency measures and ratios of the ideal Earth, moon and sun, at such a time. Their awareness of a once precise 1/108 ratio between the physical diameter of the sun and the distance between the sun and the Earth, and also of a 1/60 ratio between the actual physical circumference of the Earth, and the orbital circumference of the moon, decisively confirms this.

That the ancients therefore had full knowledge of a most harmonious past celestial configuration respecting at least the Earth, sun and moon, when such was only existent at a much earlier point in the history of the universe, implies by necessity, that they must have at one time possessed the means by which they were able to verify the physical truth of such a configuration. How else indeed would they have ever known the precise magnitudes and ratios that comprised such a system? It is not unreasonable at all then in view of this to suppose that various ancient cultures of the past, unnamed even by scholars of this age, were in possession of at least some if not all of the very laws of proportion detailed in this present work; such as would have allowed them the actual means by which to back-transform the current magnitudes associated with the Earth, moon and sun, and so discover those of a past harmonious age. The laws themselves, as known to the ancients, would indeed constitute a portion of 'high knowledge' of such a people. Moreover, one may go further to state that the very loss of these laws gave rise to the emergence of the mythical stories concerning the transformation of the noted celestial bodies. The myths over time thus replaced the exacting precision of the celestial laws, which were themselves forgotten. The re-discovery and re-statement in this present work of the laws, is thus merely the recovery of such lost knowledge. And yet, that such may occur, must lead one to consider what is without doubt the most disturbing aspect of this entire story; the great tragedy itself that actually befell the ancients, that led to the loss of their advanced knowledge.

Global Disaster & the Collapse of Civilisation

Concerning the precise nature of the catastrophe that so devastated the ancients and deprived them of their high knowledge, one cannot help but consider more deeply Plato's account of the conversation between Solon and the priest of Egypt so long ago. Here it was in this very land that Solon was actually informed by the priest that cycles of destruction were 'built in' to the very life of the Earth, such that every so often extreme events of global significance would bring even great civilisations down to the most primitive of levels. Indeed, according to the priest the events themselves came in two general forms: fire and water. What is most interesting to note though about his conversation with Solon, is that the priest actually provides a very accurate date as to when the Earth last suffered major devastation, which came in the form of a deluge or general flooding. The event was said to have occurred about 9000 years before the priest's own time, circa 9600 BC. Could this then have been the time of the collapse of high civilisation in the ancient world; the moment when an advanced culture was effectively ruined by the impact of a great flood?

According to the Egyptian priest who actually fixed the date of the event, the achievements of the ancients prior to its impact were of such an outstanding nature that they dwarfed those who followed after them. Indeed, in providing Solon with an account of his own native Greece, as existent before the deluge, the priest stated that it contained the finest men that ever existed, and that Solon himself was an actual descendent of those who survived the catastrophe that so devastated his land [2]. Moreover, of Athens in particular, it was the considered opinion of those of Egypt, that it possessed the finest constitution and was the best governed city in the whole world prior to its fall. Indeed, within the actual city itself great attention was placed upon learning, and knowledge of the divine principles of cosmology was existent during this time [3]. One could very well infer from this that the Athenians of this remote era did once possess knowledge of the laws of proportion that govern the celestial bodies of the heavens. It is at least within the bounds of possibility.

In actually evaluating the account of the ancient history of Athens given by the priest, one may perhaps be tempted to think of it as a vague recollection of past events, and that the details provided by the priest are somewhat hazy and based upon no solid foundation. However, such an argument is easily dismissed, as the priest himself specifically states that there are detailed sacred records held in Egypt stretching back 8000 years that confirm the truth of his accounts. He even invites Solon to later on examine the records at leisure for himself in order to obtain more information [4].

Periodic Cycles of Destruction & the Laws of Proportion

Due to the veracity of the account given by the priest one may be inclined to think that a flood alone was responsible for the collapse of high civilisation during the remote age of pre-history. And yet, it is just possible however that it may not have been the sole cause of the ruination of the ancient world after all. One must consider that the Egyptian priest that spoke with Solon said that there were two primary means by which the Earth suffers periodic devastation on a global scale. One was via water, a deluge clearly constituting a case of this; the other, being destruction by fire. Indeed, upon this very point, the priest specifically stated that the Greek myth of Phaethon represented exactly such an account of global devastation by means of fire.

Sensing an actual physical mechanism at work lying behind the myths, with the discovered laws of proportion, it would appear that one has indeed the very means by which to account for such destructive events as noted by the Egyptian priest. Admittedly, there is little that may be said of the deluge mentioned by the priest, in terms of generating a detailed account of its impact. However, with a shift in emphasis towards periodic global destruction by means of fire, through a comprehensive evaluation of the story of Phaethon, one is able to present in some detail a physical account of exactly how the disaster unfolded in terms of the noted Earth changes; a sort of 'running commentary' so to speak. By making use primarily of the story as related by Ovid, it can be seen that each and every one of the orbital-physical changes detailed may be accounted for by the precise manner in which the discovered laws of proportion have been found to transform the Earth over time, including the sun and the moon, which also form part of the story:

1) *Earth Orbital and Geological Disturbances*

As the story goes, almost immediately upon taking to the air in his father's chariot, Phaethon begins to deviate from the proper course. At first he travels to an extreme distance from the Earth, before then turning back to fly too close, at which point many parts of the planet are burnt up as a result, with cracks also forming over the surface penetrating through to the lower layers. Upon first reading there cannot be any real doubt as to the basic interpretation of these noted events. Essentially, what is described is a significant fluctuation in the distance separating the Earth and the sun, accompanied by major geological disruption to the Earth itself. Carefully considered, the discovered laws of proportion allow one to realise just exactly how both are physically related.

Concerning the orbit of the Earth, the relevant laws dictate that any change to the distance between the sun and the Earth, raised to the 47th power, is proportional to a change to its orbital period to the 32nd power, and that any change to the orbital period, is itself directly proportional to that of the equatorial radius of the Earth, which is itself further related to the polar radius via a 3 to 4 power law. Carefully considered one can see then exactly how a significant disruption to the mean distance separating the sun and the Earth can indeed lawfully produce a geological transformation of the Earth itself. Phaethon's initial movement away from the Earth one would associate with an extension of the Earth's orbital semi-major axis. His subsequent flight too close to the Earth would be a shortening of this distance. In accordance with the noted laws of proportion, such an orbital shift would directly result first in a general expansion of the Earth, followed next by a physical contraction. In the case of the former, one could definitely see how cracks would form over the surface of the planet. Respecting the latter, the physical contraction of the Earth would occur specifically as it moved nearer to the sun, and hence also the 'burning up' of the planet.

2) *The Sun Assumes a Position 'Lower' than that of the Moon*

In addition to Phaethon deviating from the correct path and flying too close to the Earth, there is also a statement made in Ovid's account of the story that the chariot of the sun is carried by its horses in such a manner, that the moon itself is amazed to see them lower than its own. That this could imply that the sun is moved closer to the Earth than the moon is of course completely untenable. However, a valid interpretation of this section of the story does present itself when one is taken to consider the intersecting orbital planes of the relevant celestial bodies.

Already it has been stated that the plane of the Earth's orbit about the sun currently differs by approximately 5 degrees from that of the moon's orbital plane; such being established as a direct result of the increase of the Earth year from a 360-day ideal to its present value. Indeed, a very specific law of proportion has been found to actively govern the relative transformation of the angular separation of these bodies, linked in to the orbital period of the Earth itself. In consequence, one is therefore able to understand just what may be implied by the statement that the moon is amazed by the position of the sun's chariot, even lower than its own. The disruption of the Earth orbital semi-major axis, causing further disruption to the Earth orbital period, must in its turn lawfully transform the angle of intersection of the orbital planes of both the Earth and moon. In a relative sense then, one can interpret the myth at this point to mean that the sun's apparent orbital arc as viewed from the moon, brings it closer to the Earth from an angular perspective, than the moon's own orbital arc.

3) *A Transformative, not Static Shift*

One final important point to note with regard to the evaluation of the Phaethon story is that it is critical that one realise that Kepler's general inverse square law does not apply at all with respect to the noted Earth changes. The specific reason for this is that the relevant physical changes that occur are transformative in nature and not static. Kepler's law, with its 3 to 2 power set linking the Earth semi-major axis to its orbital period is simply not operative.

Recurring Cycles of Transformation

The above analysis would appear quite conclusively then to indicate that the laws of proportion detailed throughout this work, are *the* laws that actively enforce the cycles of destruction spoken of by the Egyptian priest, that periodically over time so devastate the Earth. And yet, these are the very same laws that have indeed also been shown to be responsible for governing the full transformation of the entire Earth-moon-sun system, from an ideal configuration, assumed to exist at least on the order of many millions of years ago, to that current to this present age. That there is truth to both of these points; one is led almost naturally to draw certain distinct conclusions:

1) If the destructive events spoken of by the Egyptian priest operate under the guidance of the proportional laws, and they are cyclical in nature, then it would be most reasonable to think that the primary system transformation associated with a 5 day increase of the Earth year, is also an event that is part of a recurring cycle, as opposed to having been a one-off occurrence.

2) The transformation of the Earth from an ideal state to its present condition, being one of a series of such high impact events, would be repeated periodically one would presume, no more than once every several million years, if not less frequently, on the order of a few billion years. By contrast, the destructive cycles spoken of by the Egyptian priest, described as Earth changes that humanity has actually lived through many times, would appear to be much more frequent; on the order of every several thousand or tens of thousands of years. This is suggestive of the fact that the higher the impact of the event i.e. the greater the ratios of transformation, the lower the frequency, and vice versa.

3) The actual level of destructive intensity of the fire and water cycles identified by the priest, as judged in accordance with the proportional laws and using the Earth year as a standard, must involve a mere fractional alteration of its value; a fluctuation of the total orbital period of the Earth of much less than even a single day. For only under such conditions could one

actually conceive of humanity surviving such periodic destruction, albeit in a greatly weakened state.

4) With two distinct cycles of change isolated, it would not seem unreasonable to think that there are many more also 'embedded' within the stated laws, awaiting further discovery.

A Causal Factor

As it is all too evident that for long periods of time there is great stability of the celestial bodies, the various physical magnitudes associated with them must be maintained for the most part at an almost constant level. Not being absolutely fixed, as this would imply the universe as a whole were static, which would seem inconceivable, they must therefore be in a continuous state of transformation in accordance with the proportional laws, such that their rate of change is infinitesimally low over the vast majority of time. As a result, not only do the Earth, sun and moon, including the other planets, have the appearance of stability when viewed over the short term, the physical relations existent between the features of these noted bodies are not obvious to behold. However, when the pace of change is quickened at times of sudden upheaval; when the normal state of affairs is severely disrupted; then are the relations expressed by the laws of proportion truly seen to be manifest in nature. Given this understanding, one is thus bound to be curious as to just what may cause the onset of rapid physical transformation.

Upon this very matter it was noted earlier that the ancients held Earth-bound devastation to be the result of a disturbance of the heavenly bodies from their regular course. In light of the findings of this present work, such would appear eminently reasonable, as the discovered laws of proportion do clearly indicate that the physical transformation of any given celestial body can indeed be linked in a deterministic way to that of all others. Indeed, from the analysis of all of the planets within the solar system, they have all been found to be connected by a unified set of proportional laws. Thus, one can easily see how a change to the orbital path for example of any one planet could directly influence all others including such as the Earth, forcing a change to its own associated primary physical magnitudes. However this may be though, it does not really solve the mystery. All that is stated essentially is that disruption to the orbits of the other planets in the solar system accompanies devastation upon the Earth. The events are thus correlated only. One has said nothing about the cause of such outward effects: an actual *trigger* for abrupt celestial change.

Resonance within the Celestial Realm

In the search for a causal trigger, a most promising candidate would seem to present itself, following a more in depth study of some of the previously noted links found to exist between light, temperature, and also the natural frequencies of the heavenly bodies. Indeed, when such are examined alongside certain advances made in the physical sciences over the past two hundred years or so, one is led to conclude that the periodic transformation of the solar system is the result of what can only be described as recurring acts of 'celestial resonance'. That is to say, that resonance, a physical process widely understood by scientists to be operative within nature under a variety of conditions, is also operative with respect to the orbital and physical properties of astronomical bodies. With an understanding of just what exactly resonance is, one can grasp the basic principle at work.

The Fundamentals of Resonance

In the physical sciences it is quite well established that all outward structural forms possess, or rather are embodied, with sets of fundamental frequency combinations that are 'locked' into their very being. The particular frequencies associated with different forms depend upon their given dimensions, design configuration, and the manner in which they ultimately manifest within nature. At the most basic level, all sensible objects exist in a state of energetic electrical activity; electricity itself being fundamentally frequency based. Of course, the incredible activity taking place upon such a microscopic level is not generally perceived from a macroscopic level. As a result, objects appear as mere static volumetric forms in the everyday world of the senses; the illusion of form masking the inherent activity of the object itself. Regardless of appearance though, all objective forms existent in nature are fundamentally *bound up* by combinations of ordered energetic frequencies existing at multiple levels. Such are the very 'signature' frequencies that *define an object's state*; often referred to as its natural frequencies. It is precisely this state of affairs in nature that makes resonance possible; for just as sensible objects are bound together by ordered frequency combinations, they can be physically *undone* by them. Thus, when an object pulsating in accordance with its own natural frequency is acted upon by another object whose frequency is an equal match; there exists the potential for violent destructive oscillations that may tear apart the primary body. This indeed is the very definition of resonance itself: the disruption of a physical system by another, by means of associated frequency matching.

One may consider the case of a glass being shattered by sound. The very form of the glass in conjunction with its molecular structure determines the precise natural frequency of the object (something that must be determined by actual experimentation). When placed in a regular air-filled environment the glass may thus be impacted upon by controlled sound waves. Sound itself is of course a frequency based disturbance that propagates through air, and as such, energy itself in the form of sound waves can be transmitted through this medium. Ordinarily under everyday conditions most sound waves would not be a threat to the glass. However, if the frequency of the sound waves impinging upon the glass is a precise match to its natural frequency, the energy contained within the waves can feed into the glass to cause major disruption upon a microscopic level. If the energy is sufficient, the whole structure of the glass will be so highly agitated that the bonds of its organisation will loosen causing catastrophic failure; the result being that the glass will shatter. Thus can one grasp from example something of the destructive nature of resonance, and the potential danger that it can pose to all physical structures.

Considering then the potential cause of sudden upheaval within the celestial realm, resonant frequency matching would appear to be a very likely trigger. Indeed, resonance itself is all about frequencies; and with such celestial entities as the Earth, moon, sun, and the other planets of the solar system, actively embodying their own unique frequency configurations – that they are effectively bound up by such energy signatures – there is no question that they are vulnerable to resonance. The fact that they are large celestial bodies makes no difference. Resonance has the power to throw any physical system into great agitation, even those that manifest upon a celestial scale. One may thus posit then that certain planetary configurations – most likely involving conjunctions one would suspect – are able to produce natural resonant activity, such that periodically the whole solar system is 'quickened' and subject to a heightened state of motion, with both the orbits and the physical sizes of the bodies involved being transformed, the latter of which must result in actual global geological upheaval.

With all this said though, one should not think that the onset of resonance and the disruption it may cause to any sensible object is simply and always an act of pure devastation. Rather than leading to its ruination, it may lead ultimately to some sort of beneficial physical transformation; a reconfiguration to a higher order of being. Indeed, one may thus suggest that the entire solar system and all the bodies contained within are led to periodically undergo not just destructive changes, but also constructive changes, that may over the course of time aid the natural development of the system as a whole.

Part 2:

Advanced Nuclear Technology in the Ancient World

Chapter 13

Ancient Engineering on an Immense Scale

Megalithic Monuments from a Lost Age

Up to this point, it has been shown quite conclusively how various key numeric values of the ancient sexagesimal system can be linked not just to a series of active frequencies associated with the orbital periods of the planets, but also to a set of spatially extended wavelength values tied to the actual energetic structure and form of these bodies. This has led indeed to the physical validation of the foot, fathom, and the ideal geographical and ideal nautical mile, as important *natural* units of measure, as opposed to arbitrarily established lengths of distance. Moreover, that the ancients were cognisant of the core set of ideal frequency measures is clearly evident from the fact that they were actually built in to their calendar systems and encoded in their myths. Given this, there is a high level of confidence in the notion that the astronomical knowledge of the ancients was indeed based upon an advanced understanding of the heavens, and that sadly, only remnants of this core knowledge seem to have survived down through the ages. In the effort to truly remove all doubt though, that a highly advanced civilisation was indeed active upon the earth in some remote era, there is yet one final area of study to consider, one that may as yet prove to be totally decisive: the engineering of the ancients.

All over the world on every continent, mysterious structures built of stone are to be found, whose complexity if not size alone have commanded a sense of awe and wonder throughout the ages. Built before recorded history, they stand as silent testimony to the achievements of those from a time long since forgotten. They are giant mounds, walls, statuettes, obelisks, stone circles, pyramids; and in this present age they have been the focus of intense study, with researchers from all over the world seeking to unlock their mysteries and uncover the intention behind them. The question here and of the moment then, is could certain of these structural enigmas be evidence of advanced workmanship, and hence the product of an advanced civilisation? Indeed, if a clear linkage were to be uncovered between the key unit measures of the sexagesimal mathematical system, as already detailed, and the structural design features of certain of these ancient monuments, then this would constitute support for the idea that the people behind the advanced calendar systems established in pre-historical times, were also those who actually built the mysterious megalithic structures that date to the same lost era. Is there then any evidence of such a connection? In answering this question, there is no need to analyse a great many such megalithic monuments from prehistory. One decisive analysis is all that is required, for if just one example of the engineering of the ancients proves itself inescapably to be linked with high knowledge, the case is proven. In view of this fact, it is the intention here to select for examination a single example of the workmanship of the ancients. And there is none more fitting than perhaps the greatest and most enigmatic structure upon the planet: The Great Pyramid of Egypt.

The Great Pyramid

To date there have been many books written about the Great Pyramid. Most are highly conservative, though some are far more unorthodox. It is certainly hoped that the reader of this current work will have at least a passing familiarity with them, and consequently, most of the ideas they contain need not be reproduced here in any great detail. Suffice it to say, in brief, that the two main opposing theories concerning the Great Pyramid hold it to have been built as either a tomb for a notable Egyptian pharaoh, or as an energetic device of some sort using the highest principles of engineering and construction. To orthodox Egyptologists this latter theory is of course anathema. Be that as it may however, in the mind of this current author, the tomb theory is certainly dead. Indeed, it has been killed many times over in the past few years by a significant number of researchers. As a result, one shall not dwell upon it, but rather isolate the most up to date theories concerning the Great Pyramid and proceed from that base. With this in mind, this present chapter will seek to evaluate more deeply, and if possible reinforce the ideas that have already been put forth by Christopher Dunn, in his book *The Giza Power Plant* [1]. Before proceeding directly to his theories however, it is only proper that one provide a summary of the main facts concerning the structure of the Pyramid itself.

The Geometry and Characteristics of the Great Pyramid

The only remaining surviving wonder of the ancient world, dated (conservatively) to approximately 2500 BC, the Great Pyramid is a gigantic stone structure located in Egypt close to the city of Cairo. It's precise location is given as 29 degrees, 58 minutes and 50.952 seconds of arc north, and 31 degrees, 9 minutes and 0 seconds of arc east [2]. Of the building itself, there are a total of 4 sides, each of which have been very carefully orientated so as to be closely aligned to the geophysical points of north, south, east and west, based upon the axial pole of the Earth. The level of discrepancy on this is on average only about 3 minutes of arc in any given direction [3], which is quite an astounding feat.

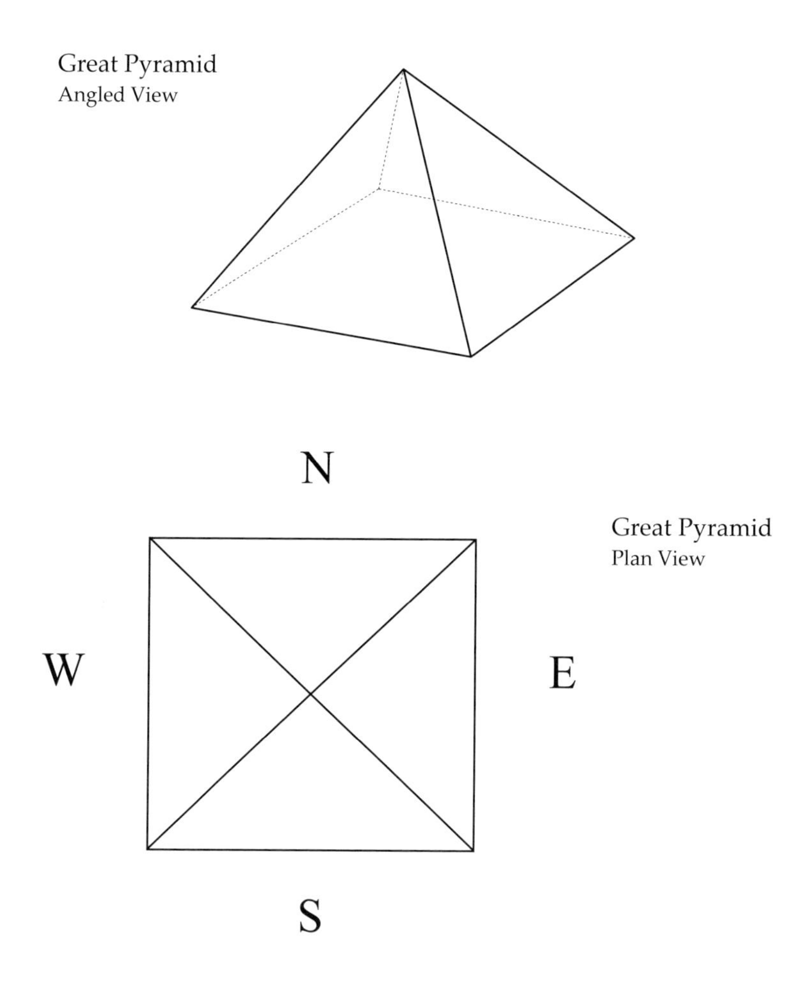

Concerning the actual dimensions of the structure, many surveys have been conducted over the past few centuries to establish its primary values. One of the most recent was conducted in 1925, from which the lengths of the four base sides, expressed in inches, were determined as follows [4]:

9065.1 North
9073.0 South
9070.5 East
9069.2 West

Calculated average: (N+S+E+W) / 4 = 9069.45 inches

Height of the pyramid: 5776 inches, + or – 7 inches [5].
(Survey by William Petrie)

With regard to the interior of the structure itself, it is for the most part composed of solid limestone blocks, approximately 2500000 in total [6]. Of extreme interest however is not so much the total number of blocks that comprise the interior, but the total number of casing stones upon the outside of the structure. Although admittedly the vast majority of casing stones have been removed from the structure over the ages, stripped off unfortunately for use in local buildings; a careful reconstruction of the outer dimensions of the Great Pyramid in line with the dimensions of the surviving casing stones, has revealed a most intriguing figure for what would have been the total number of stones to have composed the surface layer. Many researchers indeed, cite the fact that the complete number of casing stones that would have originally covered the pyramid totalled precisely 144000.

Concerning the interior of the Great Pyramid, one must certainly note that it is not composed of solid blocks of stone throughout. Rather, there are three main chambers within, with an entrance to the pyramid close to the base of the north side allowing access. Initially, upon entry, the passageway penetrates the structure at a descending angle, but after a relatively short distance, a junction is reached. At this point, one may either continue downwards into a small chamber almost centrally located underneath the pyramid in the rock foundation upon which it was built, or break off and move through an ascending passage which goes upwards into the main bulk of the structure. In following the latter, after another relatively short distance, one comes to a further junction, wherein one may either continue upwards still, or move straight ahead through a new passageway that levels out. This new passageway continues on into another relatively small chamber, generally referred to as the 'Queen's chamber', which is almost precisely upon the central axis of the pyramid, located still quite low within its interior. If one were to bypass the queen's chamber though and continue upwards, one will find that the ascending passageway which continues onwards has a sudden increase in its height. Indeed, the entire corridor from this point onwards and upwards into almost the centre of the pyramid has a continuously raised ceiling. This particular section of the pyramid is known as the 'grand gallery'. Upon termination of this section of corridor, it is to be found that once more, the corridor itself then levels off, and suffers a decrease in height as one is led to what is the third and final chamber, the so called 'King's Chamber'. This final chamber is indeed far more elaborate than the other two, and possesses a higher and more complex roof structure. Overall, the actual room itself is not located precisely in line with the central axis of the pyramid but is slightly offset. With that said however, it is placed very close to the centre of the structure as a whole. The following basic diagram details the main internal chambers:

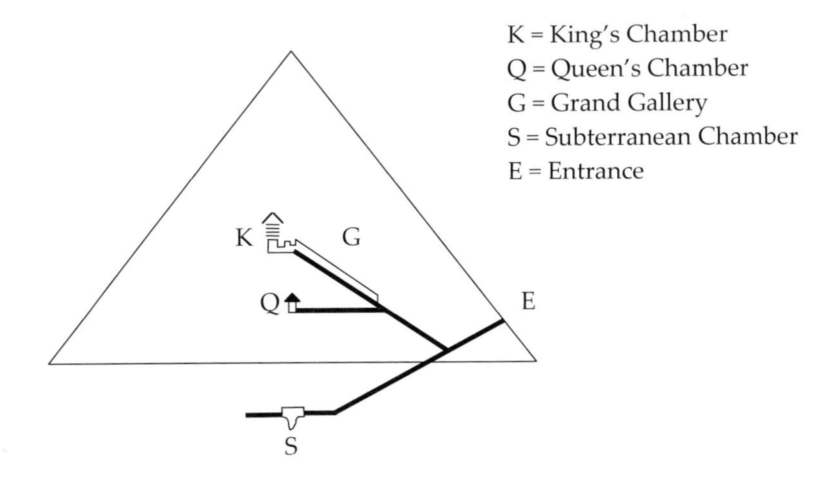

Great Pyramid Internal Chambers & Passageways

It should be noted as a point of historical interest that when the Great Pyramid was first broken into and explored during the early Middle Ages – the first time since when it was sealed – nothing has ever been found as such within its confines i.e. no mummies. Essentially, all of the corridors and chambers were found to be bare.

The Great Pyramid as a Device

With the tomb theory discarded, there have been a number of researchers over the past several years who, through their study of the structure of the Great Pyramid from an engineering/physics perspective, have been able to advance quite well the theory that it was originally built as some sort of energetically functional device. As noted above, one of the main proponents of this theory is Christopher Dunn, whose work suggests that the pyramid was built in ancient times as an instrument able to draw off energy from the Earth itself. To summarise in brief; in his book, *The Giza Power Plant,* he suggests that the materials and dimensions of the structure were carefully chosen to produce a device that would interact with the natural energy fields of the Earth, and extract useable energy through a combination of resonance and harmonics.

The idea of resonance has already briefly been touched upon in this present work in the previous chapter, where it was put forward as the cause of sudden physical catastrophes that periodically blight the Earth over long range intervals. This however, is the destructive side to resonance. In point of fact, when carefully controlled, resonance can be highly beneficial. Concerning the actual theory of resonance, to reiterate; all physical objects possess or are endowed with a set of natural frequencies, dependent upon their size and characteristic dimensions, including also their more fundamental 'electrical' make up. If one were to have an object that was continuously producing or propagating its own natural frequency, which was closely connected to another object carefully constructed to be in harmonic resonance with the first, then when set into motion, the secondary object, usually the smaller of the two, would become a coupled oscillator vibrating in sympathy to the primary structure, and be able to draw off energy from the source [7]. This essentially, is then the central idea as to how the Great Pyramid would have operated as a functional power plant according to the work of Christopher Dunn. Of course, as is detailed in his book, Dunn states, quite reasonably, that as a device, the Great Pyramid would have possessed a considerable amount of technical components and internal machinery, especially in the 'grand gallery' section of the structure, that would have been necessary for its functioning. Essentially, this equipment must then have been removed at some point, so that all that remained was the stone masonry. In effect, the pyramid power plant, after some period of operational use, was decommissioned.

In the course of his studies, Dunn presents evidence of some interesting acoustic properties found to exist within the king's chamber of the Great pyramid. Further to this, he also mentions the work of others who have examined the primary dimensions of the pyramid with the aim of demonstrating how closely they relate to the size of the Earth. Indeed, this aspect of the structure is most critical, for *if the pyramid were built to draw energy from the Earth through resonance, then its physical dimensions would need to be an integral harmonic of the Earth form in some way*. Setting aside then for the moment, the work carried out by Dunn on the resonant properties internal to the pyramid, it will be the intention here in the remainder of this section, to look more closely at the global associations of the structure; the aim being to determine if there is indeed a strong harmonic relationship between the dimensions of the pyramid and the Earth size; one that would support the idea of a resonant device.

In review of the research already conducted upon the issue of a physical correspondence between the primary dimensions of the pyramid and the Earth form, two things have been noted to date. The first is that the perimeter of the base of the structure is almost exactly half of 1 minute of arc swept out upon the plane of the equator. Mathematically, this implies that the pyramid's base perimeter is 1/43200 of the Earth's equatorial value. The accuracy of this association may be determined as follows:

Side length of Great Pyramid = 9069.45 inches

Base Perimeter = ((9069.45 × 4) / 12) = 3023.15 feet

3023.15 / 5280 = 0.572566287 statute miles

0.572566287 × 43200 = 24734.86363 miles

Compared with the equatorial circumference of the Earth as given previously:

24902.4 – 24734.86363 = 167.536363 miles

The second relation of note is the suggestion that the height of the pyramid itself has the same fractional relationship of 1/43200 to the polar radius of the Earth:

Height of Great Pyramid = 5776 inches
(5776 / 12) = 481.33333 feet

481.33333 / 5280 = 0.091161616 statute miles

0.091161616 × 43200 = 3938.1818 miles

Compared to current radius [8]: 3949.9044 – 3938.1818 = 11.7226 miles

As one can see from the expressed relations, a notable discrepancy exists in both cases. In the first instance, a very sizeable error of some 167 miles is to be had. In the second, the error is just less than 12 miles. The latter is of course then far better than the former, but yet how sure can one truly be even of its significance? Can one say on principle that either of the two connections is valid? That is to say, did the builders of the Great Pyramid deliberately intend that a ratio of 1/43200 should link the sides and the height of the pyramid to the Earth, as specified? Upon this point there are quite a number of researchers who do indeed believe this to be true. To them, the above relations are valid, and the margins of error are acceptable. However, one must concede that a great many other researchers do not accept the truth of these relations. To their minds, the level of discrepancy is too high, and the connections are entirely fictitious and imaginary. Of these two positions, this present author must hold to the latter. The relations are indeed simply not convincing enough, and lack decisive accuracy. In consequence, the suggested 1/43200 Earth links to the perimeter and height of the Great Pyramid are hereby discarded as false. Such a decision does however weaken the case for the idea that the pyramid was constructed to be in resonant harmony with the Earth; requiring that one discover some other proof of a decisive connection between the Earth form and the dimensions of the pyramid, in order to maintain the theory of a harmonic device. Can this be done?

A New Approach

If the full equatorial circumference of the Earth is not the answer, and the polar radius is also of no consequence, then what aspect of the form of the Earth may be linked in a harmonic sense to the dimensions of the Great Pyramid to prove a resonant association? The breakthrough would appear to come from a careful study of the latitude of the structure. As previously noted, the precise geodetic latitude of the Great pyramid is 29 degrees, 58 minutes, and 50.952 seconds of arc north. Upon initial viewing, admittedly, such a value does not appear to be of any great significance, especially falling just short of a seemingly 'perfect' 30 degrees north. However, earlier in this work it was demonstrated that arc length measures appear to possess far more significance than strict angular measures. As a result of this, the significance of the placement of the pyramid north of the equator may lie not in the angular measure of displacement, but rather in the actual arc length of displacement. In order to evaluate this proposal, one must of course make use of an exacting model of the Earth ellipsoid form.

Previously, it was seen how one could derive an Earth model based upon certain discovered laws of proportion. However, such a model was more indicative of the energetic frequency configuration of the planet, as opposed to its gross material boundary. Thus, to properly evaluate the position of the Great Pyramid in terms of its actual arc length displacement from the equator, the energetic model would be of no use. Instead, one must use a true ellipsoid model of the planet based upon a precise, mathematically generated 2 dimensional ellipse rotated about its full minor axis, (the details of which were elucidated in a previous chapter) as only this sort of model would capture the true 'static boundary' of the planet. Currently, the most advanced ellipsoid model of this type presently in use is the WGS84 model, under which, the value given to the physical semi-major axis (equatorial radius) of the Earth is 6378.137 kilometres; a value that is coupled with an inverse flattening measure of 298.257223563 [9]. With these two values, the totality of the entire ellipsoid form of the Earth is thus defined, and one may use advanced calculus methods to determine the arc length between any two points upon the surface of such a form. Therefore, employing the WGS84 model, the elliptical arc length measure from the equator of the Earth to the geodetic latitude of the centre of the Great Pyramid is as follows [10]:

29:58:50.952 North = 10885784.94485 feet

Suspecting that perhaps the Great Pyramid may have a harmonic affinity to this value, one may thus divide the result by the average base (side) length of the structure:

Side length of Great Pyramid = 9069.45 inches
9069.45 / 12 = 755.7875 feet

10885784.94485 / 755.7875 = 14403.234963

Immediately upon viewing, one cannot help but recall the precise number of casing stones of the pyramid: 144000. Nor can one fail to recall the value for the ideal speed of light: 144000 IGM/grid second. In view of this, it is not unreasonable to suspect then that the builders of the Great Pyramid may *wilfully have intended* that the arc length distance from the centre of the structure to the equator, should be exactly 14400 times its base side length; thus encoding the same numeric sequence of 144. Such indeed, is an intriguing possibility. And one that places significant emphasis upon the importance of the elliptical arc measure separating the pyramid from the equator. That said, were such a relation to stand alone, it is possible that one may dismiss it as mere coincidence. However, quite remarkably, an absolutely astounding further connection would appear to exist to decisively establish the extreme importance of the noted arc length; one that involves a link to the very ratio of increase of the Earth tropical year itself. In this instance, the critical connection is revealed by the simple division of what is the circular circumference of the Earth at the latitude of the Great Pyramid, by the noted elliptical arc separating the structure from the equator. The following diagram details the relevant measures:

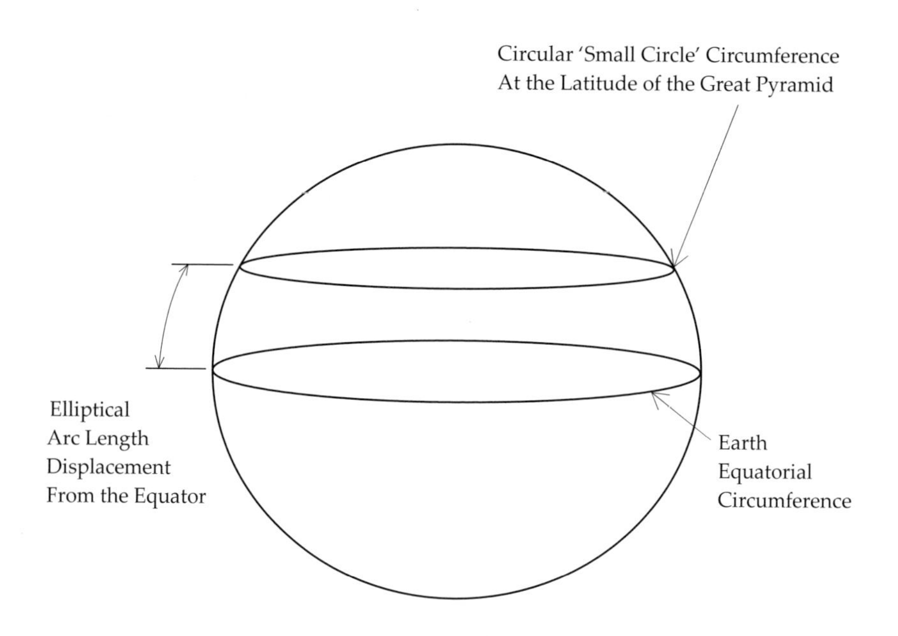

NB: The plane of the Earth's small circle circumference is parallel to the plane of the Earth's Equatorial Circumference.

Under the WGS84 Earth model, with the latitude of the Great Pyramid set at 29:58:50.952 N, the following measures hold true:

The circular 'small circle' circumference of the Earth at the latitude of the Great Pyramid = 110443685.33838 feet [11]

The elliptical arc length from the equator to the Great Pyramid = 10885784.94485 feet

Therefore: 110443685.33838 / 10885784.94485 = 10.145679516...

Compare with the expressed value of the current Earth tropical year divided by the ideal of 360 days, as noted previously:

365.2421840 / 360 = 1.014561622...

As is clearly seen, under the modern WGS84 ellipsoid model, the actual ratio between the small circle circumference of the Earth at the latitude of the Great Pyramid, and the arc length displacement of the structure from the equator, is almost exactly 10 times the ratio between the current tropical year and the ideal of 360 days. As a result, one may very well be inclined to think then, that this relation, in addition to the noted 1/14400 harmonic association, were both *simultaneously intended* by the builders of the Great Pyramid.

The Accuracy of the Noted Associations

Working under the assumption then that each of the two noted associations were indeed wilfully intended, it would be well to evaluate just how accurate they both are, set against the actual placement of the structure as given under the WGS84 Earth model:

1) On the 1/14400 Harmonic Association:

Assuming the arc length from the equator to the Great Pyramid was intended to be exactly 14400 times the base side length of the structure.

Using the average side length of the structure as previously noted, one may derive a value for its position (northwards) as follows:

(9069.45 / 12) × 14400 = 10883340 feet

Compared with its position under the WGS84 ellipsoid model, as already noted, the difference between the two figures is as follows:

10885784.94485 – 10883340 = 2444.94 feet

2) On the Earth Tropical Year Ratio Association:

Assuming the intention to build the Great Pyramid at such latitude, that the small circle circumference of the Earth at that latitude, was exactly 10.14561622… (i.e. 10 times the ratio of the Earth's tropical year increase) times the arc length from the centre of the structure to the equator.

To achieve the noted ratio with precision, under the WGS84 model, the Great Pyramid would have to be situated almost 29:58:51.44895 North precisely. If this were so, then the following would be true:

Circumference of the Earth at the latitude of the Great Pyramid =
110443505.6744 feet [12]

Arc length from the equator to the Great Pyramid =
10885835.1489 feet [13]

Therefore: 110443505.6744 / 10885835.1489 = 10.14561622…

Comparing the arc length component given above with that derived under a position of 29:58:50.952 N:

10885835.1489 – 10885784.9448 = 50.2 feet

As one can see from the analysis, the margin of error in the first association is less than half of one statute mile, and in the second, an extremely exceptional level of accuracy is to be had with an error only slightly over 50 British feet! In light of this, on a point of pure precision, the previously noted '1/43200' relations as detailed are put to the sword in comparison, and one must concede significant support for the physical validity of both the newly discovered relations; that they were wilfully intended by the builders of the Great Pyramid. That being said, when the new associations are compared directly, one cannot help but wonder at the accuracy of the 1/14400 association, and feel disappointment, along with a suspicion, that it should be far greater. How might one therefore account for this? There is little doubt it would seem, that an acceptance of the measured (survey) value for the base side length of the structure is critical. More than any other variable this would appear to be most accountable for the heightened level of error as contained within the 1/14400 relation. Consequently, if one were to determine a more truthful and exacting answer for this value, it would not be unreasonable to suspect that the accuracy of the harmonic association would be far greater. In view of this, a more extended evaluation of the size of the Great Pyramid is most definitely called for.

The Original Base Length of the Great Pyramid

From the values of each of the four sides as presented, it is quite clear that they each differ slightly. One may cite several reasons for this, including ground movement that has over several millennia disrupted the structure, the wearing of the stones, and of course, errors in measurement. As a result of such things, one could never really hope to determine the true dimensions of the pyramid no matter how extensive and repeatedly one surveyed the structure. The problem of determining the true base side length must therefore ultimately rest upon uncovering the actual *intended* dimensions of the structure. How then, might one proceed?

Initially, perceiving there to be nothing immediately remarkable in the survey results of the various side lengths, as expressed in inches, one may wonder if another choice of unit measure applied to the structure would lead to a breakthrough. Indeed, whereas almost all researchers appear to regard the pyramid survey results in inches, or even metres, as just best approximations, a general survey of the literature concerning the Great Pyramid, readily reveals that with respect to one particular unit of measure most specific to ancient Egypt, known as the Egyptian Royal Cubit, there is practically unanimous agreement over the length of the base. Indeed, it would seem that almost all scholars appear united in the view that the 4 base side lengths of the structure were all originally built to a standard of exactly 440 Egyptian Royal Cubits, and moreover, that the height was constructed to a level of precisely 280 Egyptian Royal Cubits. Thus, a great majority of researchers appear to be very confident of the exact measure of the primary dimensions of the Great Pyramid, *when expressed in terms of Egyptian Royal Cubits,* as opposed to any other basic unit of measure. However, even with an acceptance of this, the question does remain: what *exactly* is the length of an Egyptian Royal Cubit?

As a result of modern day archaeological surveys conducted in Egypt, several surviving ancient measuring rods have been uncovered, some of which now currently reside in the Museum of Turin (as of writing). Careful analysis of three of these rods has yielded the following estimates for the true length of the ancient Egyptian Royal Cubit: 20.578, 20.618 and 20.658 inches [14]. Upon viewing, the middle value leaps out almost immediately, as one recalls the length of the ideal semi-major axis of the moon: 206181.8181...IGM. With the opening number sequence of the Egyptian Royal Cubit so closely matching the moon value, one is bound to suspect a principled continuation. Moreover, with the 144 harmonic already found to be strongly linked to the Great Pyramid; that another close derivative of the primary sexagesimal set, 2061818181...would also be linked to the structure, is eminently reasonable. Consequently, one may hereby propose that the true and exact value of the Egyptian Royal Cubit, as originally devised, was a unit length equal to precisely 20.61818181...British inches (20 + 34/55 inches as an exact fraction). Of the three rods then currently held by the Turin Museum, that of 20.618 inches would appear to be far closer to the truth than either of the other two noted.

With the length of the ancient Egyptian Royal Cubit now established, to determine the original intended side length of the Great Pyramid, in British inches, one need only multiply its value by 440:

$$440 \times 20.6181818\ldots = 9072 \text{ inches}$$

Compared with the survey estimate:

$$9072 - 9069.45 = 2.55 \text{ inches difference}$$

From this calculation, it can be seen that the difference is just slightly over two and a half inches. Is this within an acceptable level of tolerance though? Interestingly enough, according to one eminent surveyor of the structure, William Petrie, in calculating the length of the sides, a mean error of 1/4000 may be all the accuracy that one could hope for [15]:

$$9069.45 \times (1 / 4000) = 2.2673625 \text{ inches}$$

Were this added to the survey value:

$$9069.45 + 2.2673625 = 9071.71736 \text{ inches}$$

Although the exact value of 9072 inches does fall just outside of Petrie's margin of error, one may still however contend that it is close enough to be accepted as true. That is, that this is the standard that the pyramid was originally built to.

Following the acceptance of a standard of 9072 inches for the original side length of the ancient structure, one may thus return then to the issue of the harmonic integration of the building with respect to its arc length displacement from the equator. Assuming that the builders of the pyramid intended perfect harmony through use of the value 14400, as is proposed, one need only multiply this value by the side length of the structure to determine the 'ideal' elliptical arc length from the equator:

$$9072 / 12 = 756 \text{ feet}$$

$$14400 \times 756 = 10886400 \text{ feet}$$

With such an exacting answer now in hand, one is thus able to refine the stated level of error in terms of the latitudinal positioning of the structure respecting the noted 1/14400 association. Setting the newly derived arc length measure against that already shown to exist under the WGS84 model, the following is to be had:

$$10886400 - 10885784.9448 = 615.055 \text{ feet}$$

One can see then that the placement of the pyramid under this new calculation is far less in error than under the earlier determination of 2444.94 feet. Indeed, the error as now shown is less than the very base length of the Great Pyramid itself.

An Arc Measure Significant in its Own Right

In continuation of the above analysis, one can also reveal a further intriguing connection. Indeed, the newly refined elliptical arc length of 10886400 feet would appear to possess significance in its own right, even without reference to the Great Pyramid's side length measure. This can be seen by way of a direct connection to the core values of the sexagesimal system itself, through the unit length of the ideal nautical mile, in conjunction with an apparent cube-root function:

10886400 / 6076.388888…= 1791.5904 INM

1791.5904 / 10 = 179.15904

179.15904 / 60 = 2.985984

Though not obvious, the value thus produced at this stage, given as 2.985984, is in fact from a series of numbers that constitute a 'cube' or 3rd power set just removed from the basic values closely allied to the primary sexagesimal progression. Such is evident from the following final transformation: Cube-root of 2.985984 = 1.44.

The Earth Ellipsoid Model of the Ancients

In view of the extended analysis above, it may be seen then that the level of error contained within the harmonic association is much reduced, even though it still does not possess the exceptional level of accuracy of the tropical year linkage; only 50 feet or so in error. That being said, there are a couple of important points that must be raised in this regard.

Firstly, one must realise that with respect to any objectively determined Earth ellipsoid model, it is quite simply impossible on principle that one could obtain *simultaneous and exacting perfection respecting both* of the noted mathematical links. A significant level of accuracy, high enough to physically validate the noted associations in one's mind, is all that one could possibly achieve; even hope for. In essence, there are certain levels of tolerance that one must posit. Indeed, taking this one step further, a second important point should also be noted. Concerning the underlying accuracy of both the harmonic 1/14400 association and that of the tropical year ratio, the analysis of each, as conducted, rests entirely upon the standard of the WGS84 Earth ellipsoid model, *which is a creation of the modern age*. And thus the evaluation of the accuracy of both associations is bound to an Earth model that undoubtedly would not have been *exactly* identical to that employed by those who built the pyramid in ancient times. Indeed, to explore this point further, it would be well to consider how the Great Pyramid stands globally under a variety of different Earth models devised over the modern era.

The Great Pyramid Positioned Under Different Earth Models

Historically, over the past two centuries there have been a multitude of Earth models that have been established, each of which has sought to specify the most exacting dimensions for an idealised ellipsoid planetary form. As a matter tradition, such models are defined by 2 particular characteristics: an equatorial radius (semi-major axis) and an inverse flattening measure. What is critical to understand here, is that with any given geodetic latitude, under different Earth models, that same latitude will possess a different arc length displacement from the equator. By way of example, one may consider the arc length distance from the Great Pyramid to the equator using the fixed geodetic latitude 29:58:50.952 North, under a variety of different historical Earth ellipsoid models [16]:

Bessel (1841): 10884713.1946 ft
Clark (1866): 10885194.3206 ft
International (1924): 10885943.2670 ft

Arc length difference from the suggested 'ideal' of 10886400:

Bessel (1841): 1686.80 ft
Clark (1866): 1205.67 ft
International (1924): 456.73 ft

Respecting the placement of the Great Pyramid then with regard to the 1/14400 association and its suggested perfect arc of 10886400 feet, it can be seen that the error rate of 615 ft obtained previously under the WGS84 model, is far more favourable than that achieved using certain older historical models from the 19th century. And yet, one can clearly see also that it is not as good as that which is obtained under such as the International (1924) model. The fact that this is so indicates quite simply, that the most subtle differences in the models can account for significant variations in the arc length measure from the pyramid to the equator. The problem one is thus faced with, in attempting to actually prove that the ancients who built the Great Pyramid, deliberately built it where they did and to the size that they did in order to express the two noted mathematical associations, is that one will never be able to truly say how accurate they were, unless one were to know precisely what Earth ellipsoid model they themselves employed. Indeed, the Great Pyramid as it currently stands at latitude 29:58:50.952 North may well actively express *both* the harmonic and the tropical year ratio associations to an exceptional level of accuracy, *assuming a certain very specific, as yet undefined Earth ellipsoid model.*

In light of the above facts, one is therefore challenged to do nothing less than reconstruct a uniquely defined physical Earth ellipsoid model of such character, that one may reasonably suppose and actively suggest, was in use by the ancients who built the pyramid, and, which when applied to the structure, yields such an astounding level of accuracy concerning the noted mathematical associations, that they are validated beyond all reasonable doubt! Upon summary of this, one could be forgiven for believing success to be impossible. And yet, strangely enough, it is indeed well within the prospect of belief. There *is* a valid, logical approach.

Combining Form & Energetic Structure

Following the tradition of the modern age, as noted, it is customary to specify the character of an ellipsoid in terms of two key variables: the semi-major axis, and an inverse flattening measure. In the attempt to discover the precise ellipsoid model suspected of being in use by the ancients, it would seem reasonable then to try to determine the specific values they would have assigned to these two variables. Beginning with the semi-major axis, just what value could the ancients have possibly used, and just how exactly would they have derived it? In exploring this, a most intriguing association existent between the 'energy template configuration' of the Earth, using the proportional laws, and the apparent gross physical form of the Earth, defined by a precise mathematically generated circle, would seem to provide the required breakthrough.

As one may recall, in a previous chapter it was detailed how one might establish a value for the current equatorial radius of the Earth using the ratio existent between the current tropical year and an ideal of 360 days:

$$3436.363636\ldots \times 1.014561622 = 3486.402666 \text{ IGM}$$

Further to this, it was also stated that this value, so derived, was a component part of what may be termed the energetic frequency template of the Earth. Also, in order that one might correctly derive the corresponding energetic frequency of the circumference of the planet, one must employ the key ratio 22/7:

$$3486.402666 \times 2 \times 22/7 = 21914.531045 \text{ IGM}$$

In contrast, if one were inclined to calculate the actual 'material' Earth equator, one would have to use a value for the true physical radius of the Earth, and employ PI instead of 22/7, as the latter would be inappropriate:

$$6378.137 \times 0.6213711922 \times 0.88 = 3487.607720 \text{ IGM}$$

$$3487.607720 \times 2 \times \text{PI} = 21913.285587 \text{ IGM}$$

Upon examination of the above sets of figures, one can see a very close correspondence between the two equatorial values, even though they are each associated with a different radius measure. The reason for this is due to a most favourable exchange between the radius values themselves, and both 22/7 and PI. Essentially, though the specified energetic radius is slightly less than the given physical radius, it is combined with 22/7, which is slightly greater than PI, which is itself linked to the physical radius. The 'trade off' is almost an exact match, as one can see:

$$3487.607720 / 3486.402666 = 1.000345644$$

$$(22/7) / \text{PI} = 1.000402499$$

Confronted with these values, the idea presents itself that one may be able to lawfully combine elements of both configurations. Indeed, through a most ingenious step, it would seem eminently reasonable to use the actual energetic circumference value as if it did indeed directly represent the Earth's physical circumference. And, furthermore, use this measure, along with PI, as a means to 'back-generate' a physical radius value for the planet:

$$3486.402666 \times (2 \times 22/7) = 21914.531045 \text{ IGM}$$

$$21914.531045 / (2 \times \text{PI}) = \mathbf{3487.805941} \text{ IGM}$$

A new, alternate value thus presents itself for the current physical equatorial radius of the Earth, differing only very slightly to that of the modern WGS84 model. Given these relations, and their highly suggestive nature, one can hereby propose then that the unit measure of 3487.805941 IGM was known to the ancients, and suggest that it did indeed form a component part of a general ellipsoid model that was in use during the time of the construction of the Great Pyramid. Admittedly, such a suggestion does not rise above the level of speculation at this point. Be that as it may, tentative acceptance does at least allow one to proceed. And to delay any evaluation of the newly derived radius until one has been able to at least obtain an inverse flattening measure with which to combine it, so that a resultant ellipsoid form can be established and tested as a complete model.

Inverse Flattening: A Most Harmonious Earth Measure

In any consideration of what may once have constituted the measure of inverse flattening under an assumed ancient Earth ellipsoid model, it would be well to review certain pertinent facts regarding the discoveries made so far.

During the course of this work it has been demonstrated on several occasions that certain key numeric sequences possess significance under a variety of different conditions. This has been especially confirmed with the evaluation of the Great Pyramid, wherein it was shown that the numeric sequence of 20.6181818... expressing the unit length of the Egyptian Royal Cubit in terms of inches, also defined the ideal semi-major axis of the moon, as given in terms of ideal geographical miles: 206181.8181...IGM. Further to this, it has also been revealed that the key ratio of increase of the Earth tropical year, 1.014561622...is of special importance when multiplied by 10, giving 10.14561622...to the very placement of the Great Pyramid upon the Earth. Such relations as these would appear to provide key insight into what measure of inverse flattening the ancients employed in their established Earth ellipsoid model.

With the extreme accuracy of the noted tropical year (×10) association defining the precise latitude at which the Great Pyramid was built, one cannot deny that the ancients who positioned the structure must have possessed an inverse flattening measure very much comparable to that of the WGS84 model of the modern age (298.257223563). But just what value exactly did they assign to this measure? If this were merely a consideration of just how accurately the ancients were able to measure the Earth, it is doubtful that one could ever truly answer this question. However, if the correct measure is in some sense an ordered numeric sequence intimately tied in to the primary values of the sexagesimal system, then the discovery of a precise answer would seem within reach. And indeed, this would appear so, for there is among all possibilities one particular value that stands out; one that has quite remarkably, already been demonstrated to be closely linked to Great Pyramid.

In the analysis of the arc length displacement of the structure it was shown previously that via the established ideal nautical mile, the Great Pyramid could be linked to the harmonic sequence of 144. But, in actually presenting this association, it was revealed that the cube or third power transformation of the harmonic 144 produced a numeric sequence of 2985984. Given this figure, were one to simply shift the placement of the decimal point in relation to the specified value, one is easily able to generate 298.5984. With an undeniable correspondence to the inverse flattening measure of the Earth under the present day WGS84 model, of 298.257223563, the bold claim is thus made that the very value of 298.5984, was indeed that which was recognised as the true measure of the inverse flattening of the Earth by those who built the Great Pyramid. This is not to suggest that they chose this value merely because it appeared 'fitting' or intriguing. But rather that they chose it because *they actively discovered during their time that it was in point of fact the correct value associated with the geophysical structure of the planet.* And as such would be combined with a precisely determined semi-major axis or equatorial radius to derive a complete and exacting ellipsoid Earth model.

Combining the results of the above analysis, it is hereby proposed then that the Earth ellipsoid model in use by the ancients who built the Great Pyramid was composed of the following values:

Semi-Major axis: **3487.805941** IGM
Inverse flattening: **298.5984**

For simplicity, the model shall hereby be referred to as the GP2007 Earth ellipsoid model, due to the fact that it was reconstructed as part of an overall evaluation of the Great Pyramid itself.

With an acceptance of the GP2007 model as being actively employed by those who built the Great Pyramid to position the structure globally to best incorporate the tropical year and harmonic associations, one would expect that its position within such a model, newly analysed, would reveal a greater level of accuracy concerning the noted associations, than present under the WGS84 model. In this one would not be disappointed. Indeed, the level of correspondence is almost practically at an optimum! The calculations are as follows:

Under the GP2007 Earth model, with the set latitude of the Great Pyramid at 29:58:50.952 N, the following holds true:

1) On the 1/14400 Harmonic Association:

Elliptical arc length from the equator to the Great Pyramid [17] =

10886476.544 feet

Compared with the theoretical ideal =

10886476.544 – 10886400 = 76.54 feet

2) On the Earth Tropical Year Ratio Association:

The circular 'small circle' circumference of the Earth at the latitude of the Great Pyramid [18] = 110449860.963 feet

Therefore, given the above elliptical arc:

110449860.963 / 10886476.544 = 10.145602253…

Moreover, to achieve the theoretical ideal of precisely ten times the Earth tropical year ratio of increase i.e. 10.14561622…, under the GP2007 model, the Great Pyramid would have to be situated 29:58:50.8423 N. If this were so, the following would be true:

Circumference of the Earth at the latitude of the Great Pyramid [19] =

110449900.626 feet

Arc length from the equator to the Great Pyramid [20] =

10886465.461 feet

Therefore: 110449900.626 / 10886465.461 = 10.14561622…

In comparing the arc length component at latitude 29:58:50.952 N with that at 29:58:50.8423 N, as given, the difference is as follows:

10886476.544 – 10886465.461 = 11.08 feet

Essentially, were the Great Pyramid to have been positioned a mere 11.08 feet south of its presently determined location, then under the GP2007 model, there would be a perfect correspondence between the ratio of increase of the Earth tropical year ×10, and the ratio between the small circle circumference of the Earth at the latitude of the pyramid, and the arc length from the structure to the equator.

An Evaluation of the Results & their Implications

In comparing the relevant Earth measures under both the WGS84 and the GP2007 models, it can quite easily be seen then respecting the two noted associations, that the latter model provides a far greater level of accuracy. Of the harmonic 1/14400 association, the difference from the 'ideal' of 10886400 feet under the WGS84 model was shown to be some 615 feet. However, under the GP2007 model this is dramatically reduced to only some 76 feet; barely a tenth of the base side length of the Great Pyramid itself. Also, of the Earth tropical year ratio (×10) correspondence, the latitudinal discrepancy in achieving the ideal ratio here is reduced from 50 feet under the WGS84 model to only 11 feet under the GP2007 model. In view of these facts, the GP2007 model is a far more exacting template upon which to have established the location of the structure to achieve the noted relations to their most optimum level. Such indeed strongly supports the overall contention that the two measures i.e. semi-major axis and inverse flattening, of the GP2007 model, are physically valid, and that this model was known to the ancients and actively employed to position the Great Pyramid. Indeed, upon this point one cannot fail to grasp the incredible implications that spring from the exceptional level of accuracy as achieved in the positioning of the stone structure. For one must understand that elliptical geometry is *inherently* more complex than simple circular geometry. The ability to even know how to calculate an arc length distance on a section of an ellipse has only been known in the modern age for a few hundred years. And yet, to have placed the Great Pyramid where they did, the ancients must have possessed exactly just such an ability, allowing them to compute not only the full dimensions of the Earth ellipsoid form, but go from an abstract set of calculations 'on paper', to the physical determination of the actual location upon the Earth wherein they intended to build the structure. Would this author be correct in thinking that one would need some sort of global satellite positioning system in order to accomplish this to accommodate an error of just a few feet? The case therefore in favour of a highly advanced civilisation existing upon the Earth in some remote age is almost sealed just with this.

Furthermore, with the deliberate incorporation of the ratio of the current tropical year and a value of exactly 360 days (multiplied by 10) in positioning the Great Pyramid, there is no better physical proof for the reality of the fact that the ancients did indeed truly believe that the Earth once possessed a 360-day year. But just why did they seek to achieve this relation in positioning the structure? The answer to this question forces one to return to the very issue of the nature of the structure, as suggested in the work of Christopher Dunn; that the pyramid was built as a coupled oscillator capable of drawing off energy from the Earth through resonance. Indeed, as has already been stated, resonance is all about frequency matching. Therefore, in view of the placement of the pyramid and its size, the builders would appear to have scientifically pre-determined that both the harmonic 1/14400 relation and the tropical year ratio (×10 multiple) incorporated into the structure, would best allow them to tap into the energetic frequencies of the Earth form and orbit.

Extended Relations: The Great Pyramid & Stonehenge

From the above analysis of the Great Pyramid, it is very clear then that an important harmonic relationship exists between the baseline length of the structure and the latitude arc displacement from the equator, as supports the idea of a resonance based device. That said however, it would seem that this particular structure of the ancient world certainly does not stand alone with such intriguing numerical associations. For indeed, at the very beginning of this work in the Introduction, it was mentioned that following an analysis of Stonehenge – by far the most famous stone circle in the whole of England – a number of researchers had come to the conclusion that it too was a technological device of some sort, just like the Great Pyramid. Now, if indeed there is truth to this, then based upon all of the work as presented so far, it would be eminently reasonable to think that it too would possess some sort of mathematical affinity to the main values of the sexagesimal system and its close derivatives, and to the Earth form. In this, one would not be disappointed.

Following a careful analysis of Stonehenge, it is revealed that it too does indeed possess a most remarkable relationship, not just to the general ellipsoid form of the Earth, but also to the Great Pyramid itself, and via the primary values of the sexagesimal system and its close derivatives. The very nature of such relations appear to indicate quite decisively that it too was built as a device, and that without any doubt, the science behind its construction was identical to that of the Great Pyramid.

The Location of Stonehenge

According to modern sources, the Stonehenge circle in England is situated at the following geodetic coordinates (deg/min/sec) [21]:

Latitude: 51:10:40.54 North
Longitude: 01:49:29.032 West

With respect to the latitude of Stonehenge, the accompanying small circle circumference parallel to the Earth's equator (GP2007 model) is given as follows [22]:

77042854.095 feet

Set against this is the equatorial circumference of the Earth itself as also under the GP2007 model [23]:

131487186.278 feet

With the ratio of the two noted values:

131487186.278 / 77042854.095 = 1.7066759509…

Following a simple set of manipulations one can see the ratio linked to a very familiar numeric sequence:

1.706675950 × 3 = 5.120027852

5.120027852 / 2 = 2.560013926
2.560013926 / 2 = 1.280006963
1.280006963 / 2 = 0.640003481
0.640003481 / 2 = 0.320001740
0.320001740 / 2 = 0.160000870

Starting out with an exact value of 1.7066666666666… (128 / 75) multiplication by 3 and successive division by 2 gives the precise sequence as follows:

5.12 → 2.56 → 1.28 → 0.64 → 0.32 → 0.16 → 0.08 → 0.04 → 0.02

It would appear then that Stonehenge was deliberately built where it was such that the ratio between the equatorial circumference of the Earth and the small circle circumference of its latitude was precisely 1.706666666666666… or 128 / 75 expressed as an exact fraction.

The Great Pyramid & Stonehenge: A Global Linkage

In addition to the above noted relation there is another most startling connection to be had; one that truly establishes beyond all doubt the significance of the placement of Stonehenge, not just with respect to latitude, but also longitude. By evaluating the direct surface arc length distance separating Stonehenge from the Great Pyramid, it can clearly be seen that the value expressed in feet, is itself linked to a close derivative of the primary sexagesimal numeric series:

With Stonehenge at: Latitude: 51:10:40.54 North
Longitude: 01:49:29.032 West

The Great Pyramid: Latitude: 29:58:50.952 North
Longitude: 31:09:00 East

The surface arc length separating the two structures (GP2007 Earth ellipsoid model) [24]:

= 11809780.017 feet

Successively divided by 9, a very familiar sequence emerges:

11809780.01 / 9 = 1312197.779
1312197.779 / 9 = 145799.7533
145799.7533 / 9 = 16199.97258
16199.97258 / 9 = 1799.996954
1799.996954 / 9 = 199.9996615

As one may surmise, ideally the numeric progression would be:

11809800 → 1312200 → 145800 → 16200 → 1800 → 200

Given a discrepancy of only 20 feet or so from an ideal arc length separation of exactly 11809800 feet, one must conclude without any real doubt, that there was a deliberate intention on the part of those who built Stonehenge to build it precisely 11809800 feet from The Great Pyramid, and that this, in conjunction the latitudinal fractional relationship of 128 / 75, was critical to its proper functioning.

The Technological Basis of the Great Pyramid & Stonehenge

Even with the above facts clearly established, one is still forced to acknowledge that though indeed both the Great Pyramid and Stonehenge appear to have been fully operational technological devices from some lost age, today they stand silent; simply the mathematical associations that link their form and features to the Earth being all that remain to indicate the existence of once mighty machines. What would decisively advance one's understanding of these ancient monuments is if one could demonstrate something of the true power that may be harnessed from the physical interaction of a man made device with the energetic Earth form.

In consideration of this, it would seem that certain critical advances in the area of nuclear physics in the present age appear to have a strong association with the harmonics built in to the Great Pyramid and Stonehenge, offering an intriguing insight into the operational nature of these ancient structures. Such then will be the primary focus of the next chapter.

Chapter 14

The Physics of a Nuclear Explosion

Of all the scientific advances of the late modern age, there are none more intriguing, more terrifying, than those that occurred within the realm of nuclear physics during the first half of the 20th century. Up until this time the physics of atomic theory was no more advanced than in the time of the ancient Greeks. Certainly there was the idea of atoms, of elemental particles so fundamental that they could not be reduced further. They were thought of simply as extremely small solid balls of matter, any notion of an internal structure at the very periphery of consideration. In the early 20th century though new and revolutionary methods were devised to probe the core of the atom; methods that allowed scientists to realise for the first time the incredible amount of energy locked within its structure, a fact that led ultimately to the development of a new type of weapon: the nuclear weapon. To understand the power of such a weapon, it is important that one first grasp something of the concept of energy per se, and of how scientists historically have expressed its measure through the use of mathematics.

The Energy Locked within the Atom

In the modern era, the development of a basic equation for energy resulted from a set of simple experiments first carried out in the 17th century involving dropping a series of weights onto a soft clay floor. Knowing that as objects fall their speed increases, scientists set out to evaluate how both the speed of impact and the mass of a falling object affected the clay i.e. the damage as produced in the form of a distinct indentation. As a result of careful study, it was found that were a given mass with an impact speed of (for example) 1 metre per second to produce a crater 1 mm deep in the clay, impacting at 2 m/s it would generate a crater 4 mm deep, at 3 m/s the crater would be 9 mm deep, and at 4 m/s an indentation 16 mm deep would be the result, and so on.

From this, it was thus discovered that one need only mathematically square the velocity of impact to reveal the (proportional) depth of the craters that would be produced. Further to this, by fixing the speed of impact and simply varying the mass of a dropped object, it was found that a mass of 1 kilogram producing an impact crater 1 mm deep (for example), would produce a crater 2 mm deep were it 2 kg, and 3 mm deep were it 3 kg, and 4 mm deep were it 4 kg, and so on. Mass was thus discovered to be directly proportional to the force of impact. Combining both results, the tests appeared to confirm then that the energy *realised* from the weights – characterised as a force of impact – was directly proportional to the mass of the weights, coupled with, i.e. multiplied by, their velocity of impact, squared. This was summed up at the time by an equation formulated by Gottfried Leibniz (1646-1716 AD):

$$E = mv^2$$

Energy = mass × velocity, *squared*

In the late 17th and early 18th centuries, the Leibniz equation was used primarily to evaluate simple objects in motion and the energy as transferred from one object to another through collision, with the impacting object retaining its fixed mass whilst having its velocity upon impact reduced (e.g. to zero). However, in the late 19th and early 20th centuries it was found that the Leibniz equation could be applied directly to a given amount of matter, in order to sum up the total amount of energy as 'contained' within its form, or mass. What led directly to this was the discovery of a new set of elements within nature; a group that was found to be strangely unstable. The name given to the first of them, as listed in the modern periodic table of elements, is Uranium.

Generally referred to as being 'heavy metals' due to their excessive weight, this newly discovered group of elements had a most striking feature previously unknown amongst all the other lighter atomic forms. They were found to be continuously engaged in a process of radioactive decay. Essentially, they were constantly 'throwing off' their matter into space by converting it into radiation. Radiation however, is in actual fact though a form of light (of very high frequency, and harmful to humans if received in excessive doses), and light itself is indeed substantial i.e. it possesses mass. Therefore, invoking the principle of conservation of energy, were a given lump of Uranium for example to cast off its own matter as radiation, which propagated away from it at the speed of light, then the measure of light emitted in any given time period would correspond to a measure of mass 'lost' from the lump of Uranium itself. Consequently, one could say that the Uranium was radiating itself away to nothing i.e. in the sense of no longer existing as a concentrated lump of matter, due to a complete dispersal of its material throughout the universe in the form of light waves. And yet, this is done at such an incredibly fine rate, that ultimately, it would take many millions of years for a given amount to completely 'dissipate'. Even so, with this understanding of what was happening with respect to Uranium, and other such elements like it, it was realised that one could apply the energy equation of Leibniz to determine the total energy of a given lump of matter. Essentially, knowing that matter unravels itself at the speed of light, the total energy of a given amount must be the mass of the matter, multiplied by the speed of light, squared. In equation form:

$$E = mc^2$$

Energy = mass × speed of light, *squared*

(NB: '**C**' denotes light speed in vacuum)

As light is such an incredibly fast speed compared to the far slower speeds that govern the basic motion of 'everyday objects', when used as a multiplier (squared) in the Leibniz equation, there is the powerful realisation that the smallest amount of matter must contain an extremely large amount of energy locked within its structure. As this fact dawned upon scientists at the beginning of the 20th century, attention inevitably turned to the question of how to tap into such energy, and this indeed *is* the story of the development of nuclear weapons.

The Basic Structure of the Atom

As a direct consequence of the discovery of radioactivity, during the first few decades of the 20th century, scientists were able to use a certain class of radioactive particles to probe the atomic form and so determine its basic structure. What they found was that atoms – of any given element – were composed mostly of empty space, with an outer periphery of negatively charged particles orbiting about a central core or nucleus, composed itself of a cluster of positively charged particles, and another type of particle found to possess no charge at all. Scientists named these particles: electrons, protons and neutrons, respectively. And found that the basic atomic structure of all known elements was composed of them. What made the primary elements different from one another was simply the number and combination of such particles that constituted their essential form. For example, the lightest known element Hydrogen is made up of just 1 proton at its core with 1 orbiting electron, whilst such as carbon is composed of 6 protons and 6 neutrons at its core, with 6 electrons in orbit. By contrast, towards the far end of the periodic table where one finds the unstable heavy metals engaged in radioactive decay, Uranium (235) has a core of 143 neutrons and 92 protons, with 92 orbiting electrons, whilst Plutonium (239) has 145 neutrons and 94 protons at its core, with 94 electrons in orbit.

In their natural state the overall charge of an atom is balanced, due to the fact that all atoms tend to possess the same number of protons within their nucleus as they do orbiting electrons. That said however, though the charge of an electron is a match to that of a proton, the actual mass of a proton is itself about 1800 times that of an electron. The mass of the neutron is also about the same as the proton, but just ever so slightly higher. Thus, the great bulk of the mass of an atom is found to be concentrated at its core in the cluster of protons and neutrons that make up the atomic nucleus. Moreover, as the elements become heavier, the number of neutrons increases within the core in such a manner that proportionally they begin to constitute significantly more of the mass of the nucleus than the protons. From a general study of the lighter elements one finds that in the nucleus of such atoms, the number of protons is equal to the number of neutrons. However, as one moves towards the heavier elements, more neutrons are to be found within the core than protons. This is very much evident in the case of the heavy metals, as detailed above, e.g. Uranium and Plutonium.

Unleashing the Energy of the Atom

Prior to the year 1945 all conventional explosives in use tapped into the merest fraction of the energy of the atom in order to produce their destructive effects. Invariably they relied on various chemical reactions and so forth to produce the heat and force of an explosion. The actual materials themselves were very bulky, and thus in the theatre of war, singularly powerful bombs built using conventional explosive materials had their limit, which was the very real consideration of how they could be practically deployed to a given target site. To comprehend this most fully one need only consider the preparations that were made by the United States in the run up to testing the first atomic device of the modern age, codenamed Trinity. Just prior to the Trinity test, which took place on the 16th of July 1945, the US military conducted a large conventional test with an explosive yield of some 108 tonnes of TNT, on the 7th of May. The purpose of this initial test was to calibrate various instruments for measuring the blast wave produced, to aid the later evaluation of the Trinity test [1]. In preparation, engineers built a special wooden platform on which to stack the boxes of TNT. In all, 4000 boxes were carefully placed upon the platform, each box weighing 50 pounds, giving a total weight of 200,000 pounds [2] – an explosive yield of 108 tonnes of TNT. In contrast to this giant pile however, the Trinity bomb itself detonated some 2 months later made use of a Plutonium core of just 13½ pounds [3], producing an explosion *equivalent* to between 20,000 to 22,000 tonnes of TNT! [4]. Needless to say, the crater of the Trinity test dwarfed that of the conventional test by some measure. But just what exactly did the scientists do to tap into the energy of the plutonium to produce such a destructive effect?

Fission & Fusion

Primarily, there are two ways in which one can harness the immense energy locked within an atomic structure, as identified by the modified Leibniz equation. Both were discovered during the 1930's through to the 1940's, and are respectively, fission and fusion. The first of these, fission, involves the splitting of atoms. The primary elements that are most suited for this (in a nuclear device) are Uranium (235) and Plutonium (239), as they each possess a very large nucleus with a far higher number of neutrons compared to protons; a fact that makes them very unstable and easily split compared to the atomic cores of other atoms. Actual fission itself is achieved by bombarding such heavy elements with a stream of free neutrons. Because such particles have no charge they are able to travel right through to the core of such atoms and by striking and 'overloading' them, cause their nucleus to split roughly into two. This creates, or rather transforms, a larger atom into two other smaller atoms more stable in nature, whilst also releasing a great amount of energy. Thus for example, a Uranium (235) atom when split may produce Krypton and Barium, plus 3 additional neutrons [5]. Moreover, the extra neutrons may well then go on to strike other nearby Uranium atoms to split them also, and thus cause a chain reaction effect, which indeed is a critical process involved in actually creating a nuclear explosion. Such fission of a sizable mass of radioactive material involving a chain reaction effect massively taps into the energy store of such atoms, as expressed by the $E = mc^2$ equation.

In contrast to fission, the second method by which one is able to tap into the immense energy contained within the atom is through fusion, which as a process is in fact the exact opposite of fission. Whereas with fission, extremely heavy elements are split to release energy causing the formation of smaller more stable atomic structures, fusion produces a release of energy by uniting extremely light elements already possessed of a highly stable configuration. Invariably, the elements best suited for fusion are Deuterium and Tritium, each being a close variation of the basic Hydrogen atom, made up itself of 1 proton and 1 orbiting electron. The sole difference between Deuterium and Tritium, and the basic Hydrogen atom, is that the former two each possess a larger nucleus due to the presence of neutrons at their core, a particle which the Hydrogen atom itself lacks. Thus, Deuterium has 1 orbiting electron with a core of 1 proton and 1 neutron, whilst Tritium has 1 electron with a core of 1 proton and 2 neutrons. The fusion of Deuterium and Tritium within a nuclear weapon creates as a by-product both helium atoms plus the release of additional neutron particles. Carefully managed, the fusion process massively taps into the atomic energy of these light elements.

High Yield Thermonuclear Weapons

Although both fission and fusion allow for the practical realisation of the enormous amounts of energy held within the atom, one should note that of the two, fusion is vastly more powerful in releasing the energy contained within a given atomic structure. That said, as a matter of necessity, in order to initiate a fusion reaction one must first trigger a fission event. This is due to the fact that fusion itself can only occur under extremely high temperatures of many millions of degrees Kelvin, and only a fission event can create such temperatures within a nuclear weapon. As a result, modern day high yield nuclear weapons, or warheads, are purpose built to combine both fission and fusion, each reaction itself taking place within a separate compartment within the overall device. An initial fission event is thus used to generate the high temperatures as are necessary to trigger a subsequent deuterium-tritium fusion within a second chamber of the warhead.

In general, devices that employ fusion are referred to as thermonuclear weapons, whereas pure fission weapons are generally referred to simply as atom bombs. Historically, the transition from one to the other came quite quickly. Following the success of Trinity in 1945, which itself was a pure fission device, the first successful test of a high yield thermonuclear device by the United States occurred only some 7 years later in 1952. To contrast the difference between the two types of weapons in terms of their power, nuclear weapons that make use solely of fission, can command explosive yields only within the kiloton (TNT) range i.e. several thousand tonnes of TNT. However, nuclear weapons that employ fusion possess no limit (theoretically) to the amount of energy that they can unleash, and in practical terms, bombs built employing this process can command explosive yields well within the megaton range i.e. several million tonnes of TNT! Indeed, the total yield of such devices is the combined energy from both the initial fission and the subsequent fusion reactions.

A Super-critical State

Generally speaking, almost all modern-day nuclear weapons make use of Plutonium (239) as the primary 'fuel' in the fission stage of the device, with actual fission itself commencing through use of a radioactive trigger built in to the device called an initiator. The initiator is itself the primary source from which neutrons are first produced to bombard the Plutonium. Once this occurs, the first atoms to split release additional neutrons; which go on to strike other atoms nearby, splitting them also. Correctly handled, in each 'generation' of the process, more and more neutrons are released than were used to split the previous set of Plutonium atoms. Consequently, both the fission of the Plutonium mass and the neutron release increase exponentially, as does the 'conversion' of the overall mass into energy, released as the explosion. However, it is important to realise that one cannot simply use any given amount of Plutonium in a nuclear weapon and simply bombard it with neutrons to cause runaway exponential fission. To begin, firstly the Plutonium within the device must be correctly configured to be in a 'super-critical' state.

All heavy metals from Uranium onwards possess a unique *critical mass value* – something that is determined by careful experimentation, which, when attained, results in the element being in such a state whereby a *sustained* fission reaction is possible, *were* the material to be hit by an influx of neutrons. The critical mass itself is in fact not purely determined by the mass of the material, but also its density and shape. Prior to the actual activation of a nuclear weapon, the key radioactive element used in the device (e.g. Plutonium) is deliberately held in a sub-critical assembly to avoid premature fission. Indeed, when sub-critical, were the element to be bombarded with an initial pulse of neutrons, they would not be able to cause a sustained fission of the whole material at all, for the reaction would not be capable of creating the release of an exponential growth of neutrons. This in point of fact is what defines the character of a sub-critical assembly of fissile material like Plutonium: a sustained chain reaction effect is not possible. Thus, the critical mass value, as it is called, is the experimentally determined boundary between a sub-critical and super-critical assembly of a fissile material. In order to cause fission, one must therefore bombard the fissile material with a stream of neutrons only once the assembly as a whole has first been transformed from a sub-critical state to a super-critical state.

Implosion

In order to cause a mass of Plutonium to become super-critical, one makes use of a process known as implosion. Essentially, a set of carefully positioned shaped charges of conventional explosive materials are placed about the Plutonium core, which itself is held in a sub-critical configuration, due either to its density being of a low order by deliberate design, or that the core has been manufactured to be hollow. By triggering the explosive charges simultaneously to activate the weapon, the Plutonium mass is rapidly compressed through implosion to high density, thus achieving a super-critical state. The initiator is then itself activated to bombard the super-critical mass with an influx of neutrons to begin the fission process. In some weapons designs, the initiator itself is actually placed inside the hollow centre of a Plutonium core, and is protected by seals, which, upon implosion are cracked to trigger the primary neutron pulse. For thermonuclear weapons, the fission event through implosion is the first of a two stage process. A second compartment of the device containing the deuterium fuel is thus carefully affected by the fission event to cause the more powerful fusion process by which even more energy is released; producing an explosion well within the megaton range, if such is desired.

Highly Classified Military Information

Although the above serves as a brief summary as to some of the inner workings of a basic nuclear device, the precise details, especially regarding the issue of implosion and of the compressional properties of the materials involved, do remain highly classified information. Indeed, the outline given above is certainly nowhere near sufficient for anybody to build a nuclear weapon, and does not address at all the immense technical challenges that must be overcome in the correct assembly of such a device. What one must realise, is that the whole process of triggering such a weapon is so carefully managed – meticulously so – in order that the internal fission-fusion reactions occur in a precise order, and with cascading physical effects, all of which occur in the merest fraction of a second, so that ultimately one is able to produce the massive release of energy and the desired destructive effect. That being said though, the precise details suppressed by the major nuclear powers concerning the exact manner in which such devices operate *do not just constitute the physical processes* ***internal*** *to the weapons themselves*. Indeed, of critical importance is the fact that *the ability to cause a nuclear explosion is dependent not just upon the inner workings of a nuclear device, but also upon* ***external*** *factors – even of a celestial nature*. And this indeed, constitutes what may be considered to be the 'true secret of the bomb'.

In the previous chapter, in examination of the Great Pyramid it was demonstrated that as a device, the dimensions of the structure were carefully chosen such that when placed upon the latitude that it was, a viable interaction was possible between the natural energy fields of the Earth, and the Great pyramid itself. Indeed, the latitude placement of the structure was found to be important in its own right, and not just with respect to the 1/14400 harmonic association noted; such being revealed by a subtle cube-root transformation. Incredible as it may seem, it would appear that modern day nuclear weapons have much in common with the Great Pyramid; operating in such a manner, *that one must conclude that they too interact with the natural energetic fields of the Earth, and that such indeed is vital for their proper functioning, as was true of the Great Pyramid*. But just how exactly is this so?

Once more one must acknowledge the pioneering work of the researcher Bruce Cathie in first cracking the secret of the bomb, much to the consternation of the major powers. As a result of his work it was found, independently of classified sources, that such devices could only be detonated upon certain key locations upon the Earth, and only at certain very precise times. This may seem somewhat bizarre, if not incredible, but the fact is, that nuclear weapons are nothing like conventional explosives. They are not simply just bigger bombs, but as explosive devices, they are altogether *qualitatively different*. Indeed, concerning a conventional explosive, one may envision such a device as a warhead placed upon an intercontinental ballistic missile (ICBM), and be of the impression that such a missile could be fired at the proverbial 'moments notice' at any given target within range, and that when hit it would suffer destruction as per the yield of the warhead. Such indeed would be true. However, were an ICBM to possess a nuclear warhead, such a device, contrary to popular opinion, could not in fact be deployed at a moments notice to a given target site at any arbitrary latitude-longitude position upon the Earth. It could only be deployed at certain select locations upon the Earth, and the actual firing of the missile would have to be very carefully timed in order that it reached its target at just the right moment, so as to be detonated under the correct geometric-harmonic conditions.

In the initial stages of the development of the bomb, it would seem that the scientists working on the [Manhattan] project realised early on that the Earth and its energy fields as a whole, *were necessary to interact with the internal processes of the device in order that an actual nuclear explosion would result.* Indeed, they would appear to have stumbled upon the fact that the Earth, as a rotating physical body with a defined equator and axis of spin, was covered over with a natural set of unique harmonic energy intervals *based upon the primary values of the very ancient sexagesimal mathematical system,* and that in order to cause a nuclear explosion, one must isolate – through actual experimentation – the correct energetic intervals that are **viable**, such that were one to place a nuclear bomb at such a site and activate it at a set time, a nuclear explosion would result. Indeed, the plain truth of the matter is that the arbitrary placement and activation of a nuclear device at any given location (on the Earth) does not in fact result in an actual explosion at all. It cannot. What would occur instead, after firing the conventional charges to produce the implosion, would be a mere localised melting of the fissile material within the device. No massive outward explosion would result, and the bomb would in fact, under such circumstances, have to be considered a 'dud'. And this would be so even if the device were manufactured without flaw and the implosion event was 'physically perfect' upon activation. In sum, *the internal processes of the device cannot on their own cause a nuclear explosion.*

Latitude Considerations & the Role of the Sun

In order to truly grasp the importance of the actual placement of a nuclear device upon the Earth, and also the timing of the activation of the weapon, one needs to understand something of just how the primary values of the sexagesimal system apply with respect to the physics of generating a nuclear explosion. One may cite the work of Bruce Cathie in particular here, as being the first 'outsider' to suggest that various nuclear test sites, were deliberately chosen for use due to their association with certain important values of the sexagesimal system. Moreover, Cathie must also be credited with the discovery that the relative position of the sun, with respect to the ground position of the bomb, is critically important at the moment of detonation, in order to actually produce the nuclear explosion [6].

In the first of these two considerations, Cathie has pointed out many times for example that two of the key test sites used by the United States (primarily during the 1950's), Eniwetak and Bikini; both small islands within the Pacific Ocean, were likely selected for nuclear testing due directly to the fact that both just happen to be about 695 nautical miles north of the equator [7]. This fact is even more noteworthy when one considers that these two islands are actually separated themselves by a distance of some 185 nautical miles. Thus indeed, is it not remarkable that out of all the islands dotted about the Pacific Ocean, the US would choose two, well separated from one another, but that just happen to straddle the very same latitude whose arc of displacement from the equator is of immense significance when expressed in terms of nautical miles? It is hardly a coincidence. Indeed, one may reinforce the point by considering two very specific tests carried out by the US in 1956 on Eniwetak island; codenamed Mohawk and Inca. Conducted during the summer months about 10 days apart, both tests were carried out at the precise latitude of 11.63000 degrees north [8]. At this geodetic latitude, the arc displacement from the equator is some 4219716.151 feet [9]. If one divides this value by the Ideal Nautical Mile unit as previously given, a very interesting result is to be had:

4219716.151 / 6076.388888…= 694.44471513

1 / 694.44471513 = 0.001439999438

One can see here once more then the key numeric sequence of 144, as indeed has been found to be linked to the Great Pyramid. In this instance then, it would seem that the latitude of these particular tests was critically important in its own right, even with the same level of exactness as with the ancient Egyptian structure.

Upon the issue of latitude per se, one may acknowledge the fact that with respect to the Earth, such is *apparently* a very 'static' consideration. That is to say that the placement of a nuclear device at a given latitude is concerned purely with correct positioning. The issue of timing the actual activation of the weapon (triggering the implosion) is far more dynamic. The reason for this, as pointed out above, is that the relative position of the sun in the sky, with respect to the location of the bomb on the ground, at the very moment of detonation, is of critical importance in order to create the conditions for a successful test. Indeed, due to both the daily rotation of the Earth on its axis and the yearly cycle about the sun, the position of the sun relative to any selected bomb site dynamically changes from moment to moment. But just how exactly is the sun 'utilised' in causing a nuclear explosion, and what is the precise nature of the relationship between its position and that of the bomb at the moment of activation?

Imagine oneself positioned at the exact centre of the Earth, and that the planet itself is a hollow sphere with a transparent surface shell, upon which is overlaid the Greenwich geodetic co-ordinate system. Were a country to conduct a nuclear test at a select location upon the surface of the Earth, one could mark out the co-ordinates of the test on the spherical shell and draw a straight line between the surface position of the bomb and the very centre of the Earth. Moreover, at the very instant of detonation of the device, one could also look to the position of the sun with respect to the Earth, and draw another straight line; this time from the centre of the sun to the centre of the Earth. Such a line would in fact pierce through the surface of the Earth at a very specific position defined by a precise set of geodetic co-ordinates, as with the location of the test site itself. One may call this point the Sun-Ground Position (SGP) for the purpose of discussion. With two Earth-bound surface co-ordinates now isolated, one could sweep out an arc length measure over the surface of the planet connecting up both points; such an arc being established upon the same plane as the lines connecting up the two noted sites, to the Earth centre. Due to the form of the planet, the arc length itself would be of the character of a complex elliptical curve, but one with a very definite measure. An example of these relations is given in the diagram below:

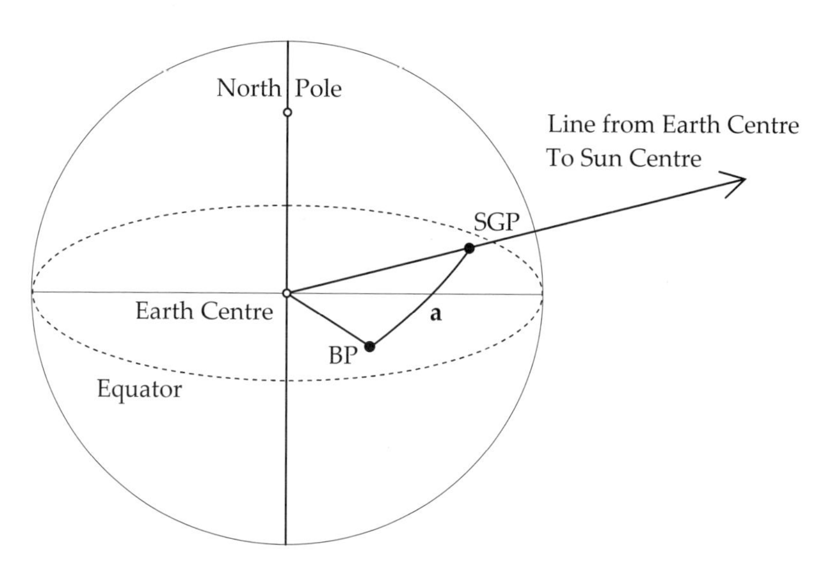

BP = Bomb Position SGP = Sun Ground Position
a = Surface Arc Length (Elliptical) Connecting BP to SGP

In his study of nuclear weapons tests, the researcher Bruce Cathie found that such a surface measure – line **'a'** in the above diagram – between the bomb position and the sun-ground position at the moment of detonation of a nuclear device was of critical importance for actually producing a successful explosion. Indeed, in the course of his work, he was able to demonstrate quite well that the measure of such values associated with nuclear events, invariably possessed an affinity to the primary values of the sexagesimal system in much the same way as pure latitude values. In his book, *The Bridge to Infinity*, he gives the example of a certain French test that was conducted in 1983 at their Mururoa atomic test site in the South Pacific. Following careful analysis of the exact location of the bomb on the island site itself, including the sun-ground position, he was able to determine that at the moment of detonation, (recorded as 5.31 am New Zealand time, 26 May) the angular separation between both positions over the surface of the Earth, was precisely 69.44444 degrees [10]; such a result, highly suggestive of the fact that the noted measure between the two positions was of immense importance.

In any evaluation of the arc of separation between the SGP and BP of a nuclear device, one cannot overstate the significance of the timing of the activation of the weapon. For in order to achieve with accuracy a desired arc of separation, such as 69.44444 degrees, as with the noted French test, quite literally, split second timing is required. Indeed, due to the Earth's axial rotation and the fact that a complete tropical day is swept out in 86400 seconds, in just 1 second of time at the plane of the equator, the sun itself i.e. the Sun Ground Position, will sweep out a longitudinal measure over the surface of the planet of some 1500 feet. And even at 23.5 degrees latitude (north or south), such a sweep will still be about 1400 feet in 1 second. Thus, were the French to have triggered their nuclear device just several minutes or so either way from the exact time that they did, the precise 69.44444 angular separation would have been lost, and indeed, in all likelihood the internal physical processes of the device would have been compromised so much so that the power of the explosion produced would have been severely reduced. Or, were the delay so great, the device would have been a 'dud' and just 'melted' on its platform.

144: A Destructive Sequence

In the analysis so far given, one is perhaps bound to wonder at the significance of the numeric sequence of 144 and its noted reciprocal 694444… strongly associated via their nuclear connections with the disruption of matter. Indeed, one may recall the ideal orbit of Ceres, determined to have been precisely 1440 solar days. For some reason such a celestial orbital frequency though would appear to have been inherently unstable, such that it could not be maintained; a fact that directly led to the shattering of the planet itself and the creation of the asteroid belt. Being the remaining major fragment, Ceres was forced to transform its orbit to a seemingly more stable frequency of 1680 days. In many guises therefore, the numeric value of 144 can be found. It has significance as a sum of solar days. Its reciprocal has meaning respecting the nautical mile unit. And indeed, it is also found to be associated with such as the Great Pyramid respecting its number of casing stones and its latitude measure from the equator.

Chapter 15

The Energy Harmonics of Nuclear Weapons

Nuclear Events & their Energetic Earth Associations

Continuing the analysis further, in order that one might reinforce the connection between the Great Pyramid and modern day nuclear weapons, a far more detailed and comprehensive evaluation of a significant number of nuclear tests is most definitely required. Using as a guide the two key points first uncovered by Cathie, that both the latitude placement of various tests, and the bomb-sun arc-length separation measure over the Earth's surface at the time of detonation, are of critical importance for the actual creation of a nuclear explosion, this present author will go on to conduct his own in-depth analysis of various recent nuclear events. In doing so, it is of course only right and proper to evaluate primarily only the most notable nuclear explosions from history, such that this author may counter at the outset any criticism of being arbitrarily selective in the choice of tests, or of being vulnerable to the assertion that 'anything can be proven with numbers'. Indeed, so that other researchers may replicate in detail the analysis that follows, the precise and exacting particulars of the methodology employed, are to be found in Appendix F.

Trinity: The Beginning of the Atomic Age

In any study of the history of nuclear weapons testing it is only fitting that one begin with the very first test. So far indeed, much has already been said of the first nuclear bomb to have been tested (by the USA) in the modern age, codenamed Trinity. Carried out on the 16th of July 1945 in the state of New Mexico, the Trinity test was able to produce an explosive yield of between 20-22 kilotons, based upon the implosion of a plutonium core of a mere 13 ½ pounds. The inner workings of the bomb aside though, in order to prove that such devices are reliant upon an energetic interaction with the Earth's fields, one would hope to demonstrate in some decisive manner an association between the positioning and/or timing of the test, and certain key numeric values of the primary sexagesimal system. Indeed, such values as would be sought, should possess a critical linkage to the basic unit lengths already identified of the Imperial system, including the foot, fathom, and the ideal geographical and nautical miles, which, as demonstrated upon numerous occasions throughout this work, do appear to represent a key set of natural energetic wavelength measures associated with the Earth, both under its ideal and its present configuration.

The Positioning of the Device

From a review of the relevant sources, the actual co-ordinates of the Trinity test as known to history, reveal the precise position of the device at the moment of detonation to have been [1]:

Latitude: 33°40'36" North
Longitude: 106°28'31" West

Focusing initially upon the latitude, it may further be noted that the actual elliptical arc length swept out to the geodetic angle of the bomb from the equatorial plane, is determined to be [2]:

12230495.539 feet

Is such an arc length displacement of any real significance though? Most assuredly it is. By way of a relatively simple mathematical transformation, it can be seen that this value is strongly associated with a key sequence of numbers linked directly to the destructive harmonic value of 144. Moreover, the very nature of the association is such that it is heavily tied in to the transformations linked to the Great Pyramid itself. As one may recall from a previous chapter, it was shown that the latitude displacement from the equator of this ancient monument was important in its own right, and that such could be demonstrated most fully by expressing the measure in terms of Ideal Nautical Miles: 1791.5904 INM. Moreover, by successively dividing this measure by 10 and then 60, one could further produce a value of 2.985984. The key thing of note here, as was pointed out earlier, is that this particular value is of a series one step removed from one of the major sexagesimal numerical variations involving the 144 harmonic, by a 3 power transformation. Thus indeed, the very cube-root of 2.985984 is exactly 1.44. What is most striking then about the Trinity latitude arc, is that the value of its equatorial displacement, *expressed in feet,* is tied in to the very same numeric sequence, also via a 3 power transformation:

Cube-root of 12230495.539 = 230.3994039

From this initial answer one is immediately led into a numeric set whose progressions directly involve the 144 harmonic value, simply through use of the number 2 as a basic divisor:

230.3994039 / 2 = 115.1997019
115.1997019 / 2 = 57.59985098
57.59985098 / 2 = 28.79992549
28.79992549 / 2 = **14.39996274**

Ideal = **14.4** precisely

Working in reverse from an exact value of 14.4, one can determine a suggested ideal latitude displacement for the Trinity test:

230.4 Cubed = 12230590.464 feet

Difference: 12230590.464 – 12230495.539 = 94.924 ft

Given the above associations it is hereby suggested then that the scientists working to produce the bomb deliberately selected the latitude of the test site with the express intention that its arc length displacement from the equator was exactly 12230590.464 ft; being the cube of 230.4. By doing so, they thus recognised the critical importance of the Imperial foot unit as being a very significant energetic measure linked to the Earth's form and fields. It was required by necessity, so that the bomb would be able to interact with the frequency interval present at this latitude, defined by such a unit measure.

What's in a Name?

As an additional point of interest, it would not seem beyond the realm of possibility that the very name assigned to the first test, Trinity, was chosen due directly to the linkage between the value 230.4 and the arc length equatorial displacement of the test (in feet) via the noted 3rd power relation. For indeed does not Trinity = 3? Admittedly, one cannot be completely certain here due to the fact that there are several different stories floating about to account for the name of the test. Of course, the man actually responsible for its choice is indeed well known to history, as Robert Oppenheimer himself, who led the scientific team that actually built the Trinity device. By his own admission though, he is on record as saying that he was not entirely certain at the time why he chose the name, but that it may have been due to a poem that he had recently read. Such though is the 'official' story. And one cannot help but wonder. For if it truly was due to the mathematical relations stated above, there is no way that he would have openly acknowledged this at the time, as it would have constituted a very important piece of technical information regarding the true operational nature of the device.

The Arc of Separation between the Sun-Ground Position & the Bomb Position

With the latitude measure of the Trinity test thus shown to be important in its own right, what then of the arc length over the surface of the Earth separating the bomb from the sun ground position? At the precise moment of detonation, the value of this particular measure was [3]:

35643162.809 feet

Can one demonstrate that such a value is of special significance though? It would appear so, as is revealed by a careful analysis of the ratio between the Trinity latitude measure and the bomb-sun arc. The ratio itself is given as follows:

35643162.809 / 12230495.539 = 2.9142860723

What is most intriguing about this ratio is that via a simple manipulation involving certain key prime numbers – of a low order – one can reveal a very strong association to the principal values of the sexagesimal series:

2.9142860723 × 7 = 20.400002506
20.400002506 / 17 = 1.2000001474
And: 1.2000001474 × 12 = 14.400001769

Ideal = **14.4**

As can be seen then, through use of the prime numbers 7 and 17, a ready association is to be had to the main sexagesimal values. Accepting the principled truth of such relations, one may thus go on to refine the given value for the bomb-sun arc length, as was done with respect to the Trinity latitude value. This is achieved as follows:

((14.4 / 12) × 17) / 7 = 2.91428571428… (102 / 35)

2.91428571428 × 12230590.464 = 35643435.066… feet

Difference between model value and refined:

35643435.066 – 35643162.809 = 272.256 feet

What is implied by this result is that the positioning of the Trinity test; and the very timing of the activation of the weapon – which determines the length of the SGP–BP arc measure – were both very well planned in advance. Indeed, with the latter in particular being a dynamic measure determined by the rotation of the Earth, in pure practical terms, the implosion of the device had to be executed quite literally with split second timing, as the arc length separation between the device and the sun ground position would have only fleetingly held its value with respect to the latitude arc (achieving the ratio 102 / 35), in order that the bomb might realise a resonant association to the Earth.

One must acknowledge of course that the prime number associations as detailed above may seem quite tentative at this stage. However, as will be revealed in due course, the further analysis of additional nuclear events of prominence does appear to strongly reinforce the physical validity of such relations.

The Wartime Bombs

Following closely upon the success of the Trinity device, the US, in order to quickly force the Japanese to accept terms of unconditional surrender, made the decision to use atomic weaponry against them. Two specific cities were targeted: Hiroshima and Nagasaki. And, upon destruction of the latter the US did indeed bring the Second World War to a swift end. The relevant details of each target site are as follows [4]:

Target: *Hiroshima.* Yield (TNT): 15,000 Tonnes
Date: 05 August 1945. Time: 23:16:02 (GMT):
Geodetic Bomb Co-ordinates (Deg):
Lat: 34.3914 N Long: 132.4581 E
Geodetic Sun-Ground Position Co-ordinates (Deg):
Lat: 16.98844797621 N Long: 167.54666666666 W
Arc Length: Equator to Latitude of bomb (feet): = 12490608.033
Arc Length: Bomb Site to Sun-Ground Position (feet): = 20442969.562

Target: *Nagasaki.* Yield (TNT): 21,000 Tonnes
Date: 09 August 1945. Time: 01:58:00 (GMT):
Geodetic Bomb Co-ordinates (Deg):
Lat: 32.7702 N Long: 129.8657 E
Geodetic Sun-Ground Position Co-ordinates (Deg):
Lat: 16.11401239077 N Long: 151.87166666666 E
Arc Length: Equator to Latitude of bomb (feet): = 11900649.121
Arc Length: Bomb Site to Sun-Ground Position (feet): = 9461198.149

In the Trinity analysis it was seen that the ratio between the arc of the bomb to sun ground position, and the latitude arc of the device, was 102 / 35 (2.9142857142857…). And also, that the value of the ratio could be reduced to the primary sexagesimal series via the use of two prime numbers: 7 and 17. Careful study reveals that similar associations appear to be had respecting the arc measures of yet other nuclear events, even by combining a variety of arc measures between different tests. In many instances, such relations to the primary sexagesimal values can either be very direct, or, somewhat more complex. One example of the former can readily be had from an examination of the ratio between the bomb-sun arc measures for both Hiroshima and Nagasaki:

20442969.562 / 9461198.149 = 2.160716775

Ideal = **2.16** precisely (as an exact fraction 54 / 25)

Although the above value is slightly higher than an exact 2.16, one would only need to further separate the arc measures themselves by an additional 3000 feet to achieve 2.16 precisely. And indeed, as these values are dynamically determined primarily by the rotation of the Earth, such a discrepancy (of 3000 feet) would disappear to nothing in less than 3 seconds of additional time separating both explosions. Given this, one may be very confident that the scientists behind the devices had explicitly planned in advance that the Nagasaki bomb-sun arc was to have been 25/54 of the length of the Hiroshima arc. They had determined that the relevant arc measure for Hiroshima possessed a viable energetic frequency such as would be required to cause a nuclear explosion, and thus that a fractional harmonic of 25/54 of this arc measure, was also viable.

The Grapple Series

With the war at an end, in the decades that followed, quite a number of countries sought to develop their own nuclear arsenals. One such country was the United Kingdom. Beginning in 1952 with their first fission bomb, by the end of the decade they had managed to successfully detonate a series of high yield thermonuclear weapons. The primary test site for the largest of these devices was Christmas Island, located just north of the equator in the Pacific Ocean. During late 1957 through to 1958 a total of 6 nuclear devices were exploded on this island. The details of two of these tests, codenamed Grapple Y and Grapple Z/Halliard 1, are given below. And, with respect to them, one may note the fact that Grapple Y stands to this day as the highest yield nuclear weapon ever tested by the United Kingdom, at some 3 megatons, with Grapple Z/Halliard 1 being the 4th highest in terms of its yield, at 800 kilotons.

Shot: *Grapple Y* (United Kingdom). Yield (TNT): 3,000,000 Tonnes
Date: 28 April 1958 Time: 19:05:00 (GMT)
Geodetic Bomb Co-ordinates (Deg):
Lat: 1.67 N Long: 157.25 W
Geodetic Sun-Ground Position Co-ordinates (Deg):
Lat: 14.26293782508 N Long: 106.88 W
Arc Length: Equator to Latitude of Test (feet): = 605845.063
Arc Length: Test Site to Sun-Ground Position (feet): = 18734896.299

Shot: *Grapple Z/Halliard 1* (UK). Yield (TNT): 800,000 Tonnes
Date: 11 September 1958 Time: 17:49:00 (GMT)
Geodetic Bomb Co-ordinates (Deg):
Lat: 1.67 N Long: 157.25 W
Geodetic Sun-Ground Position Co-ordinates (Deg):
Lat: 4.58890257277 N Long: 88.08333333333 W
Arc Length: Equator to Latitude of Test (feet): = 605845.063
Arc Length: Test Site to Sun-Ground Position (feet): = 25238794.623

What is so remarkable about these tests are the incredible harmonic-type associations they possess with respect to the wartime bombs dropped on Hiroshima and Nagasaki. An evaluation of the latitude arc of the Hiroshima bomb and the SGP–BP arc of Grapple Y is quite simply astonishing:

18734896.299 / 12490608.033 = 1.499918678

Ideal = **1.5** (3 / 2 as a perfect fraction)

Detonated some 13 years apart at completely different times of year and day, to achieve a 3/2 fractional relationship with such extreme accuracy; it would be impossible to believe that such was due to chance. Rather, one must accept that it was intended. The scientists working on the UK test deliberately triggered the device at exactly the right instant such that the bomb-sun arc length at the moment of detonation was 3/2 of the Hiroshima latitude value. Moreover, one may further link the Hiroshima bomb-sun arc to that of Grapple Z / Halliard 1, to uncover yet another startling connection:

20442969.562 / 25238794.623 = 0.8099820085

Ideal = **0.81** (81 / 100)

By simply multiplying the above theoretical ideal of 0.81 by 16 (2 × 2 × 2 × 2) and then dividing by 9 (3 × 3) one easily achieves the destructive harmonic of 1.44:

0.81 × (16 / 9) = **1.44**

Further to this, one may also note a distinct relationship between the given values for the latitude arc and the bomb-sun arc of Grapple Z /Halliard 1 itself; the ratio between them of undeniable significance:

605845.063 / 25238794.623 = 0.0240045165
0.0240045165 × 6 = 0.144027099

Ideal = **0.144**

A Refinement of the Values

From the above analysis of the noted Grapple series and the wartime bombs of Hiroshima and Nagasaki, it is quite evident that a distinct set of ratios link up the various arc length measures associated with these tests; those identified so far including 102/35 (2.9142857…), 54/25 (2.16), 3/2 (1.5), 81/100 (0.81) and 3/125 (0.024). Moreover, with the high accuracy of the noted ratios, one may be extremely confident that such precise ratios stated are in principle absolutely correct, in that they were intentionally sought. And yet, one must concede that closely scrutinised, the actual arc length measures upon which the ratios are based do not appear in and of themselves to be too remarkable. However, as a result of the Trinity analysis, one would very much expect that in principle they should be, and by way of some sort of distinct association to the various numerical progressions of the sexagesimal series. The values thus, as 'stand alone' arc measures, ought to be susceptible to refinement. And this in fact does appear to be the case.

To begin, from a careful examination of the latitude arc of the Trinity test, and a particular association to the Hiroshima bomb-SGP arc, one is able to draw a mathematical link as would seem to allow for a precise refinement of the latter value, and thus all other values as are tied to it through the various ratios given. The specific connection is found to involve a combination of the basic prime numbers 7 and 13, representing a slight variation on 7 and 17, as shown previously to be associated with the Trinity arc measures. Presented in stages, the mathematical linkage is as follows:

Trinity latitude, refined = 12230590.464 feet
12230590.464 × (13 / 7) = 22713953.718…

From this initial calculation, one may note a very simple fractional relation to the Hiroshima bomb-SGP arc:

20442969.562 / 22713953.718 = 0.900018104…

With this now established, multiplying the answer by 2 successively reveals a most notable affinity to the primary sexagesimal values:

0.900018104 × 2 = 1.800036208
1.800036208 × 2 = 3.600072416
3.600072416 × 2 = 7.200144832
7.200144832 × 2 = 14.40028966

Ideal = **14.4**

A precise evaluation of the above values thus leads one to determine that an exacting ratio between the Trinity latitude and the Hiroshima bomb-SGP arc, as would have been *intended* by the scientists, so as to achieve a successful explosive nuclear reaction, would be 117 / 70. And thus, the Hiroshima bomb-SGP arc is given as:

$12230590.464 \times (117 / 70) =$

$20442558.346971428571428571\ldots$ feet

Further to this, in consideration also of the latitude arc measure of the Nagasaki target site, one is able to reveal a direct link to the numeric values closely tied to the primary sexagesimal sequence, by way of the following:

$11900649.121 / (9 \times 9 \times 9 \times 9 \times 9 \times 9 \times 9) = 2.4881300969$
$2.4881300969 / 8 = 0.3110162621$
$0.3110162621 / (6 \times 6 \times 6) = 0.0014398901$

Ideal = **0.00144**

This result would appear to suggest then that Nagasaki as a city was very carefully selected as a target due to the energetic associations of its latitude placement from the equator. And indeed, such would have been of critical importance over and above any real desire to choose a target that was of true strategic value.

In view then of the associations so far given, one may refine the arc length measures noted, using the refined Hiroshima bomb-SGP arc measure as a standard, in conjunction with the suggested ideal ratios. With the Nagasaki latitude measure possessing its own 'standalone' association, as noted:

Nagasaki (bomb-sun arc) = $20442558.346 / 2.16$
$= 9464147.382857142857142857\ldots$ feet

Nagasaki (latitude arc) = $0.00144 \times (9^7 \times 6^3 \times 8) =$
$= 11901557.42208$ feet

Grapple Z/Halliard 1 (bomb-sun arc) = $20442558.346 / 0.81$
$= 25237726.354285714285714285 7\ldots$ feet

Grapple Z/Halliard 1 & Grapple Y (latitude arcs)
= 25237726.354 × 0.024 = 605705.4325028571428571428 57... feet

In addition to the above, and to complete the series in terms of the refinement of the latitude arc of the Hiroshima bomb, and the SGP–BP arc of Grapple Y, one may introduce yet a further nuclear test of interest. One that is indeed of immense historical significance; being none other than the very first thermonuclear device ever to have been tested in the modern age.

Less than one month after the very first British fission test in 1952, the United States of America, already well progressed with its own weapons program, exploded the first ever thermonuclear device employing the principle of fusion. The site used was Eniwetak; the test itself codenamed Mike:

Shot: *Mike* (United States). Yield (TNT): 10,400,000 Tonnes
Date: 31 October 1952 Time: 19:14:59.4 (GMT)
Geodetic Bomb Co-ordinates (Deg):
Lat: 11.2372 N Long: 162.1964 E
Geodetic Sun-Ground Position Co-ordinates (Deg):
Lat: 14.37430409918 S Long: 112.835 W
Arc Length: Equator to Latitude of Test (feet): = 4077159.503
Arc Length: Test Site to Sun-Ground Position (feet): = 32123600.768

A careful analysis of the Mike bomb-sun arc would seem to reveal a very close association to both the latitude arc of the Hiroshima device and the bomb-sun arc of Grapple Y, the character of which is suggestive of a possible means by which all three values might be refined relative to one another; the critical connection involving the basic prime numbers 7 and 11.

Consider the refined arc length of the Hiroshima bomb-SGP arc set against the noted Mike bomb-SGP arc:

20442558.346 / 32123600.768 = 0.6363719464

Studying carefully the numeric sequence, one would doubtless be of the mind that ideally, it would progress in terms of a recurring series of 0.6363636363...

Indeed, one could hardly doubt this, as such a value is the expression of the basic fraction of 7 / 11:

0.636363636363... = 7 / 11

Accepting the refined Hiroshima bomb-SGP arc and this specific ratio, one may thus refine the Mike bomb-SGP arc value as follows:

20442558.346 × (11 / 7) = 32124020.259... feet

With the aid of this new value for Mike, one is able next to refine the Grapple Y bomb-sun arc; the required breakthrough to be had from a simple examination of the numeric sequence of the basic ratio between both values:

18734896.299 / 32124020.259 = 0.5832052198
0.5832052198 / (9 × 9) = 0.0072000644
0.0072000644 × 2 = 0.01440012888...

Ideal = **0.0144**

With reference to the above figures, one may recall the previously stated ideal Venus semi-major axis of 58320000 IGM, including also the ratio between this and the Earth semi-major axis, 1 / 0.72. There would thus seem to be no real doubt that the intended ratio between the two relevant bomb measures was exactly 0.5832. And thus, the ideal Grapple Y bomb-SGP arc:

32124020.259 × 0.5832 = 18734728.615... feet

With the refinement of the Hiroshima latitude:

18734728.615 / 1.5 = 12489819.076... feet

With this final value given, the entire set of refined arc measures as so far stated may be compared against the model values as given; derived through use of various nuclear test archives, a selected Earth ellipsoid model, and also an astronomical program for modelling the past positions of the sun at the times of the tests (See Appendix F for full particulars). The differences are expressed in feet as follows:

Hiroshima (bomb-sun arc) = 411.215
Hiroshima (latitude arc) = 788.956
Nagasaki (bomb-sun arc) = 2949.233
Nagasaki (latitude arc) = 908.300
Grapple Z/Halliard 1 (bomb-sun arc) = 1068.269
Grapple Y & Z/Halliard 1 (latitude arc) = 139.630
Grapple Y (bomb-sun arc) = 167.683
Mike (bomb-sun arc) = 419.491

From the above one can easily see then that the refined values differ from the derived (archive-astronomical) model values by no more than several hundred feet in most cases. The primary exception is the Nagasaki bomb-sun arc which possesses a discrepancy of about 2950 feet. Of course, due to the dynamic nature of this arc length and the fact that it is primarily determined by the rotation of the Earth on its axis, in real terms the suggested error would in point of fact be no more than 3 seconds of time. That being said, in the analysis that will shortly follow, one will see a set of further nuclear associations presented, that will reinforce to an even greater degree the suggestion that the refined Nagasaki bomb-sun arc is correct.

King & Bravo

After the successful test of Mike, the first thermonuclear device ever exploded by the US, the very next test was King, conducted some 15 days later on the same site: Eniwetak Atoll. This subsequent test was not however, as one might suppose, a second thermonuclear test. King was in point of fact an extremely high yield pure fission device, being the equivalent of 500 kilotons (i.e. 500,000 tonnes) of TNT. Indeed, the very significance of King is that of all of the tests carried out by the US, King stands as the highest yield pure fission device that they have ever tested. That being said, the US did of course continue on to further develop their thermonuclear (fusion) devices during the 1950s. And, it was only 2 years later in 1954 that shot Bravo was successfully detonated, which has its place in history as being the highest yield device ever tested by the US in the whole of their weapons program, period. The recorded yield of this particular bomb was an incredible 15 megatons of TNT (15,000,000 tonnes)! The primary details of King and Bravo are as follows:

Shot: *King* (United States). Yield (TNT): 500,000 Tonnes
Date: 15 November 1952 Time: 23:30:00 (GMT)
Geodetic Bomb Co-ordinates (Deg):
Lat: 11.5622 N Long: 162.3525 E
Geodetic Sun-Ground Position Co-ordinates (Deg):
Lat: 18.78649565341 S Long: 176.315 W
Arc Length: Equator to Latitude of Test (feet): = 4195109.611
Arc Length: Test Site to Sun-Ground Position (feet): = 13428155.350

Shot: *Bravo* (United States). Yield (TNT): 15,000,000 Tonnes
Date: 28 February 1954 Time: 18:45:00 (GMT)
Geodetic Bomb Co-ordinates (Deg):
Lat: 11.6908 N Long: 165.2736 E
Geodetic Sun-Ground Position Co-ordinates (Deg):
Lat: 8.006922846136 S Long: 98.09166666666 W
Arc Length: Equator to Latitude of Test (feet): = 4241782.290
Arc Length: Test Site to Sun-Ground Position (feet): = 35807863.081

In an analysis of the bomb-sun arc measures of both tests one can immediately see the significance of the ratio as existent between the two values:

35807863.081 / 13428155.350 = 2.6666256195

With: 2.6666256216 × 3 × 3 × 3 × 2 = 143.99778345

Ideal = **144**

Reversing the mathematics and working backwards from an exact 144 value, the precise ratio in question that would appear to apply to the respective bomb-sun arc measures is 8 / 3, or 2.6666666666… Considering that both tests were conducted at significantly different points upon the Atoll, and at very different times of day (and year), that such an exacting fractional relationship exists between the tests; there is no way that such was due to chance. Rather it was due to physical necessity, in order that the devices were able to actually function properly. A key wavelength (arc) measure was determined to be energetically viable respecting King, with a most exacting mathematical fraction of this arc subsequently employed for the Bravo test, as would also achieve a successful explosion. Moreover, given the precise fraction of 8 / 3, one may note that such a ratio reinforces the physical validity of certain arc length values as have already been presented; specifically, the bomb-sun arc of Grapple Z/Halliard 1 set against that of the Nagasaki blast:

25237726.354 / 9464147.382… = 2.66666666…

In seeking to refine the arc measures associated with both Bravo and King, it is to be found that once more, the basic prime number 11 is critical as a transformative value, in revealing just how their noted arc length measures are linked in to the basic sexagesimal series. In beginning with Bravo, one can thus see the following:

Bravo: bomb-SGP arc = 35807863.081

35807863.081 / 11 = 3255260.280
3255260.280 × 8 × 8 × 8 = 1666693263.43
1 / 1666693263.43 = 0.0000000005999904253171

Ideal = **0.0000000006**

Bravo: latitude arc = 4241782.290

4241782.290 × 11 = 46659605.196
46659605.196 / (6 × 6 × 6 × 6 × 6) = 6000.46363118

Ideal = **6000**

Using the specified ideal values as indicated, by reversing the math, one is thus able to generate a set of refined values for the Bravo test.

Bravo: bomb-SGP arc = 35807291.66666666... feet

Bravo: latitude arc = 4241454.54545454... feet

From the above analysis, it is easily seen that both arc measures of the Bravo test are intimately tied in to the basic sexagesimal system, and to the primary number 6 that is its base. And indeed, it would be a mere formality to further link them to the 144 'destructive value' via additional manipulations. But, this point aside for the moment, one can, with the above refinements, thus proceed onwards to also refine the basic arc measures associated with King, beginning with its bomb-SGP arc:

35807291.66666666... × (3 / 8) = 13427734.375 feet

With the bomb-SGP arc thus refined, one is able to draw attention to a further value of interest that would appear to link it up in a simple manner to the latitude value of King. Consider thus the following:

13427734.375 / 4195109.611 = 3.200806562

With this result, one cannot ignore the idea that the intended ratio between both King arc measures was exactly 3.2, which itself is a value of obvious significance. Consequently, one is able to produce a refinement of the King latitude arc as follows:

13427734.375 / 3.2 = 4196166.992 feet

And further to this, one is able to return to an arc value that has up to this moment been left unrefined, but which now, as a result of the above determined value can be dealt with: the Mike latitude arc. By making use of the King latitude one is able to draw out what would appear to be the decisive connecting ratio:

4077159.503 / 4196166.992 = 0.97163900
0.97163900 × 4 = 3.88655600
3.88655600 / (3 × 3 × 3) = 0.1439465186
Ideal = **0.144**

Reversing the math; a refined value for the Mike latitude:

4196166.992 × 0.972 = 4078674.316 feet

Concerning all of the now refined arc measures of this section, one can judge them against the model answers, and thus compare (in feet) the discrepancies:

Bravo (bomb-sun arc) = 571.414
Bravo (latitude arc) = 327.745
King (bomb-sun arc) = 420.975
King (latitude arc) = 1057.380
Mike (latitude arc) = 1514.813

Russia & France: The Giants

In continuing the analysis, one may find further examples that stress the importance of the chosen arc measures associated with nuclear testing, from the evaluation of what are on record as the largest atmospheric tests ever conducted, by both Russia and France. The primary details are as follows [5]:

Shot: *Tsar Bomba* (Russia). Yield (TNT): 58,000,000 Tonnes
Date: 30 October 1961 Time: 08:33:27.8 (GMT)
Geodetic Bomb Co-ordinates (Deg):
Lat: 73.8 N Long: 53.5 E
Geodetic Sun-Ground Position Co-ordinates (Deg):
Lat: 13.83754538227 S Long: 47.56166666666 E
Arc Length: Equator to Latitude of Test (feet): = 26880677.302
Arc Length: Test Site to Sun-Ground Position (feet): = 31932231.750

Shot: *Canopus* (France). Yield (TNT): 2,600,000 Tonnes
Date: 24 August 1968 Time: 18:30:00.5 (GMT)
Geodetic Bomb Co-ordinates (Deg):
Lat: 22.22800 S Long: 138.64400 W
Geodetic Sun-Ground Position Co-ordinates (Deg):
Lat: 10.97134533948 N Long: 96.95666666666 W
Arc Length: Equator to Latitude of Test (feet): = 8067740.589
Arc Length: Test Site to Sun-Ground Position (feet): = 19182655.110

Beginning first with an analysis of the Tsar Bomba latitude, one can see an extremely simple association to the main sexagesimal values of interest, initially just through use of the prime number 7:

26880677.302 / 7 = 3840096.757
3840096.757 × (3 / 8) = 1440036.284

Ideal = **1440000**

From this, one is able to back generate a latitude arc value that one would assume the Russians were employing:

(1440000 / (3 / 8)) × 7 = 26880000 feet

Further to this, by way of a very simple transformation, one is able to derive a refined bomb-SGP arc associated with the Tsar Bomba, making use of the prime number 11. In this instance, the number 11 is applied to the existent ratio between both arc measures linked to the test to reveal the decisive connection:

31932231.750 / 26880000 = 1.187955050…
1.187955050 / 11 = 0.107995913
0.107995913 × (4 / 3) = 0.143994551

Ideal = **0.144**

With the truth of these associations, one is able to refine the bomb-SGP arc of Tsar Bomba as follows:

(0.144 / (4 / 3)) × 11 × 26880000 = 31933440 feet

Moving on to the French test, Canopus, here one is able to derive a value for its bomb-SGP arc length based upon a major fraction of the bomb-SGP arc of the US test, King: 13427734.375 feet:

13427734.375 / 19182655.110 = 0.69999352529

Ideal = **0.7**   (7 / 10)

With the refinement of the Canopus bomb-SGP arc:

13427734.375 / 0.7 = 19182477.67857142857142…feet

Next, on the issue of the pure latitude of the Canopus test, one can see a standalone connection to the allied numerical progressions of the basic sexagesimal system. In this instance, a basic combination of the prime numbers 13 and 11 is the key:

(13 / 11) / 8067740.589 = 0.000000146486884237

0.000000146486884237 × 8 × 8 × 8 × 8 × 8 =
0.00480008222271
3 × 0.00480008222271 = 0.014400246668

Ideal = **0.0144**

The refinement to be had for the Canopus latitude arc measure from these associations is as follows:

(13 / 11) / (0.0144 / (3 × 8 × 8 × 8 × 8 × 8))

= 8067878.7878787878... feet

Concerning then the refined arc measures of both Tsar Bomba and Canopus, one can thus judge them against the model answers, and compare (in feet) the discrepancies:

Tsar Bomba (latitude arc) = 677.302 feet
Tsar Bomba (bomb-sun arc) = 1208.249 feet
Canopus (latitude arc) = 138.198 feet
Canopus (bomb-sun arc) = 177.431 feet

Trinity, Hurricane & Pajara

In addition to the high yield tests carried out by the major powers, it would be well to consider also the very first test carried out by the United Kingdom, codenamed Hurricane, and in particular how its arc measures relate to the Trinity test, including a certain other test carried out by the US much later, called Pajara.

Shot: *Hurricane* (United Kingdom). Yield (TNT): 25000 Tonnes
Date: 03 October 1952. Time: 00:59:24 (GMT)
Geodetic Bomb Co-ordinates (Deg):
Lat: 20.4 S Long: 115.57 E
Geodetic Sun-Ground Position Co-ordinates (Deg):
Lat: 3.893760642537 S Long: 162.43833333333 E
Arc Length: Equator to Latitude of Test (feet): = 7403797.423
Arc Length: Test Site to Sun-Ground Position (feet) = 17693500.883

As with all the major powers still in existence at the close of World War II, the UK had a unique opportunity to study the early tests of the US; an obvious aid to their own nuclear ambitions. Given this, that they would seek to 'play it safe' with their own initial series of tests by basing them closely upon the early configurations of the US would not seem unreasonable. And indeed, this would very much appear to have been the case with respect to Hurricane.

At the very moment of detonation (of Hurricane), the arc length over the surface of the Earth from the test site itself to the sun ground position, was

17693500.883 feet. Upon evaluation, such a measure appears to possess an intriguing connection to the latitude displacement of Trinity, the very first US test. This fact is clearly evident from their ratio:

12230590.464 / 17693500.883 = 0.691247625

Examined closely, one can see that the first 4 numbers following the decimal point are an exact match to the previously suggested ideal orbital period of Mars: 691.2 solar days. As a result, by a simple set of transformations, one may derive a value very close to 144:

0.691247625 / 8 = 0.086405953

0.086405953 / 6 = 0.014400992

With a starting value of exactly 0.6912, producing:

0.6912 / (8 × 6) = **0.0144**

The closeness of the ratio is enough for one to actively suggest that it was intended that the bomb-sun arc of Hurricane was 1 / 0.6912 (625 / 432) of the Trinity latitude. Accepting this, a refinement can be made, and the level of discrepancy revealed:

12230590.464 × (1 / 0.6912) = 17694720 feet

17694720 – 17693500.883 = 1219.116 feet

Given the dynamic nature of this measure, a difference of just over 1200 feet would disappear to practically nothing within 1 second of time. As a result, one may have confidence in the refined measure. And indeed, such confidence is greatly reinforced by consideration of yet one further nuclear test, codename: Pajara.

Shot: *Pajara* (United States). Yield (TNT): 5000 Tonnes
Date: 12 December 1973. Time: 19:00:00 (GMT)
Geodetic Bomb Co-ordinates (Deg):
Lat: 36.99149 N Long: 116.02413 W
Geodetic Sun-Ground Position Co-ordinates (Deg):
Lat: 23.25030024078 S Long: 106.53666666666 W
Arc Length: Equator to Latitude of Test (feet): = 13436898.148
Arc Length: Test Site to Sun-Ground Position (feet) = 22118750.551

With the obvious expense and logistical challenges of conducting nuclear tests within the Pacific, the United States shifted towards conducting tests primarily within its own territory. To this end, a new test site was established in the state of Nevada, called simply the Nevada Test Site or NTS for short. Pajara was one of a great many tests carried out at this site, possessing a most striking series of harmonic associations. The first one of note is the ratio between the latitude and the bomb-sun arc of the test:

13436898.148 / 22118750.551 = 0.60748902239

The value as is shown would appear to be directly associated with a key variation of the sexagesimal series, revealed by a very simple set of transformations:

0.60748902239 / (9 × 9 × 9) = 0.00083331827488

(1 / 0.00083331827488) × 12 = 14400.26021

Were the ratio exactly 0.6075 (243 / 400) then a precise 14400 value would be returned. Moreover, a careful analysis of the latitude arc of the test, combined with such an exacting ratio, allows for the refinement of both the latitude and bomb-sun arc, as follows:

13436898.148 × 2 = 26873796.297

26873796.297 / 9 = 2985977.366

Cube-root of 2985977.366 = 143.99989336449

Precise ideal = **144**

Once more a cube-root transformation is evident. And, by reversing the mathematics from an exact 144 answer, one may derive a suggested ideal arc measure that the scientists conducting the Pajara test sought to achieve:

$(144^3 \times 9) / 2 = 13436928$ feet

With: 13436928 – 13436898.148 = 29.851 feet

Combining this newly derived latitude arc with a precise ratio of 243 / 400, a refined value for the bomb-sun arc can be determined:

13436928 / 0.6075 = 22118400 feet

With: 22118750.551 – 22118400 = 350.551 feet

With the above refinements thus made, one may return to the matter of a suggested association between Pajara and Hurricane. A careful evaluation of the bomb-sun arc measures associated with each test reveals a very strong harmonic association. Using first the original values as stated, and then the refined values for both tests, one can easily see the truth of this:

17693500.883 / 22118750.551 = 0.79993220425

17694720 / 22118400 = 0.8 (4 / 5 as an exact fraction)

The association is practically undeniable: a basic 4 / 5 connection between the two arc measures. And yet another clear demonstration that nuclear tests conducted decades apart and in completely different locations around the world do appear to be precisely linked by various key fractional harmonic intervals, the exactness of which cannot be put down to chance or coincidence.

Chapter 16

The True Operational Nature of Nuclear Devices

The Physics of Generating a Nuclear Explosion

With the analysis of the selected nuclear explosions now complete, and with an indisputable linkage proven between such events and the primary values of the sexagesimal system, the question that must inevitably follow, is just what exactly does this imply about the operational nature of nuclear weapons per se? For indeed, just what is the real significance of the mathematical affinity between the two noted types of arc length measures and the sexagesimal values, and why do such associations seem to be so essential? The answers to these questions would appear to be had from a closer evaluation of the transformation of the very solar system itself as a whole, and in particular the creation of the asteroid belt, as initially discussed in a previous chapter.

As one may recall, the law that governed the transformation of the fifth planet of the solar system, indicated that it suffered an extremely radical increase to its orbital period; more so than any of the other planets that were themselves all transformed in unison. The result of this quite literally fractured what was a unified body, leading to a general increase in the orbital frequency of what became its most prominent fragment, Ceres, from exactly 1440 days per year to some 1680 days. Critically however, the suggested cause of such a catastrophic event was given as resonance. And indeed, resonance was also put forward as the primary agent responsible for periodic planetary upheaval, as related in the myths of the ancients. For it is through such a mechanism, that one could readily account for the 'quickening' of a physical system, and of how a structure could be so suddenly thrown into agitation, even to the point of breaking up, were it pushed beyond a certain vital threshold.

A Resonant Trigger to Unlock the Energy of the Atom

Given what is known about resonance as a physical principle, in addition to the highly suggestive associations between the celestial frequencies of the planets and also the Imperial arc length measures associated with nuclear events, it would seem that precisely just such a principle would appear to be operative respecting the actual detonation of nuclear weapons. Indeed, this would account very well for the unusual conditions under which they are apparently triggered. But how exactly though does the geographical location of a nuclear device and the sun ground position at the very moment of detonation work to produce a massive explosion through resonance?

The Earth as a Musical Instrument

What is often taken for granted when one thinks of the Earth, is the truly dynamic nature of its existent condition. It is not simply that of a static material body spinning through space. And its energy fields are not unstructured. With a clear axis of spin, the planet is in effect continually 'operational' very much like an actively sounding string that is plucked on a musical instrument; one whose signal strength is constantly replenished. Indeed, as is well known respecting such instruments, any given string is bounded at its ends to its main body, with the ends themselves forming a set of fixed nodal points of zero motion. When actively plucked, the string sounds out its pre-tuned frequency, being split up into its fundamental harmonics whose intervals play out over its length. In exactly the same way, one can thus suggest that the physical axis of the Earth, terminating at its two poles, thus defines the nodal points of the 'Earth instrument'. And that just as with a musical instrument, various key harmonic intervals also play out over the curved surface of the planet, indicative of the frequencies embodied within its dynamic form. One may thus postulate the existence of a structured energy grid matrix orientated specifically to the Earth's physical ellipsoid form, superimposed over the entire planet such as would set up a series of key harmonic intervals with respect to both latitude and longitude; the intervals themselves based up the primary noted values of the sexagesimal system. Concerning the exact intervals of just such a grid, one would strongly suspect that the Ideal Geographical Mile of 6000 feet would be among its major basic components. Of course, one would certainly expect also a series of complex sub-harmonic intervals embedded even within these unit measures. And one must not discount either the presence of still greater harmonic intervals covering the Earth, splitting up the planet into very basic fractional units, such

½, 2/3, 1/3, ¼, 1/5, etc. The suggested pattern then, is of a complex three dimensional grid with a range of energetic intervals playing out over the full surface of the Earth, extending into space, and even penetrating through to its interior.

Favourable Earth Harmonies

If the Earth as a physical entity is then a musical instrument, albeit upon a celestial scale, then the very science of music applies to its being. Given this, an important point must be stated concerning this discipline. With regard to the very development of musical theory, it is well to note that the primary scales of tone and frequency in use today, and the harmonic structures they contain, *were* ***discovered*** *as a result of actual experimentation*. Indeed, music itself, as with mathematics, is not a static subject, and never has been. Rather, it is one wherein new discoveries and 'proofs' are continuously sought through a combination of experimentation and precise analytical reasoning. With this understanding, one can definitely isolate the character of the work carried out by those scientists who developed the first nuclear weapons.

In the rapid search to unlock the seemingly incredible energy stored within the basic atomic structure, the scientists working on the (Manhattan) project turned their attention to the forces of the macroscopic realm, as a means to disrupt those of the microscopic. They focused upon the Earth itself, and recognising it as a powerful instrument, sought to discover the primary wavelength intervals and natural frequencies of its energetic form: the very signature of its musical harmonies, so to speak. The whole process though was one of discovery. It was an investigation that ultimately was aimed at determining the very real spatial intervals in terms of latitude and longitude that cover over the Earth, and which have the potential via resonance, to disrupt the atomic structure of a nuclear device placed precisely at such locations. And it is from this, that the true nature of such devices is revealed. Quite simply, *a nuclear weapon is nothing more than an extremely intricate musical instrument manufactured to produce upon activation a particular energetic signature of such frequency and character, so as to be in sympathetic resonance with the localised energetic Earth frequencies* ***naturally convergent at certain precise points upon the globe****, which indeed have been pre-determined experimentally to offer a* ***viable interaction*** *with such a device, in order to achieve a nuclear explosion.* Consequently, the sites that one might choose to destroy upon the Earth are very select, and the mere arbitrary choice of a given location in terms of latitude and longitude will not suffice to create such an explosion. And indeed, were a nuclear device to be activated at a randomly selected point, it would in effect be a dud. One can see then that the major test sites of Bikini, Eniwetak, Christmas Island, and Nevada, were all carefully pre-determined to possess viable harmonic signatures that lent themselves to generating extreme nuclear reactions.

The Harmonics of the Sun

In all of the discussion concerning the structure of the Earth energy grid, and the importance of selecting only those locations embedded within its matrix that possess the necessary harmonics in order that one might create a nuclear explosion, one must not forget the significance of the sun, and the timing of the activation of a nuclear weapon. For as one will doubtless recall, the arc length over the surface of the Earth between the bomb position and the sun ground position at the moment of detonation, is itself of critical importance for producing a successful nuclear event. This indeed provides a dynamic aspect to the whole process. But just what exactly is the role of the sun in the creation of a nuclear explosion? The answer would seem to be had from a more in-depth consideration of the orbital period of the Earth as a whole.

In any evaluation of the full length of the Earth tropical year as given in solar days, one must realise that such a measure is the *celestial pulsation rate or frequency of the orbit.* And all frequency based phenomena are cyclical, alternating between extremes and a point of equilibrium. With respect in particular to the rotation of the Earth on its axis, the sun itself is the (very visible) marker denoting precisely where any given point upon the planet stands with regard to a single complete (daily) cycle, or pulsation. Thus for example, if the sun were directly above a given point upon the surface of the Earth i.e. at its highest point in the sky at noon, then that surface position could be said to be at maximum amplitude respecting the sun. From this initial configuration, following the completion of a quarter of one single day, it could then be said to be at equilibrium. Moreover, following yet another quarter, it would achieve minimum amplitude. And from this, the next successive quarter would return it back to equilibrium, with the final quarter once more returning it to maximum amplitude. Such a progression would establish a basic pattern that could be graphed as follows:

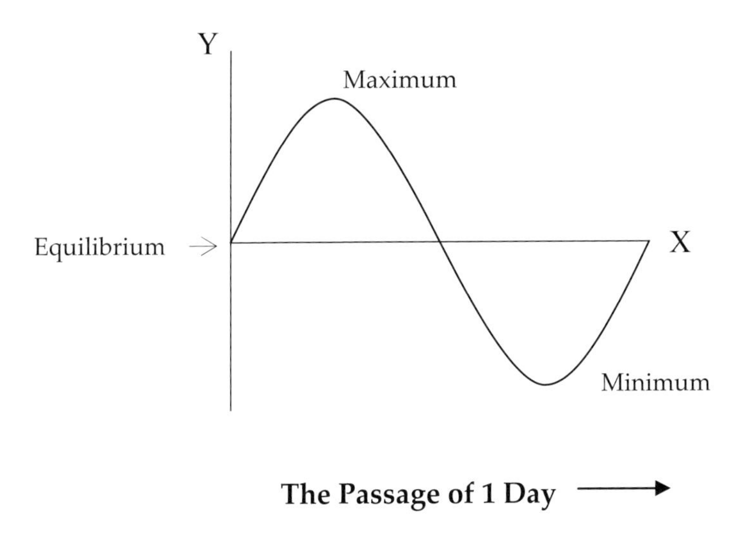

The Passage of 1 Day →

Concerning the completion then of each successive solar day, any given point upon the Earth will experience such an alternating relationship with respect to the position of the sun. Indeed, at the very moment of detonation of a nuclear device upon the surface of the Earth, the instantaneous arc length separating the sun-ground position from the bomb position *must by necessity be a **viable** energetic measure*. Such unit measures are therefore the dynamic equivalent of what merely *appear* to be nothing more than 'static' arc length measures up from the equator. Both indeed possess the same energetic reality, such a point accounting for the fact that both types of arc measures are readily interchangeable.

Ideal Earth Measures Apply

One critical point to emphasise respecting the dynamic arc measures associated with the sun is that ideal Earth measures do apply with regard to the creation of a nuclear event. Thus indeed, as previously stressed, with respect to the increase of the Earth tropical year, the fundamental time for the completion of one (solar) day remained the same i.e. 24 hours, or 1440 minutes. All that changed was the number of days contained within the Earth orbit. Moreover, the full ideal Earth circumference as shown to be linked to the time measure of 1440 minutes is exactly 21600 ideal geographical miles, or 129600000 feet. *Essentially, one unit distance of 129600000 feet is the full wavelength measure accompanying one complete solar day cycle 1440 minutes in duration. And all nuclear events as artificially engineered must employ an interlocking combination of various arc length measures as are viable fractions of the primary wavelength distance that is 129600000 feet.*

Given then the extremely dynamic nature of such relations, one is led to a very specific conclusion respecting the timing of the implosion of a given nuclear device. With it noted that only those locations that have been harmonically verified as globally viable can be chosen for destruction, in light of the evaluation of the SGP-BP arc measures and their importance in determining the timing of the activation of such weapons, it would seem clear then that *any given viable Earth location, is only* ***actively viable*** *for a brief moment in time when there is established a* ***natural harmonic convergence*** *of the energetic frequencies of the Earth's fields, including those of its orbit, the latter being the very solar day cycle itself and its various sub-harmonics.* It is therefore only under precisely these conditions that there exists the *potential* to create such a destructive effect.

Breaching the Threshold

At the very moment when a given position becomes *actively viable*, powerful energetic forces momentarily converge in harmony upon the location itself. And yet critically however, such forces are held just beyond a certain vital threshold, which ordinarily prevents them from 'spilling out' into the world. They silently pass through one another practically unnoticed; a mere fleeting configuration. Indeed, a person might walk through such an energetic convergence as it happens and not even notice. Of course, to those who understand the physics of nuclear reactions, such events, as are capable of being accurately predicted, can be exploited for destructive purposes. And the nuclear weapon is the very device for achieving precisely this.

Upon activation at a pre-determined spot undergoing natural harmonic convergence, a powerful energetic signal is produced in sympathetic resonance with the briefly united Earth harmonies. At such a moment the alignment of the frequencies causes a 'bridge' to occur between those produced by the device, and those of the Earth, as are associated with the sun (ground position). As a result, there is a critical breach of the aforementioned vital physical threshold. And the powerful energetic forces of the Earth-sun harmonies pierce through this natural veil to connect with the signature frequencies as generated by the imploded nuclear device itself. 'Rushing' into it, they quickly unravel its atomic structure at the most fundamental level. In effect then, the nuclear weapon does not in point of fact actually cause an explosion through its own internal processes or 'power', for indeed it lacks the very capability. Rather, *the massive outward explosion that results is due to the resonant destruction of the nuclear weapon itself by a set of* ***external*** *celestial harmonies that converge upon it at the critical moment.*

Modelling Harmonic Convergence

With an extremely accurate Earth ellipsoid model, combined with a well refined astronomical model of the Earth's yearly passage about the sun, it is thus possible for war planners to determine precisely in terms of both space and time, the future locations, *as they arise,* that could be subject to nuclear destruction. Were one thus to produce a detailed computer model in 3-dimensions of the Earth's ellipsoid form, programmed with all of the experimentally verified harmonic configurations viable for nuclear destruction, one would be able to forecast the exact global co-ordinates for any given Earth-harmonic event, including the precise times at which they would so briefly occur, and down to the nearest second, *even years in advance!* With such a model in hand, a given nation would be able to plan a nuclear strike upon an enemy, achieving an optimum level of deployment of its nuclear arsenal.

Of course, if two countries were to go to war, both of which were nuclear powers, and both of which possessed the same sort of model, then there would be no surprises at all in terms of when and where each country would strike. The military forces of both would know precisely where to attack their enemy and when, and also exactly the times and locations at which their enemy would strike them, as within their own territory. Indeed, quite literally dependent upon which day of the year it was, subtly different targets and times would be chosen, reflecting the dynamic nature of the emergence of various harmonic convergence events, as are themselves determined by the characteristic elements of the Earth orbit. Moreover, in a basic strike – counter strike scenario, one would expect that the bulk of the 'action' would initially take place over the course of one full axial rotation of the Earth i.e. the sweep of one complete solar day. And that the exchange would certainly not be overwhelming with everything 'all over' within 30 minutes or so. Rather, it would be an odd sort of exchange, lasting the full duration of a single solar day; destructive of course, yet globally, within certain limits.

Chapter 17

Nuclear Weapons: Exotic Effects

In view of the analysis given in the previous chapter, it is quite clear then that nuclear weapons are nothing like conventional weapons in terms of their operational nature. For indeed, in order to generate a nuclear explosion, external factors are required; the correct celestial harmonic configurations as are necessary to interact with a nuclear device at the moment of activation, to produce the explosion. And yet, in still further consideration of such devices, it would appear that they are far more exotic than even the evidence so far presented suggests; the vast power of these weapons producing certain highly anomalous additional physical side effects.

Nuclear Explosions & Nuclear Aftershocks

It goes without saying that the energy released from nuclear devices is immense, especially from fusion type weapons. And significant disruption to the geophysical state of the Earth at the site of a blast is a given. One may note also regarding the effects of such weapons, the fact that the physical disruption caused is very much comparable to that generated by sizable natural earthquakes. Indeed, upon this very point, throughout the world and for some time now, not only have monitoring stations been set up to record earthquake activity to aid scientific research, monitoring stations have been established also with the specific purpose of ensuring that various nuclear test ban treaties are actively upheld, through detection of suspicious geophysical disturbances indicative of unlawful nuclear weapons testing.

As a direct result of nuclear experimentation, scientists have been able to develop scales of measure to express the relationship between nuclear yield values and seismic disturbance. In the case of the former, as already noted, nuclear yield values are usually given in terms of TNT. By contrast, the measurement of earthquakes is through use of the Richter scale. For a basic comparison of nuclear yield values set against their Richter scale equivalent, one may consult the table below [1]:

Richter Magnitude	*TNT for Seismic Energy Yield*	*Example (approximate)*
-1.5	6 ounces	Breaking a rock on a lab table
1.0	30 pounds	Large Blast at a Construction Site
1.5	320 pounds	
2.0	1 ton	Large Quarry or Mine Blast
2.5	4.6 tons	
3.0	29 tons	
3.5	73 tons	
4.0	1,000 tons	Small Nuclear Weapon
4.5	5,100 tons	Average Tornado (total energy)
5.0	32,000 tons	
5.5	80,000 tons	Little Skull Mtn., NV Quake, 1992
6.0	1 million tons	Double Spring Flat NV Quake 1994
6.5	5 million tons	Northridge, CA Quake, 1994
7.0	32 million tons	Hyogo-Ken Nanbu, Japan Quake 1995. largest Thermonuclear Weapon
7.5	160 million tons	Landers, CA Quake, 1992
8.0	1 billion tons	San Francisco, CA Quake, 1906
8.5	5 billion tons	Anchorage, AK Quake, 1964
9.0	32 billion tons	Chilean Quake, 1960
10.0	1 trillion tons	(San-Andreas type fault circling Earth)
12.0	160 trillion tons	(Fault Earth in half through centre, OR Earth's daily receipt of solar energy)

Of the above figures, one may note that each point of increase on the Richter scale is indicative of an earthquake approximately 32 times greater or more powerful than the last point that came before e.g. a level 7.0 earthquake is about 32 times more powerful than a level 6.0 quake, with a level 8.0 earthquake being about 32 times more powerful than a level 7.0 quake.

Of Primary Events & Aftershocks

From the study of both nuclear weapons and earthquakes it is well established that both share one common attribute. They are both responsible for what geologists call 'aftershocks'; secondary seismic events that follow on closely after the initial primary event, but usually significantly time delayed, quite often by several days. Of course, though one might call an aftershock caused by a natural earthquake event also natural, one could hardly label a seismic aftershock following the test of a nuclear weapon as natural, even if there was no actual physical device to trigger it.

From a scientific viewpoint, the exacting nature of the causal linkage between any given primary event – nuclear or natural, and its secondary aftershock, is a matter of immense interest. Just how exactly does a nuclear weapon (or earthquake event) actually trigger a time delayed secondary aftershock event? In the attempt to answer this question, there is one particular pair of events well worthy of consideration.

Engineering a Nuclear Aftershock

In the history of the US nuclear weapons program, it is to be noted that the highest yield underground test ever conducted by the US, codenamed Cannikin, was in 1971 on the island of Amchitka; a part of the Aleutian chain. Indeed, only 3 nuclear tests in total were ever carried out on this island. In order, they began with Longshot in 1965, Milrow in 1969, and then Cannikin in 1971. Being the last of the three tests, Cannikin was by far the highest yield, with the actual placement of the device upon activation being a small underground chamber 50 feet in diameter at a depth of some 5875 feet [2]. The primary details of Cannikin are as follows [3]:

Shot: *Cannikin* (United States). Yield (TNT): 5,000,000 Tonnes
Date: 06 November 1971 Time: 22:00:00.06 (GMT)
Geodetic Bomb Co-ordinates (Deg):
Lat: 51.47190 N Long: 179.10690 E
Geodetic Sun-Ground Position Co-ordinates (Deg):
Lat: 16.09596624633 S Long: 154.09 W
Arc Length: Equator to Latitude of Test (feet): = 18715833.051
Arc Length: Test Site to Sun-Ground Position (feet): = 26001130.562

Of the test explosion itself, it can be seen from the table cited above that the yield of Cannikin, at some 5 megatons (5 million tonnes of TNT), is the equivalent of a 6.5 level earthquake as measured by the Richter scale. Such indeed may be externally verified as the actual blast itself was detected at the time by various monitoring stations, and currently finds itself (as of writing) on the official Australian Government Earthquake database ('Natural Hazards Mapping') [4]:

6th Nov-1971	Time: 22:00:00 (GMT)
Latitude: 51.47 N	Longitude: 179.11 E
Earthquake level: 6.6 Magnitude	

In addition to the above listing, one may note also that in consulting the very same official source just two days after the Cannikin test, the following event is detailed [5]:

8th Nov-1971	Time: 11:54:12 (GMT)
Latitude: 51.472 N	Longitude: 179.107 E
Earthquake level: 4.9 Magnitude	

One can clearly see in this second event, co-ordinates that match up most exactingly to none other than those of the primary Cannikin test, as occurred itself only some two days previously. In consulting a nuclear test database though for details of further US tests carried out on Amchitka following on from Cannikin, there are none listed throughout the whole of the rest of the year – even up to the present day; a fact which clearly confirms that the second event was not the result of any new nuclear test as such. It could only have been a 'nuclear aftershock' following the initial Cannikin event. Much reduced in power the secondary event was capable only of a Richter scale disturbance of 4.9, equivalent to a yield of just less than 32000 tonnes of TNT, or about 2 Hiroshima bombs. At a full 5000000 tonnes of TNT, the primary Cannikin test, was thus far greater than the aftershock that occurred at the same site some two days later.

Cannikin Evaluated

In an initial study of the primary values of the Cannikin test, simple division of the bomb-sun arc by the latitude arc measure reveals a most notable association:

18715833.051 / 26001130.562 = 0.71980843318

With: 0.71980843318 × 2 = 1.43961686636

Naturally, from this, one is bound to suspect that the true measures sought at the moment of the activation of the bomb were related by a precise value of 0.72. And indeed, that by far the most probable arc length measures as would combine to produce this simple ratio would be 18720000 and 26000000, for by their selection, one can easily note that 26000000 divided by 2 produces 13000000, and that with the basic number 13 used itself to divide 18720000, one is able to produce 144000. However tempting such relations appear though, it would seem that such refinements are bound to be in error. And the reason for this is that the US could never have placed the device on the isle of Amchitka at precisely 18720000 feet from the equator, at the specified longitude, due to the reality of the geography of the island itself. The basic layout of the island is that of a thin stretch of land about 42 miles in length by only about 2-4 miles in width, orientated approximately 45 degrees with respect to the Earth's equator. Were the Cannikin device to have been placed at 18720000 feet from the equator upon the noted longitude, it would have been positioned in the sea!

Clearly then the Cannikin device (and its aftershock) was not positioned at the exact latitude of 18720000 feet north. It had to have been positioned slightly to the south of this full measure, but in accordance with just what governing ratio or harmonic? The answer is to be had with reference to an arc length measure associated with a test already reviewed in a previous chapter: Pajara (Test Ground: Nevada). At the moment of the activation of Pajara, the bomb-sun arc length measure possessed a value of some 22118750.551 feet. It would appear that using the basic ratio of 13 / 11, one can connect up the latitude value of Cannikin to this value, and to an exceptional level of accuracy:

18715833.051 × (13 / 11) = 22118711.788

In addition to this, and as the further analysis of Pajara revealed, the intended bomb-sun arc was determined to have been 22118400 feet precisely. Using this measure, a refinement of the Cannikin latitude may be had as follows:

$$22118400 \times (11 / 13) = 18715569.230 \text{ feet}$$

With the latitude arc appropriately refined then, what of the bomb-sun arc of Cannikin? Here one is tempted to think that this arc was indeed as initially suspected, of a precise value of 26000000 feet at the very moment of detonation; the reason being simply that such a value does possess an undeniable standalone association to the 144 numeric through use of the prime number 13: $(26000000 / 13) \times 72 = 144000000$. Of course, in consequence, the (first) noted 0.72 ratio would clearly appear not to have been evident at the moment of the test; the actual harmonics being more subtle.

The Cannikin Nuclear Aftershock

In view of the fact then that a significant earthquake struck the exact same site as the Cannikin test, only some two days following the blast, a definite causal link between the two events must exist; but of just what nature? Quite remarkably, the answer would seem to be had from the inspired decision to evaluate the Cannikin aftershock in exactly the same way as any given nuclear event. And thus, reproduced in like format, the primary values of the aftershock that lend themselves to such an analysis are as follows [6]:

Shot: *Cannikin Aftershock* (US). Yield (TNT): ~32,000 Tonnes
Date: 08 November 1971 Time: 11:54:12 (GMT)
Geodetic Epicentre Co-ordinates (Deg):
Lat: 51.472 N Long: 179.107 E
Geodetic Sun-Ground Position Co-ordinates (Deg):
Lat: 16.56608684877 S Long: 2.61833333333 W
Arc Length: Equator to Latitude of Quake (feet): = 18715869.553
Arc Length: Quake Site to Sun-Ground Position (feet): = 52915427.507

As is seen from the above data, the site of the Cannikin aftershock constitutes the actual epicentre (EC) of the earthquake, as occurred at practically the exact same spot as the initial nuclear test. Consequently, the arc length from the equator up to the site of the quake is practically of the same length as the Cannikin event some two days previously. Furthermore, at the time of the aftershock, just as with the nuclear events already evaluated, there is a distinct arc length measure from the sun ground position (SGP) to the epicentre of the earthquake, which is also noted.

Using the values as already refined from the Cannikin test, one is able to quickly bring to light a most intriguing numerical association between the latitude arc of the site and the epicentre-sun ground position arc of the aftershock; one that makes use of the prime number 11, already noted to be of significance, and a 'base-6' progression, which would appear to directly tie in the aftershock to the 144 harmonic sequence:

52915427.507 / 18715569.230 = 2.827348014646
2.827348014646 × 11 = 31.100828161107
31.100828161107 / (6 × 6 × 6) = 0.143985315560

Ideal = **0.144**

From the above, if one were to reverse the procedure starting out from an ideal of 0.144, then a refined value for the EC-SGP arc of the aftershock is determined as follows:

0.144 × 6 × 6 × 6 = 31.104
31.104 / 11 = 2.82763636363... (3888 / 1375)
(3888 / 1375) × 18715569.230 = **52920824.123** feet

Difference: 52920824.123 – 52915427.507 = 5396.615 feet

As a note, though the error may seem slightly greater than has been seen before, it is well to consider that earthquakes as such can and usually do occur over the course of several seconds, if not up to half a minute or so on some occasions for the most powerful. Thus, in modelling the dynamics of such events on par with known nuclear test explosions, it should be no surprise to find a slightly higher level of inaccuracy in the figures.

That said however, it is also very important to realise that for any given nuclear event or earthquake, what is actively derived from a refinement of the raw values associated with such events, are simply the ideal configurations as would allow for a 'maximum energetic reaction'. The true celestial configurations as are present at the moment of a nuclear or earthquake event are of course bound to be imperfect as such, in that they do differ from the ideal noted configurations determined from a further careful evaluation of the 'raw' arc lengths as modelled. However, the specified ratios and arc lengths noted in all the examples are indeed close enough to an optimum or ideal set as does allow for an energetic nuclear-seismic disturbance to occur. Moreover, it would be eminently reasonable to conclude that the closer the celestial configurations actually are to their ideals, the greater the power of the nuclear-seismic events.

The True Nature of Nuclear Aftershocks

As the mathematical associations appear to indicate respecting the above, it would seem then that the primary Cannikin test and its noted aftershock are clearly linked via the harmonics of the basic sexagesimal system through their associated arc measures, with an undeniable 'nuclear signature' evident in both cases. Just as the test weapon required a very precise celestial configuration to be present at the exact moment of activation to achieve success, the dynamics of both the sun and the Earth also appear to have governed the very occurrence of the subsequent nuclear aftershock. But what exactly do such relations say about how earthquake events of this type are caused per se, and what mechanism lies behind them?

If one were to ask a geologist to account for an aftershock event following on shortly after a primary event, one that was especially close to the primary, they would doubtless tend to answer by saying that the primary itself caused a weakening of the ground about the site, such that the occurrence of natural tectonic activity a few days later, combined with the weakened substructure, breached a critical threshold thereby causing a secondary seismic disturbance; though usually one of far less power. In saying this it would seem though that such geologists would be greatly mistaken, as a careful study of the Cannikin relations appears to reveal.

Based upon the analysis it would appear undeniable that the Cannikin test *caused* the subsequent aftershock. Also, the very fact that the aftershock precisely coincided with a celestial configuration possessed of a distinctive nuclear signature, forces one to conclude that the earthquake itself was in no way random in its timing. A true deterministic causal link between the two events is clearly evident, but of what nature exactly? The answer, leads one to realise that the true operational nature of nuclear weapons as energetic devices, is far more exotic than one might imagine. Indeed, so far unmentioned until now, is the fact that the extreme intensity of the reaction that occurs when a nuclear weapon is triggered, *causes a rip in the very fabric of space; one that results also in a major disruption to the very 'flow rate' of time itself at the site of the blast,* affecting even the material components of the bomb.

From a previous chapter, it was explained as per the work of Walter Russell, that all physical objects are in a state of constant interaction with the universal 'zero-vacuum' of stillness, and that cyclically all things emerge from the universal vacuum, taking up form and motion, only to withdraw once more into its structure. The very *manifestation of matter* is thus fundamentally tied in to the pulsation rate of its continual emergence from, and withdrawal into, the primary vacuum. And what physicists understand to be 'time' is nothing more than simply *the 'rate of existence' of a given object*. The key point here to note though is that different objects pulsate or exist in time at different rates. Thus, in terms of such as action at the atomic level, matter may in the space of a single second go through a million cycles of distinct manifestation, emerging from the zero-universe to take on form, and withdrawing back into the very same, but with each manifestation being ever so slightly different, giving the object the appearance of having been both shifted in space i.e. moved, and of being altered slightly e.g. transformed; evolved (or in the case of humans, aged).

In the course of the 'everyday universe', one may speak of the natural activity of things wherein one would assume no surprises in terms of the general predictability of events in space and time. However, with regard to the force and character of nuclear reactions of the type that occur at the centre of a bomb upon activation, one must understand that such reactions have the power to massively distort the nuclear material (radioactive core) of the device, so much so as to interfere with the regular time cycles that govern its recurring existence within the physical universe. To be clear; what actually happens when a nuclear device is exploded, is that a significant fraction of the power *and matter* of the explosion *is not released as part of the initial blast,* but instead withdraws into the zero-vacuum, and in such a manner that its physical-frequency base, or time i.e. the rate of existence of its regular manifestation, is severely disrupted. And consequently, it does not automatically re-emerge from the zero vacuum straight away as it normally would. Its next re-emergence into the physical universe rather will be based upon a much longer time cycle i.e. it will, in effect, be significantly *time delayed*. But, when it does emerge, the actual character of its next manifestation will be none other, than an earthquake!

By way of example therefore, one could say in estimate, that the true power of the Cannikin event upon activation of the device was some 5032000 tonnes (TNT). However, at the very time of its activation, the initial explosion as manifest was only on the order of about 5000000 tonnes. The remainder, constituting a blast force of some 32000 tonnes, occurring precisely 2274.199 minutes later, was in fact a secondary nuclear explosion literally *transmitted into the future*; being the violent yet ordered re-emergence of a fraction of the matter of the bomb that had withdrawn into the universal vacuum at the moment of the primary event. Its re-entry into the universe after a significant time delay some two days later thus produced nothing less than a further lower yield nuclear explosion – an event easily mistaken for a natural earthquake.

The Transmission of Nuclear Forces through Space & Time

In response to the above points, and with full acceptance of the fact that both Cannikin and its aftershock were nuclear in nature, the next question that is bound to arise is, to what extent is it possible to actually control the transmission of a nuclear explosion through space and time? Essentially, is it possible to generate a nuclear blast at one point, with full predictive knowledge as to exactly where and when its secondary aftershock will emerge? Certainly, in the case of Cannikin the precise timing of the aftershock was fully determined in advance. Once the initial nuclear test had been conducted it was essentially guaranteed that a secondary event would occur in the future precisely when it did and where it did. But did the scientists at the time know the specifics of the aftershock time and location? It is doubtless the case that following all of the nuclear testing over the past several decades that the dominant military powers on the Earth have built up a sizable body of information on the patterns involved both with respect to engineering a primary nuclear weapons blast, and also secondary reactions. In consideration of this, one has to wonder then at just how far they have been able to progress with this science. Thus one may inquire, have they as yet developed the ability to control exactly how much of the power of an initial explosion can be manifest in the primary blast, and how much can be transmitted through the zero vacuum to a secondary site? Indeed, to go further, have they been able to actively develop the technical means to generate in safety a high magnitude nuclear explosion, and transmit it in totality from a primary site or facility to an altogether completely different location upon the Earth? The answer to both questions would seem quite remarkably to be a resounding yes!

Although rarely discussed in the mainstream press, over the course of recent years a significant amount of information has indeed managed to find its way out of the secret world of 'black projects' and into the proverbial light of day. And this is especially true with regard to knowledge gained from nuclear experimentation by the so called 'great powers'. The achievements, as have emerged and been set into print, though limited, clearly indicate that advances in the physical sciences over the past few decades have been truly astounding. And yet, so very many have been (ruthlessly) suppressed. Be that as it may, confirmation of the fact from reliable and credible sources that certain powers have indeed attained the ability to wilfully transmit nuclear explosions through space and time, and in a highly controlled manner, has emerged. One may cite in particular Thomas E. Bearden PhD (U.S. Army, retired) who has written many important works concerning a variety of suppressed scientific-technological advances that have been made over the past few decades; specifically with regard to certain new (revolutionary) principles of nature as discovered by Russia, that have led them to the direct weaponisation of the controlled transmission of nuclear forces through space-time. Such weapons as are able to achieve this astounding feat are generally referred to by Bearden as energetics or psychotronic based weapons. Indeed, Bearden goes so far as to name specific Russian facilities as have been built to transmit a fully generated nuclear explosion from one place to another; made possible by tuners that can be planted essentially anywhere in the universe to 'receive' the incoming explosion. A facility located at Semipalatinsk is named by Bearden as just such a place where this capability exists, which he identifies specifically as a "third-generation hyperspace nuclear howitzer" [7]. Moreover, in providing yet further details of the actual process of transmission, Bearden states that the [nuclear] explosions themselves do not in fact travel through or in ordinary space at all, but instead they move through hyperspace [8]. Such an established fact is of critical importance here; being strongly tied to the findings of this present work.

To advance therefore on the work of Bearden, which would indeed appear quite harmonious with what has already been said of nuclear weapons previously, the precise manner in which a nuclear explosion would appear to be transmitted from one place to another, would seem to be via a complex series of highly stable hyperspatial pathways, which first allow for its withdrawal into the zero vacuum (of Russell), and then for its ordered re-emergence into the world at another location and/or time. And yet critically, the very ability to transmit a nuclear explosion through hyperspace still would clearly appear to rest upon the essential dynamics of various Earth-sun celestial configurations; fundamentally required by necessity in the first place for the very resonant destruction of a nuclear device. That this appears to be so would seem to be had from a very intriguing statement by Bearden concerning what he refers to as a "strange weakness" possessed by nuclear howitzers [9]. To quote him in full:

"...the superweapons [nuclear howitzers] have a strange weakness and failure mode possessed by no other kind of weapon. The hyperspace amplifiers in the psychotronic weapons depend on a large number of stages, say sixty to one hundred, and the hyperspatial flux in all these stages must be very stable and isotropic. At odd intervals the flux coming in from hyperspace suddenly becomes turbulent; this destroys the smooth coherence in the multiple stages of the amplifiers, and thus the basic amplification mode. Thus *all* the psychotronic weapons will suddenly cease working at the same time, and this "all fail" mode may last from fifteen minutes to even three weeks in an extreme case. Thus, the weapons are erratic and not dependable." [10]:

What this statement would appear to be hinting at is the fact that the pathways via which a nuclear explosion may be transmitted through space and time are in a dynamic state of flux, and that only when the correct alignment of the pathways is present is one able to actually generate a nuclear explosion and successfully send it to a desired location in an ordered manner. If this point is combined with the findings of this present work, one is thus bound to suggest that the stable pathways able to transmit such explosions are dynamically established by the celestial configurations of the Earth and sun. And that only at certain propitious times are they momentarily aligned to allow for the transmission. This would certainly explain such long delays when it is not possible to send an explosion. Major shifts in the axial orientation of the Earth to the sun over the course of a year may be the key factor, but also, one may suggest another variable so far unmentioned in this regard: the moon. Already it has been noted that the disc of the moon in terms of its size is a match to that of the sun, and that the full orbit is about 29.5 days. The fact that Bearden mentions a time period of about 3 weeks is very interesting, as this is only slightly less than the moon synodic month. Could it be that the orbital configuration of the moon is also a decisive factor in the equations; either an aid or disrupting influence to the stability of the hyperspatial pathways? Certainly this point is well worthy of further exploration, and indeed will be touched upon in the next chapter.

In light of the above information then, the ability to transmit a nuclear explosion in totality through space and time would appear to have been well mastered by certain military powers. Making use of various celestial configurations which dynamically occur over the course of a given year, and as momentarily establish an ordered set of natural hyperspatial pathway alignments between two points, one is able to both actively generate a nuclear explosion at one location, and wilfully transmit it to another.

Chapter 18

The Nuclear Physics of Earthquakes

In view of the analysis given in previous chapters it is clear then that nuclear weapons are nothing like conventional weapons in terms of their operational nature. As indeed explosive reactions of the former type may only be generated at certain precise points upon the Earth, relative to various fleeting harmonies of certain physically energetic celestial configurations, as are necessary to interact with a nuclear device via the principle of resonance at the moment of activation. And further to this, any secondary nuclear aftershocks as may arise from a primary event appear themselves to be determined in both space and time by a subsequent set of celestial relations. Such future harmonies, acting as a doorway to allow entry into the physical world of a set portion of the explosive matter of the primary event, transmitted up to that point through a series of well ordered hyperspatial pathways 'underneath' what would seem to be a certain critical energy threshold of the 'ordinary' physical realm.

The Geophysical Science of Earthquakes

In all that has so far been said concerning the importance of various celestial configurations as would appear necessary to allow for a nuclear explosion, one must realise that were it not for somebody taking actual advantage of such configurations to actively generate a nuclear reaction, then the fleeting harmonies would on their own appear to pass by unnoticed, being of their own 'power' insufficient to cross that most critical energy threshold as would cause a real disturbance in this present physical realm.

However, in view of the noted fact that a nuclear weapon is indeed capable of causing subsequent earthquake type events as are labelled aftershocks, one must wonder if genuine earthquake events may in fact themselves be caused by none other than *a natural breach* of the aforementioned critical energetic threshold. And thus that the physical science governing nuclear weapons is also the natural physical science governing the occurrence of earthquakes, the latter of which have of course been occurring throughout the history of the world before any modern-day weapons programs were ever established. If this is so, then one would expect that the same harmonic-type relations found to be present upon the detonation of a nuclear weapon would also be present at the moment of a natural earthquake.

The Celestial Relations of Earthquakes

In order that one might test the theory that earthquakes are actively caused or triggered by various celestial patterns, which produce a highly targeted seismic disturbance at some geophysical location upon the Earth, one would need to select for evaluation a set of well recognised or prominent earthquakes from the recent period, for which there is reliable data in terms of actual location and time of occurrence. Also, to be as certain as one can be that one is choosing for study genuine earthquakes, and not covert nuclear weapons events, it would be appropriate to choose only those earthquakes of the recent past as possessed a magnitude well in excess of the power of what are generally recognised as the most powerful nuclear tests ever to have been conducted. On this very point one may note that the highest yield nuclear weapons test ever in the history of testing (excepting covert tests that one could only speculate about) was on the order of some 58 megatons by Soviet Russia in 1961 (codename Tsar Bomba. See chapter 15 for a complete analysis) [1], producing a Richter scale disturbance of just slightly over 7.0. Clearly then, to test whether the occurrence of earthquakes rely upon the same sorts of celestial patterns as nuclear weapons, it would be wise to study a series of earthquakes whose magnitudes are well above this level.

The Seismic Events of December 2004

Of such events one may note two in particular that occurred at the close of 2004, the epicentres of which were located deep under the oceans. The first occurred just to the south-west of New Zealand on the 23rd of December 2004, and had the power of an 8.1 magnitude earthquake, equivalent to some 1000 megatons of TNT. The second event which followed closely after only some 3 days later was in the Indian Ocean just north of the equator, constituting a 9.0 earthquake equal to about 32000 megatons of TNT [2]:

23rd Dec-2004	Time: 14:59:04 (GMT)
Latitude: 50.24 S	Longitude: 160.13 E

Earthquake level: 8.1 Magnitude

26th Dec-2004	Time: 00:58:50 (GMT)
Latitude: 3.298 N	Longitude: 95.779 E

Earthquake level: 9.0 Magnitude

Of the above listed events, the first did not receive too much global attention. The second that took place on the 26th of December 2004 most certainly did however; the reason being that the power of the seismic disturbance was so great that it caused a massive disruption to the surrounding ocean, resulting in the generation of a whole series of tsunamis – freak high amplitude waves. Following several hours after the initial event, whose epicentre was just off the coast of Indonesia, the tsunamis swept through the low lying regions throughout the Indian Basin. In all, the earthquake had a magnitude higher than any other to have hit the region in over 40 years, and the devastation was so great that some 227898 people were confirmed dead shortly after the impact of the tsunamis [3].

Nuclear Evaluation of the 2004 Earthquakes

With the time and location of each of the noted earthquakes given, a precise determination of the celestial patterns existent in the heavens at the moment of both events can thus be had, allowing one to derive the key values necessary for a nuclear analysis. However, in an extension of the previous analyses, in addition to pure latitude values and their associated SGP arc measures; in light of the note made by T. E. Bearden concerning the 'strange weakness' of the energetics weapons as developed by Russia, it would seem highly appropriate to examine also the associated Moon Ground Position (MGP) arc measures of the noted earthquakes, so as to determine if they too are of any significance [4]:

Shot: *Ocean Event*. Yield (TNT): ~1000,000,000 Tonnes
Date: 23 December 2004 Time: 14:59:04 (GMT)
Geodetic Epicentre Co-ordinates (Deg):
Lat: 50.24 S Long: 160.13 E
Geodetic Sun-Ground Position Co-ordinates (Deg):
Lat: 23.56410147126 S Long: 44.92833333333 W
Geodetic Moon-Ground Position Co-ordinates (Deg):
Lat: 22.50506572015 N Long: 99.17833333333 E
Arc Length: Equator to Latitude of Quake (feet): = 18266215.786
Arc Length: Quake to Sun-Ground Position (feet): = 37617448.593
Arc Length: Quake to Moon-Ground Position (feet): = 32921928.760

Shot: *Ocean Event*. Yield (TNT): ~32,000,000,000 Tonnes
Date: 26 December 2004 Time: 00:58:50 (GMT)
Geodetic Epicentre Co-ordinates (Deg):
Lat: 3.298 N Long: 95.779 E
Geodetic Sun-Ground Position Co-ordinates (Deg):
Lat: 23.50115225058 S Long: 165.42833333333 E
Geodetic Moon-Ground Position Co-ordinates (Deg):
Lat: 27.86553040614 N Long: 21.61666666666 W
Arc Length: Equator to Latitude of Quake (feet): = 1196450.142
Arc Length: Quake to Sun-Ground Position (feet): = 26574940.685
Arc Length: Quake to Moon-Ground Position (feet): = 41005188.158

From a careful study of the above events, several highly significant associations are revealed. Not only would it appear that the two ocean events are strongly linked to the Cannikin test and its low yield aftershock, they are also found to be linked to other known nuclear tests previously evaluated. Indeed, the various arc measures associated particularly with the 23rd of December 2004 event are found to have a most striking affinity to various ratios and fractions evident in some of the most prominent nuclear tests; so much so that one is bound to conclude that both 2004 ocean events were in fact governed by the same physical science as nuclear weapons.

From a previous chapter, one may recall the key arc length measures associated with Trinity and Hurricane, including also the ratio between the pure latitude placement of Trinity and the bomb-SGP arc of Hurricane, upon activation:

12230590.464 / 17694720 = **0.6912**

What is most striking about the Ocean Event of December the 23rd 2004 is that this same ratio is evident respecting its SGP arc length and that of the previously noted Cannikin test:

Cannikin: Bomb-SGP arc = 26001130.562 feet
Ocean '23' Event: EC-SGP arc = 37617448.593 feet

26001130.562 / 37617448.593 = 0.691198673361

And, to link to the destructive harmonic of 144:

0.691198673361 / 8 = 0.086399834170
0.086399834170 / 6 = 0.014399972361

Ideal = **0.0144**

In addition to the above, noting the previously suggested ideal bomb -SGP arc of Cannikin, 26000000 feet; in conjunction with an exact 0.6912 value, an ideal Ocean '23' event EC-SGP arc can be had as follows:

26000000 / 0.6912 = 37615740.740…feet

Still further, additional relations may also be noted. Indeed, a focus upon the Moon values demonstrates that they too would appear to be intimately associated with both of the noted earthquake events. Using simply the number 9 (3 × 3) as a basic multiplier, and also divisor, it is easily seen that the noted MGP arc lengths are indeed decisively linked to the 144 numeric sequence of interest. Consider first the Ocean '23' MGP arc relations:

32921928.760 × 9 = 296297358.844
296297358.844 × 9 = 2666676229.601

2666676229.601 × 9 = 24000086066.411
24000086066.411 × 6 = 144000516398.467

Ideal = **144000000000**

Also, with regard to the MGP arc of the Ocean '26' event:

41005188.158 / 9 = 4556132.0176
4556132.0176 / 9 = 506236.89084
506236.89084 / 9 = 56248.543427
56248.543427 / 9 = 6249.8381585
6249.8381585 / 9 = 694.42646206

1 / 694.42646206 = 0.0014400372

Ideal = **0.00144**

In respect of the above one may carefully refine the noted arc length measures to produce the following:

Ocean '23': EC-SGP arc = 37615740.740740…feet
Ocean '23': EC-MGP arc = 32921810.699588 feet

Ocean '26': EC-MGP arc = 41006250 feet

Given these relations and their accuracy, one may thus continue on to detail another most forceful relation, between the noted SGP and MGP arc lengths specific to just the Ocean '26' event, that indeed allows for the further refinement of the remaining arc measure; that is the Ocean '26' EC-SGP arc. Consider the following:

26574940.685 / 41006250 = 0.6480704937
0.6480704937 / 9 = 0.0720078326
0.0720078326 × 2 = 0.1440156652

Ideal = **0.144**

Assuming the truth of the noted 41006250 refinement, by reversing the mathematics from the above ideal of 0.144, a refined value for the Ocean '26' EC-SGP arc can be determined as follows:

$$((0.144 / 2) \times 9) \times 41006250 = 26572050 \text{ feet}$$

In addition to the above, there is also a most notable relation to be had involving both the pure latitude arc and the EC-SGP arc of the Ocean '26' event; their simple division revealing a value of obvious significance:

$$1196450.142 / 26572050 = 0.04502664047$$

Ideal = **0.045**

With: $0.045 \times 2 \times 2 \times 2 \times 2 \times 2 =$ **1.44**

With an ideal of 0.045 thus established in principle, which further leads on to the 1.44 value, then due to the refinement already made of the Ocean '26' EC-SGP arc length, a refined value for the pure latitude arc of the event may also be had:

$$26572050 \times 0.045 = 1195742.25 \text{ feet}$$

With the latitude thus refined for the Ocean 26 event, one may turn to consider also that of the Ocean 23 event. And note the following:

$$18266215.786 / 41006250 = 0.445449554321$$

$$0.445449554321 \times (11 / 7) = 0.699992156791$$

Ideal = **0.7**

Working backwards from the ideal of 0.7 one can generate a refined arc length measure for the latitude of the Ocean 23 event as follows:

$$0.7 / (11 / 7) \times 41006250 = 18266420.45454545\ldots \text{ feet}$$

In this refinement, one can see once more the very familiar numbers 11 and 7, as have been demonstrated on numerous occasions to be intimately associated with nuclear events.

The Venus Linkage

Before moving immediately to certain conclusions as may be drawn from the analysis as presented so far, one is asked to consider yet one final relation associated with the Ocean '26' event, that involves a further celestial body. In addition to the celestial ground position points of the Sun and the Moon as linked to the epicentre of the '26' event, the arc length measures associated with the planet Venus also would seem to be of great significance. Summarised in the familiar format as above, for the moment of the earthquake itself, they were as follows [5]:

Shot: *Ocean Event.* Yield (TNT): ~32,000,000,000 Tonnes
Date: 26 December 2004 Time: 00:58:50 (GMT)
Geodetic Epicentre Co-ordinates (Deg):
Lat: 3.298 N Long: 95.779 E
Geodetic Venus-Ground Position Co-ordinates (Deg):
Lat: 21.34561489299 S Long: 140.685 E
Arc Length: Quake to Venus-Ground Position (feet): = 18375197.503

Previously it was shown how the number 7 is intimately linked to certain transformations of the arc measures of various nuclear tests. This very number applied twice to the Venus EC-VGP arc, yields an immensely significant association:

$$18375197.503 / (7 \times 7) = 375004.030$$

NB: The inverse of the number 375, produces the very familiar sequence of: 26666666666…:

Thus: 0.375 = 3 / 8 & 2.6666666666…. = 8 / 3

Multiplied by 6, the modified Venus arc produces the same numeric sequence as the previously stated ideal Venus orbital period, of 225 solar days:

375004.030 × 6 = 2250024.184
Also: 2250024.184 × 8 × 8 = 144001547.785

Ideal = **144000000**

Working from the specified ideal, one may thus derive a refined arc value for the Ocean '26' EC-VGP as follows:

375000 × 7 × 7 = 18375000 feet

In bringing the analysis of this section to a close, it would be well to state the difference between all of the refined values and those as modelled. And thus will the reader see just how accurate all of the noted refinements are:

Ocean '23': Latitude arc = 204.667
EC-SGP arc = 1707.852
EC-MGP arc = 118.060

Ocean '26': Latitude arc = 707.892
EC-SGP arc = 2890.685
EC-MGP arc = 1061.841
EC-VGP arc = 197.503

All values as given are in feet

A General Theory of Earthquakes

In view of the above it would appear then that there is an undeniable significance to the arc length measures linked to both the Sun and Moon covering over the Earth's surface at the very moment of the December 2004 earthquake events, including those of Venus linked specifically to the Ocean '26' earthquake. The complex harmonies evident at the time of each event are seen to possess a most striking affinity to the mathematics of the sexagesimal system, including allied numerical progressions. And thus with high confidence one may conclude, as indeed this present author does, that the celestial relations of the heavenly bodies most important to the Earth – the Sun and Moon at the very least – in their dynamic configurations, are directly responsible in a true causative sense for producing highly targeted seismic disturbances upon the surface of the planet (or just beneath) in the form of earthquakes. And thus, the physical science as actively governs the natural occurrence of such seismic activity, is the very same as that which is employed to cause high powered nuclear explosions, as may themselves result in extreme seismic activity [6].

Matter-Energy Transfer between the Celestial Bodies

Extending the above analysis to its ultimate conclusion, one is led to accept that the principle celestial bodies that make up the very solar system, each body with respect to every other, must be continuously engaged in a *highly dynamic matter-energy exchange in the course of their very being*. And that such constitutes an active mechanism of the solar system. Indeed, following the evidence, one must point to the functional hyperspatial pathways, discovered to lie outside of 'normal' space and beyond a certain critical energetic threshold, as being the key to facilitating the entire process. Various propitious celestial configurations as may thus be achieved in the heavens are able to briefly establish certain extremely specific links between set target locations, via such pathways. And that the precise alignments as occur, are able to momentarily establish an 'instantaneous space-time' linkage; and this between points that in terms of normal 3-dimensional space may in fact be separated by an immense distance. As to the character of such a link, one would doubtless assume it to be based upon a 'path of least resistance' type principle; *one that actively transfers* ***matter*** *from one given location to another*; the dynamically established hyperspatial pathways of the sub-domain region acting as a conduit.

From the above description one can thus perceive just how earthquakes are caused. With respect to the Earth as a target planet so to speak, the celestial patterns of the primary bodies of the solar system, when able to achieve certain very specific configurations respecting the Earth, establish a distinct energetic connection via the hyperspatial sub-domain region. Once achieved, a set portion of matter (and thus energy) is 'released' from (a given) source body; being actively channelled by the momentarily aligned hyperspatial pathways that link it to the Earth. An earthquake on or near to the surface of the Earth as would then strike, is nothing more than the *physical matter* emerging from a hyperspatial tunnel as transmitted from another celestial body; its 'reception' violently disrupting the Earth's own physical structure, due to it being forced into the same space as that currently occupied by the existing matter of the planet.

Extending the analysis, one may indeed point to the Earth itself as an astronomical body, that may in conjunction with others, including such as the sun and moon for example, form a part of a fleeting configuration that sees it transmit a portion of its own matter to another celestial body, causing itself a highly targeted earthquake to manifest upon its surface. Carefully thought through, one can thus picture an extremely dynamic celestial system at work, with all of the physical bodies of the solar system being active participants in an ongoing energetic transference of matter and energy; the ever so subtle configurations of the heavens actively regulating both the timing and the flow rate of the transfer of power between each of the bodies.

The End of an Age: Signs in the Sky

With a general theory as to the true cause of earthquakes, one is able to return to a subject first raised in a previous chapter concerning the global devastation said to have occurred sometime circa 9600 BC as per the writings of Plato. As one may recall, the Egyptian priest from whom Solon received his information stated that the Earth suffered periodic destruction over long range intervals, and that the planets themselves were the key component in regulating such upheaval. With this in mind, from what has now been revealed as to the physical principles that govern both natural earthquakes and engineered nuclear explosions, it would seem most certain that the underlying science as given is exactly that which must govern also the destructive events spoken of by the priest. Both fire and water (deluge), as said to blight the Earth at various set times, must be triggered by certain very precise and exacting celestial patterns as achieved in the heavens. And indeed, one may actually cite the work of a very specific individual from antiquity who noted what he considered to be *the* precise configurations as are responsible for actually causing global destruction through fire and water:

Berossos, who interpreted the prophecies of Bel, attributes these disasters (the end of the world and its aftermath) to the movements of the planets. He is so certain of this that he can determine a date for the Conflagration and the Great Flood. He maintains that the earth will burn whenever all the planets, which now have different orbits, converge in Cancer and are so arranged in the same path that a straight line can pass through their orbs, and that there will be a further great flood, when the same planets so converge in Capricorn. For under the sign of Cancer occurs the change to summer, under Capricorn the change to winter. They are signs of great power, occurring when there are movements in the change of season [7].

As can be seen then, Berossos identifies a set of high precision celestial conjunctions involving the Earth at certain of its seasonal markers, and what one would assume to be some or possibly all of the then known planets, as actually being responsible for triggering sudden global destruction. Such a proposal, carefully considered, is indeed thoroughly in line with the findings of this present work.

From the evaluation of the 2004 earthquakes it was revealed how a set of exacting and harmonious, though nonetheless complex configurations, may cause highly targeted high magnitude quakes to strike the Earth. In these sort of cases though, for all the harmony present (from both the arc values and ratios), the various celestial bodies involved are not actually in conjunction with one another. Rather, there are distinct ground-based arc measures separating the epicentre of the earthquakes from the noted celestial bodies e.g. the Sun, Moon, or Venus 'ground positions'. Considering the writings of Berossos in view of this, it would seem that such subtle patterns as these are only capable then of causing earthquakes of far less power than when high precision planetary conjunctions are achieved in the heavens. For in this latter instance, it is implied that the forces unleashed, are such as to be able to literally throw an entire planet into great agitation – the Earth, along with, one would assume all of the other celestial bodies involved in the alignment. The resonant forces activated at the precise moment of such conjunctions, with simultaneously a stable hyperspatial link dynamically established between the planets, are able to create a matter-energy exchange on such an immense scale as to cause entire worlds to tremble. Those of the esoteric traditions would recognise such events as marking the end of one world age and the beginning of a new one.

Chapter 19

The Philadelphia Experiment

In view of the analysis given in the previous chapter, it is quite clear then that the physical science behind both earthquakes and nuclear weapons is extremely sophisticated and most exotic. Indeed, that both noted types of events rely upon the celestial configurations of the heavens is quite an astounding finding. That being said however, in consideration of both it is quite evident that they involve the unleashing of what are essentially destructive forces either naturally or by wilful design, creating also secondary events of the same type. Thus can one cite the fact that an earthquake may cause a delayed aftershock, being a fraction of the matter-energy of the initial event transmitted through time to a precisely defined location that may indeed be well removed from the epicentre of the primary event. One could say the same even of nuclear weapons; that they too are able in the same way to cause secondary events that may be time delayed, spatially displaced, or both. Further to this though, with a general theory of earthquakes established and recognition given to the fact that matter is energetically being transmitted between different celestial bodies via the configurations of the heavens in an ongoing capacity, one is forced to the conclusion that the wilful detonation of a nuclear device at some point on the Earth, may in fact cause an aftershock not necessarily bound to the Earth at all, but to a highly targeted location on an altogether completely different celestial body; quite possibly the Moon, or indeed any other body within the entire solar system, including the sun!

Given then the physical science underlying the hyperspatial transmission of matter as outlined, the thought is therefore bound to arise if such a capability could be made use of for transportation purposes; that one could actually engineer a device to establish what would in essence be a functional portal or doorway linking in to a whole series of stable hyperspatial pathways, to enable one to transfer a material object or life form to another location both safely and instantaneously, or even with a highly controlled time delay? In exploring this very possibility, the examination of certain wartime achievements of the scientists of the modern age, over and above just their work on nuclear weapons, would appear most apt. And in this, one may cite what is known to history – having passed almost into legend – as the so called "Philadelphia Experiment".

Highly Classified Wartime Research

Conducted by the United States military in 1943 at the height of the war with the Axis powers, the Philadelphia Experiment was the codename given for a secret project to develop a technological means by which to counter radar detection. Such facts as indeed have become known to the general public concerning this covert program have emerged primarily through a series of leaks down through the years. As a result, and as one would imagine, many books have thus been written concerning the purported experiment, exploring 'every angle'. In brief then, from what is known; the scientists at work on the project, in order that they might gain an advantage over enemy forces, sought a way to baffle incoming radar signals as would ordinarily detect armoured vessels, by cloaking them inside an extremely intense electromagnetic field of some sort. In theory, it was expected that this would result in the incoming signals being scattered at odd angles, absorbed, or refracted by some measure, all of which would prevent detection. In order to test the theory then, a seagoing vessel was used: a Destroyer class ship, DE173. Manned only by a skeleton crew, the ship was outfitted with specially constructed equipment in order to generate the necessary energetic fields as would completely envelope the entire vessel. According to sources, experiments were conducted both upon the Delaware River and also out to sea sometime between the 20th of July through to the 20th of August [1]. With regard to the first initial trial, the Destroyer was manoeuvred into position at a special point upon the river; a second observational vessel situated alongside. Once in place, and with everything set, the specially constructed electronic equipment upon the ship was activated and the field established. What happened next, according to reports, was both astounding and also catastrophic.

According to a witness who observed the DE173 from the second ship, as the field built up to full strength the vessel rapidly became invisible to human eyes, with only the outline of the ship's hull in the water able to be seen [2]. The whole mass of the Destroyer as enclosed within the intense electromagnetic field was apparently fully cloaked, and not just with respect to radar, but also ordinary sight. Initially then, up to this point, the experiment would have appeared a success beyond all expectations. However, after the field was switched off, the full impact of what had actually happened quickly dispelled the notion that the experiment had been successful. For when the ship did reappear, some of the crew did not return with it. They were quite simply gone! Yet certain other members of the crew who did come back with the ship were dead; being strangely merged in with the steel bulkheads of the vessel itself [3]. For those who did manage to return alive though, accounts given by the men of what happened whilst within the field indicate a thoroughly bizarre experience. While stood upon the deck of the ship, the other crew members as sighted, including even themselves, appeared vague in form, and looked to be walking upon nothing [4]. Further to this though, and worst of all for the men who took part in the experiment, was the fact that even when they were no longer within the field and had returned to their regular duties, or taken shore leave for that matter, the effects of the field could still seemingly overcome them at any future time, and without any warning. Reports thus exist of men just disappearing right in front of witnesses, never to be seen again. Or of men freezing on the spot, experiencing a strange mind-body disconnect, with an acute sense of awareness of the passage of time [5]. And indeed, were a man to so freeze, if it occurred for more than a day in ordinary time, when they finally came out of it, they would usually go completely insane; stark raving mad, running and gibbering. Following on from the experiment, almost half of the officers and crew suffered this very fate [6].

One could go on in much greater detail concerning the issue of the experiences of the crew. However, with regard to this present work, the main focus here is upon the ship itself, and of one further and highly unusual reported incident as occurred to the test vessel at the time of the actual experiment. From its dock in Philadelphia, at the very point at which the field was first activated, the ship was seen to completely disappear for a few minutes, only to reappear at its other dock, located in Norfolk, Newport News, in the Portsmouth area. And yet following just these few minutes, it then disappeared once more only to return again back to its initial position at the Philadelphia dock. Indeed, it is said that this event as witnessed was even noted in newspapers at the time [7]. Considering that *the actual distance separating the two docks is some 220 miles*, one is hardly likely to be surprised at this, for indeed, how could such an event not be newsworthy!

Given then the quite remarkable double teleportation event, one is of course bound to suspect that the precise positions of the ship at both ends of the transfer, including any associated arc length measures, would possess a certain harmonic affinity to those as have been shown previously as necessary to allow for nuclear explosions. In order to test this however, one would need to know the precise geodetic locations of both points. Fortunately, the information as is required for such an evaluation is to be had; and again it comes from none other than Bruce Cathie. Having spent many years researching the Philadelphia Experiment Cathie made several contacts including the author of what many consider to be the definitive book upon the entire subject, Charles Berlitz [8]. Suspecting a possible connection between the two sites as would have been involved in the double teleportation event, and also the Earth energy grid matrix that he had uncovered, Cathie attempted to determine for himself precisely just what the co-ordinates were for each site. And in this he appears to have very carefully examined the evidence as presented in the book by Berlitz, *The Philadelphia Experiment: Project Invisibility*.

Initially, from the study of several accounts, each purporting to detail the actual position of the Destroyer DE173 upon activation of the field, Cathie focused upon the words of one particular source who indicated that it [one would assume the initial test] occurred upon the Delaware River [9]. Of course, the river itself is quite long and one could suggest many points upon it wherein the experiment may have taken place. Indeed, from still other sources, references are made to the actual Philadelphia Naval shipyard along the Delaware River as being the key site. Be that as it may however, as a result of careful thought, Cathie hit upon a brilliant connection, as seemed to settle the issue quite decisively. In a full account of the matter as provided in his book *The Bridge to Infinity: Harmonic 371244*, he reasoned that the experiment may well have had some association with the Philadelphia College of Science, whose latitude placement was shown via his own studies to coincide with the harmonics of the speed of light [10]. The actual latitude placement of the College of Science is just on the edge of the Schuylkill River relatively close by to the naval shipyard itself. Sensing therefore the importance of the latitude of the college and being cognisant of the Delaware River connection, Cathie reasoned that the actual position of the DE173 Destroyer at the moment of the experiment was up the Delaware River just off the site of Penn's Landing at the very same latitude as the College of Science [11]. With this in hand he then proceeded further to establish the co-ordinates of the Norfolk docks, where the ship was said to have been transported to. Cited directly from Cathie's book *The Bridge to Infinity*, the co-ordinates of both are given as follows [12]:

Penn's Landing Area (Delaware River):
Latitude: 39 deg, 56 min, 35.77 sec, North
Longitude: 75 deg, 08 min, 55.8 sec, West

Norfolk Naval Docks:
Latitude: 36 deg, 55 min, 08.634 sec, North
Longitude: 76 deg, 19 min, 45 sec, West

After presenting the co-ordinates in his book, Cathie then sets out to evaluate them in terms of his system of grid harmonics. However, in proceeding here, it is the intention of this present author to seek to evaluate the positions and their associations in line with the findings as are contained in this present work, thereby offering the reader an alternative evaluation to that of Cathie. For an in-depth comparison of Cathie's associations set against those of this present author, as will shortly follow, one may consult Appendix I.

The Nuclear Harmonics of the Philadelphia Experiment

Initially, upon the issue of a suspected linkage between the site of the experiment and the Philadelphia College of Science, one may recall the fact as previously verified, that the latitude arc measure of a structure from the equator is important in its own right, if certain energetic effects involving space-time disruption/manipulation are to be achieved. And indeed, the latitude placement of the college, theorised to be that also of the experiment itself, is found to be of immense significance when evaluated under the WGS84 ellipsoid Earth model, as the following reveals:

Philadelphia College of Science / Site of Experiment:

Latitude (geodetic): 39 deg, 56 min, 35.77 sec, North
Latitude (arc length) [13]: = 14511909.464 feet

If one simply takes the latitude value as given in British feet, and divides it successively by the number 6, it becomes apparent that the numeric sequence is of great significance:

14511909.464 / (6 × 6 × 6 × 6 × 6 × 6 × 6 × 6 × 6) = 1.44000270

In reversing the procedure, starting out from 1.44 exactly and then multiplying by 6 successively, one may derive a precise theoretical value for the latitude of the experiment:

14511882.24 feet

With: 14511909.464 – 14511882.24 = 27.224 feet

In view of the above associations it is very likely then that not only was the Destroyer at the noted latitude when the experiment was conducted, but that quite possibly also the College of Science, being situated where it was, allowed researchers located there to be able to conduct preliminary experiments in space-time though on a much smaller scale, in preparation for the actual test.

In addition to the above, one can also demonstrate a direct connection between the co-ordinates of both noted sites, once again by making use of the WGS84 Earth ellipsoid model. In this instance the numerical associations are of even greater simplicity:

Direct elliptical arc length between both sites [14]:

= 1152082.005 feet

And the simple division of this value by 8:

1152082.005 / 8 = 144010.250

Were the arc length exactly 1152000 feet, then such a division by 8 would result in a precise figure of 144000. That said however, the actual length of the Destroyer vessel itself was 306 feet [15], and thus one is well within a reasonable margin of error with only an 82 feet discrepancy. With this in evidence there would seem little doubt then that the alleged ordered transportation through hyperspace of the DE173 between the two noted geodetic points is confirmed.

Matter Transportation between Worlds

Given the fact then that it is physically possible to transport matter in a relatively controlled manner through hyperspace, and not just in terms of a purely destructive transmission, as in the case of nuclear weapons blasts or earthquakes, one must conclude, in light of the dynamic physical principles found to govern the process, that the controlled transportation of matter between a set of target locations *on entirely separate celestial bodies* is achievable.

That being said however, one would of course look for evidence of such a feat; that the various powers upon the Earth have indeed been able to actively transport an object through hyperspace to another celestial body. But does it exist? At first thought one might think that it would be impossible to prove either way. However, quite remarkably there is indeed intriguing evidence to be had, which is tied in strongly to the Philadelphia Experiment, as already noted.

At the height of the Second World War, with its experiments in radar invisibility the US was of course looking to gain a major strategic advantage over what were then its enemies. And with a plan to retake the whole of Europe from German forces, establishing first a beachhead in France, it goes without saying that such an operation would rely upon a massive naval force. One can clearly see therefore the focus of the US in developing a means to cloak military ships at sea from radar; something that could potentially be of immense value in allowing a naval fleet to arrive off the coast of France undetected. To assume however that the US would only have been interested in experimentation involving naval vessels would seem rather naïve. Indeed, one would certainly expect them to have extended their research to include such as tanks and airplanes; being armoured vehicles also susceptible to radar detection. With respect in particular to the latter of these, evidence does present itself to confirm such suspicions.

Some two decades ago, as from the time of writing this, a remarkable story appeared in the press making front page headlines the world over. Originating in Russia, it told of the detection by satellite of nothing less than a U.S. World War II bomber on the surface of the Moon! According to one Dr. Stanislav Makeyev who spoke of the discovery at the time, a set of high resolution images revealed the plane to be situated in the middle of a crater and almost entirely intact, with what looked like only minor damage from meteor impacts. Moreover, the images in question were said to be no hoax either, but entirely genuine [16]. Of course, at the time of the story, there was no official acknowledgement as to its authenticity, everyone of concern simply declining to comment upon the matter. Somewhat 'conveniently' though, the bomber apparently then went 'missing' only a day or two later following discovery. As a result the whole incident did come to be seen as a hoax by the wider public. However, there are many serious researchers who do indeed take the story seriously and believe that the bomber was detected, and its disappearance engineered very quickly just after discovery. This present author is certainly one of them, as the story is perfectly in line with the hyperspatial physics detailed as would appear to govern both earthquakes and nuclear reactions. And thus, of the two noted incidents, concerning both the Philadelphia Experiment and the World War II bomber, it would appear that in the case of the former, teleportation of the destroyer to another location somewhere upon the Earth was achieved. In the case of the latter, the bomber was teleported to an entirely separate celestial body.

In a further comment on both events, it is highly likely that the actual hyperspatial transportation of both the ship and the plane was an unforeseen side-effect of the tests, just as with the crew of the destroyer becoming merged in with the ship when the field was turned off, or indeed of them disappearing altogether. Initially then, it would seem that the US just stumbled upon such exotic effects by accident. That being said however, once they did, it is certain that they did not ignore them, but continued to study such phenomena very carefully, and in secret. Despite a somewhat disastrous start, they were bound to go on, no doubt driven by the slogan "If we don't do it, somebody else will." For certainly, who could doubt the strategic advantage to be gained from the technological mastery of such exotic physics? In terms of sending whole fleets of armoured vehicles past the defences of an enemy in an instant, transporting them to their targets before anyone could respond, or indeed, as developed by Russia, the ability to directly transmit controlled nuclear explosions from one point on the globe to another; which military power would not want such capabilities?

Such blatant military applications aside though, the extent to which one could scientifically master hyperspatial travel, is also the extent to which one would truly be able to explore the outer reaches of space (one would hope peaceably). Indeed, the possibilities for discovery are potentially limitless through use of the hyperspatial pathways of the sub-domain region. Of course, one must wonder in view of this just how close the human race is to truly mastering the science required for travel in this manner beyond the Earth, both to known bodies within the present solar system and even outside of it. In all likelihood, the full truth of the matter is that the very means to safely transport both inanimate objects and life through hyperspace to other bodies and back again has already been mastered, but that the technology has unfortunately been kept secret. And thus, to go further, it is not at all unreasonable to suspect that various powers upon the Earth – and not necessarily governments – have at present already established bases on the Moon, and also the planets Venus and Mars. And that movement between such bases and the Earth is carried out via instantaneous hyperspatial travel [17]. Or perhaps even controlled, time delayed hyperspatial travel.

Chapter 20

Re-engineering the Lost Science of the Ancients

In further consideration of the issue of matter transportation using hyper-spatial pathways, one may be inclined to wonder then if there are any distinct facilities that have been constructed upon the Earth as would contain the necessary technological equipment that would allow for such exotic movement. Of course, one must have in mind the concept of portals of some sort. In trying to narrow down the possibilities, based upon the fact that the physics involved with this type of travel requires the active use of various celestial harmonics tied in to nuclear reactions, one would expect any fixed facilities to be globally sited as would encode such values in some way. A basic analysis of the various geodetic positions of what one may consider to be 'facilities of interest', specifically upon account of their global placement, would appear to offer up a few intriguing possibilities. Indeed, a careful review of certain very notable military-intelligence installations would appear highly fruitful. And in this regard, there are two particular facilities deserving special attention. Both owned and operated by the United States, they are: The Pine Gap complex located in Australia, and the Menwith Hill facility in England.

Pine Gap

Located almost at the exact centre of the geophysical landmass of Australia, the so called Pine Gap Base was first established in 1966 by the United States with consent from the Australian government. Officially it is called the Joint Defence Space Research Facility [1]. And in the beginning it was established to research space defence technology. In terms of its basic layout, the site consists of a large central rectangular building complex surrounded by several radomes (big 'golf balls') that house satellite receiving dishes. In operation, the site is almost exclusively manned by U.S. personnel, the staff totaling just over one thousand. Moreover, very few people inside the Australian government itself appear to know exactly what goes on at the base. It is indeed a 'top secret' facility.

From what little information has emerged concerning the Pine Gap complex, it is said that it contains a drilled hole some 5 miles deep; the largest found throughout the whole of Australia. Of the hole itself, some suggest that it may act as an antenna to transmit extremely low frequency electromagnetic signals to submarines [2]. In addition to this though, the facility is also linked in to signals intelligence, and of the interception of global telecommunications for U.S. national security purposes. And yet, more than even this however, there are reports that exist of a whole host of extremely intriguing activity said to have taken place at the base, witnessed by numerous people over the course of several years, as would make this a site of extraordinary significance. The activity in question includes sightings of UFO's, and of strange beams fired up into the distant sky from the complex. In consideration of such reported incidents, and most definitely bearing in mind the connection as first noted by Bruce Cathie, of a link between the flight patterns of UFO's and the basic sexagesimal system, it is most interesting to note the latitude placement of the Pine Gap facility.

As noted from surveys, the primary latitude of the Pine Gap base is generally given as some 23.801 degrees south (23:48:3.6 S). Evaluated using the GP2007 Earth ellipsoid model, the arc length distance from the equator to the base, in feet, is given as follows [3]:

23 deg, 48 min, 3.6 sec (south) = 8639817.200 feet

Converting the specified measure into IGM units, the arc length displacement of Pine Gap from the equator:

8639817.200 / 6000 = 1439.96953 IGM

Clearly one can see that the value as returned is almost exactly 1440 IGM, and indeed, were the initial value expressed in feet as 8640000, this would be the exact answer as generated. The discrepancy as one may note from such an ideal is very small indeed:

8640000 – 8639817.200 = 182.800 feet

That this relation is clearly existent; combined with all of the others as given throughout the course of this work, it is powerful evidence of a connection between the UFO phenomenon and the Pine Gap facility.

Menwith Hill

Located in northern England and built during the early 1950's, the Menwith Hill site in terms of its basic layout is almost identical to the U.S. Pine Gap facility, consisting of a large central rectangular building complex surrounded by a scattering of radomes. As one might thus imagine, the primary purpose of the facility is very much akin to that of the Pine Gap base: signals intelligence; monitoring/ intercepting electronic communications, including phones, faxes, E-mails etc. Such being said however, it is worth noting that the Menwith Hill site is on a far greater scale than Pine Gap, being in fact the largest such station engaged in electronic monitoring in the world [4]. All size considerations aside though, just as with the Australian base, the on-site personnel are almost exclusively U.S nationals. Menwith Hill is thus clearly run by the U.S. for their own strategic interests, all the while technically the land is owned by the British Government i.e. it is the territory of Great Britain.

Noting then that the actual complex is itself similar to the U.S. Pine Gap base in Australia, one would very much expect that the Menwith Hill site would also have been positioned in line with various key harmonic values related to basic Imperial measures, and also the sexagesimal system. And in this one would not be mistaken. From an evaluation of the placement of the site, the pure latitude alone, as of some 54.00755 degrees north (54:00:27.18) is highly suggestive. Evaluated under the GP2007 Earth ellipsoid model, the arc length from the apparent centre of the complex to the equator is as follows [5]:

54 deg, 0 min, 27.18 sec (north) = 19642800.527 feet

Simple division by the basic fraction 11 / 7, whose significance has been demonstrated on numerous occasions, immediately transforms the arc length into a very recognisable value that can easily be tied in to the 144 sequence, just as with Pine Gap:

19642800.527 / (11 / 7) = 12499963.972
12499963.972 / (6 × 6) = 347221.221
347221.221 × 2 = 694442.442
1 / 694442.442 = 0.00000144000415

Ideal: **0.00000144**

Were one to work backwards from the noted ideal to generate the latitude arc, the precise answer obtained would be 19642857.142... feet; differing from the survey value by only some 56 feet.

A Global Linkage

As one may recall from Chapter 13, in the study of both Stonehenge and the Great Pyramid, it was revealed that in addition to both structures being linked to the Earth form harmonically in terms of the primary values of the sexagesimal system, the direct arc length separation between them was also of significance. The value itself as noted, 11809800 feet, was found to be directly tied in to a major variation of the primary sexagesimal set. As a result of this, it is very apparent therefore that those responsible for both structures wilfully sought to energetically link them via the natural harmonic intervals of the Earth form. That this is so proves most interesting in light of an evaluation of both Pine Gap and Menwith Hill.

Already it has been demonstrated that the pure latitude values of the two U.S. facilities are of great significance. What is of even greater interest however is how they relate to one another. As one will doubtless recall, from what is known of both facilities, their basic function is very similar, primarily involving communications and signals intelligence. For the purpose of world-wide surveillance covering the globe, one would certainly suspect some sort of linkage between each of the two sites. In this, one would not be mistaken. Given the geodetic positions of both sites, the arc of separation between them proves to be of immense significance:

Pine Gap:	Latitude: 23:48:3.6 South Longitude: 133:44:12.75 East
Menwith Hill:	Latitude: 54:00:27.18 North Longitude: 01:41:18.42 West

Arc Length Separation GP2007 [6] = 49310146.768 feet

What is of great interest with respect to the above value is that it would appear to be a highly significant fraction of the equatorial circumference of the Earth; one that has been shown previously to be of importance to various nuclear tests:

Earth Equator (GP2007) [7] = 131487186.278 feet

131487186.278 / 49310146.768 = 2.666534068

One will doubtless not forget the ratio evident between the bomb-SGP arc lengths of Bravo and King:

35807863.081 / 13428155.350 = 2.6666256195

Ideal refinement: 2.666666666... or
8 / 3 expressed as a basic fraction

Given the above associations, it is not unreasonable to suggest then that the United States deliberately sought to construct both facilities such that the elliptical arc of separation between them was as near as possible to being exactly 3 / 8 of the full equatorial circumference of the Earth. One would expect therefore that this separation would allow for a unique energetic linkage between both sites, as would be conducive to achieving certain physical effects far more exotic than the US government would be willing to admit to.

A Link to the Ancient Matrix

With a global connection thus established between Pine Gap and the Menwith Hill site, one is able to see a modern energetic link, much the same as would appear to have existed in ancient times between The Great Pyramid and Stonehenge. That said, the associations as are to be had are not as separated in time as one might imagine. Indeed, careful analysis of the Menwith Hill facility would appear, quite remarkably, to link it to none other than the ancient site of the Great Pyramid itself, and through the same harmonic associations as have become all too familiar in the course of this work. The critical link in question is revealed from a simple study of the direct arc length separation between the two sites:

Menwith Hill → Great Pyramid:
Arc Length Separation (GP2007) [8] = 12301598.213 feet

Using the number 9 to successively divide the arc length separation value, it becomes quite clear that one is moving through a major progression allied to the basic sexagesimal series:

12301598.213 / (9 × 9 × 9 × 9 × 9) = 208.3286459
208.3286459 / 3 = 69.44288197
1 / 69.44288197 = 0.014400324

Ideal = **0.0144**

Reversing the procedure; starting out with an exact value of 0.0144, an ideal arc length separation between Menwith Hill and the Great Pyramid may be obtained:

Ideal Arc = 12301875 feet

Discrepancy: 12301875 – 12301598.212 = 276.786 feet

Given the fact that the main building complex at Menwith Hill is itself several hundred feet in length and also of a sizable width, a careful review of the precise separation will reveal to the reader that the noted error is in fact well 'contained' within the base itself.

With an ideal measure of separation now to be had between the Great Pyramid and Stonehenge, and also the Great Pyramid and Menwith Hill, it is most interesting to note the fractional association of both arc measures:

Stonehenge → Great Pyramid: 11809800 feet
Menwith Hill → Great Pyramid: 12301875 feet

11809800 / 12301875 = 0.96
With: 0.96 × (3 / 2) = **1.44**

There can be no doubt as to the significance of the noted values and the general conclusion that must follow. Respecting both Pine Gap and Menwith Hill, it is very clear that the scientists of the modern age are actively trying to replicate the lost technology of those from the last great civilisation to have flourished upon the Earth. And to this end, they have sought to develop a series of modern energetic facilities and integrate them harmonically into the Earth's natural energy grid matrix; even aligning them to some of those of the very ancient past.

An Off World Connection

As intriguing as the above associations would appear to be, the very scope of this line of inquiry can be expanded still further, into the celestial realm itself. For indeed, as one may recall from the natural physics of earthquakes, including the top secret experiments of the US during the Second World War involving radar invisibility; the ability to actively transmit matter in a controlled fashion from one celestial body to another is distinctly possible. Now, with the study of both The Great Pyramid and Stonehenge, in addition to the Pine Gap and Menwith Hill facilities, due to the global associations of the structures, one is bound to suspect that in some way an active energetic link must connect each pairing. And that such links would exist via the hyperspatial pathways of the sub-domain region. With this line of enquiry taken further, one cannot help but suspect then that such structures of the ancient world, including the facilities of the modern age, may have been built to achieve communication not just via electronic signals, but via actual matter transportation i.e. physical travel. With an acceptance of this, it is just one more step, to concede the point that controlled hyperspatial travel between different celestial bodies is entirely possible, were one to establish an off world facility upon another body that possessed an energetic pairing to one already established upon the Earth itself.

In accordance with the principles of hyperspatial physics as previously discussed, were indeed some civilisation to establish a technical facility on another world in order to achieve hyperspatial interplanetary travel, then such a facility would by necessity have to be carefully placed to be harmonically viable; being in line with all the relevant values of the sexagesimal system as would give it the required energetic signature. With this established in theory, one is bound to enquire then, if indeed there is any evidence to suggest that such facilities have ever been built upon other worlds. Indeed there is.

The Pyramids of Mars

For the greater part of recorded human history, with the exception of the moon, all of the other celestial bodies of the solar system as observed by man, have been seen merely as dots of light in the sky. It was only with the advent of telescopes in the modern era that astronomers began to be able to observe specific surface features on various bodies. And indeed, it is only very recently, with the aid of space probes, that extremely detailed images have been taken of the surface of the major planets of the solar system. Of all of the planets though that have been subject to such close scrutiny, there is one planet in particular that has seemingly captured the imagination over and above all others when one is led to enquire, is there or has there ever been life on any of the other planets in the solar system besides the Earth? The planet in question is Mars.

In the late 19th century, from a detailed study of the surface of Mars using telescopes, certain astronomers of the time observed what they believed to be a series of canals covering over the planet. Indeed, it was further theorised that they were artificially built for the purpose of irrigation by an unknown civilisation that dwelt upon the planet. The theory held even into the early decades of the 20th century. However, with the advent of more detailed images of the surface of the planet, the canals were shown essentially to have been a trick of the light, and of no real substance. Indeed, with the era of space probes, the whole idea of a surface canal system was totally falsified, as new close-up images covering over the entire planet finally put the theory to rest once and for all. However, with one issue settled, another emerged; one that rejuvenated the idea of life on Mars to a whole new level.

In 1975 the US space agency NASA launched two probes to Mars, called Viking 1 and Viking 2. Consisting each of a lander and an orbiter, they arrived in the vicinity of Mars in 1976, their purpose being to search for signs of life. The lander craft, filled with various instruments, took ground samples and also examined atmospheric composition [9], whilst the orbiters mapped the entire surface of the planet down to a resolution of 490-980 ft, with certain selected areas to a resolution of 26 ft. [10]. After analysis of the samples and surface readings, no signs of life were detected. However, when the pictures taken of the surface of Mars by the orbiters were examined, they caused a sensation when it was found that one appeared to contain an image of a humanoid face pointing out into space!

Discovered in the Cydonia region of Mars, at a latitude of some 41 degrees north and 9.4 degrees west, the height of the face, as it simply came to be known, was found to be at least 1650 feet, with a base size some 1.6 miles long and 1.2 miles wide [11]. And yet however, the face itself was not the only feature on the surface of Mars to draw interest, for it was also discovered that just to the south-west of the face were a distinct group of 12 pyramid type structures, including, over to the east and just offset from the main group, an additional five sided pyramid of immense size. Examined by many researchers over the years since their discovery, the main pyramid grouping is often referred to as 'the city', with the giant pyramid offset to the east and south of the face, generally referred to as the D & M Pyramid, after the initials of those who discovered it, DiPietro and Molenaar. Indeed, with further reference to the D & M pyramid in particular, the base dimensions of this gigantic structure were determined to be very much on par with 'the face', being some 1.6 by 2 miles across [12].

Following careful study of these intriguing new structures found upon the surface of Mars, it was clear to all from the images as captured their form, that they were far from being in a state of 'geometrical perfection'; most especially the pyramid structures, as their edges looked distinctly 'weather worn' or eroded. And indeed, as a result of this very fact, there is intense controversy surrounding them, which has existed right from the time when the pictures were first released. For just as with the theory of the canals of Mars noted earlier, there are a great many people, especially from the world of 'officialdom', that claim that the structural formations are entirely natural and in no way artificial, being merely a trick of the light. In truth, the debate as to the authenticity of the formations themselves continues to rage even to this day, with researchers on both sides of the divide trying to prove their case in one way or another.

Concerning the whole issue of the debate, it is interesting to note that the arguments as have been put forward in support of the idea that the Martian structures are artificial, do not rest entirely upon the issue of the perceived geometrical sharpness of the stone formations, as imaged. Rather, quite a number of researchers have sought to examine the noted structures with the aim of deriving some sort of geometric or mathematical association as would prove intentional placement. As such a line of inquiry is very much allied to this present work it would certainly be well to examine then some of the research done by others in this area, to see if indeed a decisive linkage does exist.

In a review of the literature concerning the main structures of interest in the Cydonia region, it is to be found that with respect to uncovering some mathematical basis for their placement, most researchers have tended to focus upon the D & M pyramid, both in terms of the physical structure itself and in terms of its global position. Such indeed would seem to make sense, as 'the face' itself for example, is clearly not a distinct geometric structure as such, but rather artistic. And therefore, were any mathematical associations to be discovered, especially of a geometrical basis, it is far more likely that they would be linked to a physical structure of the shape of such as a pyramid.

Focusing then upon the D & M pyramid, has anyone to date been able to discover a mathematical association of significance to decisively lend weight to the contention that it truly is an artificial structure, as opposed to relying solely upon just the photo images in support of this idea? In a review of the research so far, several ideas have been set forth, most notably linked to the latitude placement of the pyramid, which according to the surveys of the Cydonia region of Mars, is some 40.868 degrees north of the Martian equator. Now, upon viewing this figure, as expressed in degrees, quite a number of researchers have sought to determine if as a value, it possesses some sort of numerological significance. One theory as set forth, clearly suggests that it does, based upon an angular tie-in to the ratio of two well recognised mathematical constants: PI and *e*. The proof itself is reproduced as follows [13]:

Where: PI = 3.14159265
e = 2.71828182

2.71828182 / 3.14159265 = 0.86525597

Converted from a ratio (arc tan) to obtain an angle:

0.86525597 = 40.8681937 degrees

One can see from this then an answer returned that does match up closely to the angular latitude of the D & M Pyramid. Further to this though there has also been proposed an interesting variation upon this particular mathematical linkage. For at least one researcher has further noted that the value for *e* may be replaced by the ratio as exists between the surface area of a tetrahedron form, and that of a sphere as circumscribes it, which is determined to be

2.72069. Now, if one were to use this variation upon the above, then an alternative value for the latitude angle of the pyramid would be given [14]:

2.72069 / 3.14159265 = 0.8660225252

Converted from a ratio (arc tan) to obtain an angle:

0.8660225252 = 40.893300 degrees

From this alternative ratio one can see then a slightly greater latitude value as given for the D & M pyramid. And indeed, one may well wonder at such a shift to the north, and just how acceptable it may be. Upon this point there are two things of note. The first of which, is that upon the matter of the longitude/latitude grid as established over the surface of Mars, the accuracy or precision of the grid is on the order of 0.017 degrees, and the suggested alternative latitude of 40.893 degrees is well within this margin [15], were one to take the value of 40.868 degrees as the primary reference point. Also, is the already noted fact that is the very size of the D & M pyramid. A latitude difference of some 40.868 to 40.893 degrees is equal to about 4840 feet over the surface of Mars. With the base size of the pyramid itself between 1.6 and 2.0 miles (8448 and 10560 feet respectively), the noted shift is contained within the dimensions of the structure.

A Decisive Latitude Association

One can see then of the above, two ideas put forward suggesting an intentional placement of the D & M pyramid at a latitude linked-in to various well known mathematical constants. The essential idea thus being that for some reason, certain universal values of this type were deliberately employed to site the structure upon the face of Mars. The key question therefore, is are such ratios as these valid in any true physical sense, or are they mere mathematical fantasies?

In the view of this present author, based upon the research as so far presented in this current work, both connections as detailed would appear to be without foundation, and thus false. And the main reason for this, which has indeed been elaborated on earlier with reference to the positioning of the Great Pyramid, is that angular measures do not appear to possess true physical validity at all as such; being merely abstract in their nature. Rather, in truth it is only actual arc length measures as cover over the physical form of a body, and that are in accordance with British Imperial Units, which possess true validity. The very physics of both nuclear weapons and earthquakes attest to this fact.

If indeed then both of the above mathematical associations are false, is there still to be found one that is of true significance and that does imply intentional placement of the D & M Pyramid upon Mars, thus supporting the contention that it is an artificial structure? It would seem so, with the critical link to be had as proves this point most decisively being of the same type as shown previously to apply to Stonehenge.

As one may recall, with respect to the latitude of Stonehenge it was shown that it appears to have been based upon a very specific ratio; that between the equator of the Earth and the associated small circle circumference of the noted structure. Reproduced in full, the essential relations as given previously, are as follows:

Equator of Earth (GP2007) = 131487186.278 feet

Small Circle Circumference of Stonehenge (GP2007)
= 77042854.087 feet

And the ratio of the two noted values:

131487186.278 / 77042854.087 = 1.7066759511…

Following this, an assumed ideal of 1.7066666666… (128 / 75) was thus put forward as the intended ratio, from which one could link up the stone circle to the following basic mathematical progression:

1.7066666666… × 3 = 5.12

Successively divided by 2:

5.12 → 2.56 → 1.28 → 0.64 → 0.32 → 0.16 → 0.08 → 0.04 → 0.02 → 0.01

With it an undisputed fact that Stonehenge is a man-made structure, to find such a high precision mathematical ratio of 128 / 75 to be the very basis of its latitude, and linked to such a significant numerical progression; were one to find a relationship of exactly the same type respecting the pyramid on Mars, this would greatly increase support for the idea that it too was artificial in nature.

Following careful study of the D & M pyramid, a numerical relationship of immense significance, involving exactly the same components, is also to be found. Just prior to actually presenting the proof though, including the decisive ratio in question, a certain refinement of what is assumed to be the true latitude of the pyramid must be considered, for the original stated value of 40.868 degrees would seem to be quite inadequate. Indeed, strangely enough, it would appear that a latitude value very close to precisely 40.893 degrees, shown to be derived via the noted mathematical association above – that this present author considers to be entirely without foundation – does in fact, quite by chance it would seem, yield the correct answer almost dead on. It would thus appear most apt in presenting the proof, to make use of this particular latitude value for the placement of the D & M pyramid, noting that it has previously been shown to be within an acceptable margin of error:

The Mars Ellipsoid Form:

Equatorial Radius [16] = 1857.611548444 IGM
Inverse Flattening [17] = 154.40915337461204

Under this ellipsoid model, the equatorial circumference of Mars, as converted into feet, is given as follows [18]:

70030305.529 feet

With a latitude 40.893 degrees north of the equator, the small circle circumference of the D & M Pyramid is itself given as follows [19]:

50421665.597 feet

And the ratio: 50421665.597 / 70030305.529 = 0.719997795…

With: 0.719997795 × 2 = 1.43999559

There can be no real doubt at all as to the significance of the answer returned. The suggestion is thus, that the D & M Pyramid clearly is an artificial structure and that the builders specifically intended that its latitude placement would be determined by a precise ratio of 0.72 (18 / 25) between the associated small circle circumference of the pyramid, and the full equatorial circumference of the planet (Mars). Moreover, for the sake of accuracy, were the pyramid to be placed a mere 35 feet to the south, at latitude 40.892818148 degrees north, then a precise value of 0.72 would be obtained. Such a shift would yield a latitude arc length as follows [20]:

7875535.563 feet

In addition to the above associations there is also one of further note that links the D & M pyramid on Mars to the Great Pyramid itself upon the Earth, resting upon a relationship between the arc length latitude values up from the equator of each structure:

10886476.544 / 7875535.563 = 1.382315711

And: 1.382315711 / (2 × 2 × 2 × 2) = 0.08639473195

0.08639473195 / 6 = 0.01439912199

Ideal = **0.0144**

The values as given from a practical sense are of course not exactly perfect, but then, as explained previously with reference to the Great Pyramid, it would be impossible to achieve a ratio of some exacting harmony respecting the equatorial and small circle circumference of a planet, *in addition* to an equally harmonious stand alone arc length latitude value (of a noted structure), with an objectively determined ellipsoid planetary model. One may also note as well that the actual discrepancy level between both mathematical associations is so low considering the size of the D & M Pyramid, that just as with the Great Pyramid, one is able to accept that on a practical-energetic level, both are within tolerance levels. And thus, assuming the Mars ellipsoid model as established by modern science to be true, then within the allowed precision of that model, if one assumed perfect positioning of the D & M Pyramid in terms of the 0.72 value, then the 1.3824 value respecting the Great Pyramid upon the Earth would not be achieved with exacting precision; being in error by some 480 feet. Whereas vice versa, were one to hold that the 1.3824 ratio were perfect, then the 0.72 value would itself be slightly inexact, and by the same measure. All in all though, considering the size of the D & M Pyramid is some 8448 by 10560 feet, such a difference cannot be of any real consequence.

The Great Pyramid / D & M Pyramid: An Energetic Pairing?

With the above analysis one can state with confidence that the ratios as given are indeed of real energetic significance, and that they do tip the balance in favour of the theory that the Pyramid on Mars is an artificial structure purposely built, and in accordance with the same physical science as the Great Pyramid upon the Earth. Indeed, to be absolutely clear upon this point, with regard to both of these structures, in light of the global positioning upon their respective celestial bodies, and the affinity they each possess to the arc length measures employed to generate nuclear explosions, inescapably one must conclude, that the Great Pyramid and the D & M Pyramid were both (in some previous lost age) functional high technology devices based upon advanced nuclear principles. Of course, one is in no way suggesting here that either pyramid was an actual nuclear weapon, as such indeed would possess only a 'one-time only' usage capability. Rather, what does seem likely is that both were built to harness the natural energy of their respective planets through resonance, drawing it off in a highly controlled manner.

Upon the issue of the intended purpose of both the D & M Pyramid and the Great Pyramid, one must imagine that what would hold true for one, would also hold true for the other. And indeed, based upon the essentials of hyperspatial physics one is hard pressed to discount the possibility that both such stone structures may well have formed a distinct energetic pairing, such that communication between them both would be possible under the correct physical conditions. Of course, there is indeed a great amount of speculation regarding both of the noted structures in question. The accepted age of the Great Pyramid is certainly far from a settled matter. And also, of the D & M Pyramid on Mars, or even the other noted structures, it is certainly conceivable that they could be many millions of years old. At this very moment it is simply impossible to know for certain based upon the current literature. That being said however, based upon the energetic signatures of both noted pyramids, including the physical science as would appear to govern hyperspatial travel between celestial bodies, *were* one to establish an energetic pairing of structures as would facilitate such travel, then an established configuration as present in the case of these two pyramids, is how it *would* be done.

One could thus imagine then that both pyramids would be in possession of various internal chambers as would be some sort of focus for the energies collected by the pyramids themselves. In the very specific case of the Great Pyramid for example, one would be inclined thus to point to the 'King's Chamber' in this regard, with the idea that an equivalent chamber would exist inside the D & M Pyramid. And that both could be energetically tuned with the correct harmonies, as would open up or establish a set of portals to connect up to a series of ordered hyperspatial pathways, as would allow for transportation, of such as a person or material object, for example. It is conceivable; somewhat speculative perhaps, but nonetheless very conceivable.

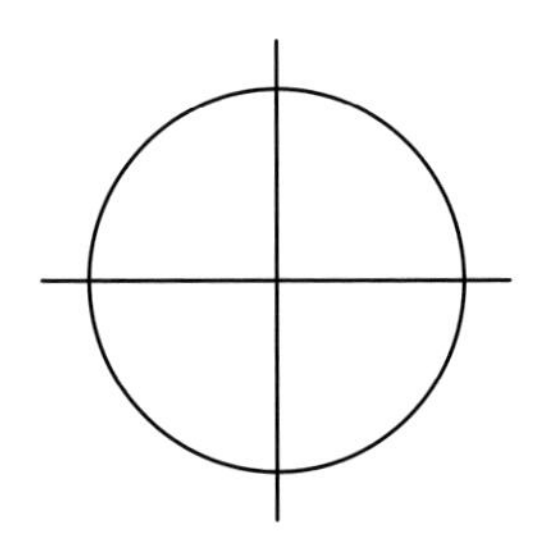

In Conclusion:

Nothing New Under the Sun

With all the facts as presented, and all the analysis and evaluation; the myths of the ancients, their calendar systems, and their immense megalithic structures; as remarkable as it may seem, the evidence put forth decisively demonstrates that at some point prior to known recorded history, there was a civilisation active upon the face of the Earth at least as sophisticated as that of the current age, having reached the level of nuclear physics, and quite possibly gone even further still. Indeed, from the reconstruction of the mathematics and calendars of this lost civilisation, it is certainly clear that they based their units of measure, of both time and distance, upon an extremely profound understanding of the natural cycles tied in to the celestial bodies of the immediate solar system. The mythical stories thus passed down from ancient times are in effect a 'record' of their knowledge, even indicative of the basis of their systems of measure. Moreover, such monuments as the Great Pyramid stand – even to this day – as the active manifestation of their lost knowledge, being perhaps a far more significant record of their achievements than is contained in their stories.

When all is considered, there is little doubt then that those of the modern age are simply rediscovering the lost knowledge of the ancients; attempting to advance once more to almost godlike status. And indeed, upon this very point, from all the evidence presented, it would seem that the very advancement of any civilisation, is only possible through the discovery of those uniquely valid physical units of measure that capture the true essence of the universe; for indeed, the search for knowledge of the deepest mysteries of the universe by necessity involves the very discovery of its measure. With the key to unifying all the essential characteristics of the universe, including man, being inextricably tied in to the ongoing discovery of that one universal system of measures that binds all sensible and spiritual phenomena in the deepest and most profound manner. Were the path of study to be thus; such an approach could not fail to lead one to the most sublime insights the universe has to offer, revealing the solution to the greatest of mysteries; those not of any one age, but those of all ages; even those that have not yet come into being.

Appendix A – A Summary of the Discovered Laws

The Earth, Moon and Sun: The Governing laws

The physical laws of proportion governing the transformation of the Earth, Moon and Sun (also, precession and the speed of light):

$$TY \propto PE_{(e)}$$

$$TY^4 \propto CE^3_{(m)}$$

$$PE^4_{(e)} \propto CE^3_{(m)}$$

$$TY^{12} \propto \frac{1}{SM^{11}}$$

$$CE^9_{(m)} \propto \frac{1}{SM^{11}}$$

$$LS^{11} \propto PE^{9}_{(e)}$$

$$TY^{9} \propto LS^{11}$$

$$CE^{47}_{(e)} \propto TY^{32}$$

$$SMA^{3}_{(p)} \propto SMI^{4}_{(p)}$$

$$TY \propto \frac{1}{OI_{(m)}}$$

$$PE_{(s)} \propto PE_{(e)}$$

$$PE^{9}_{(s)} \propto LS^{11}$$

$$SM^{9} \propto PE_{(m)}^{32}$$

$$TY^{3} \propto P^{5}$$

KEY

In each instance, the celestial characteristics as given are representative of a set of ratios:

present magnitude / past magnitude

They are designated as follows:

TY = Tropical Year
PE (e) = Physical Equator (Earth)
PE (s) = Physical Equator (sun)
PE (m) = Physical Equator (moon)
CE (m) = Celestial Equator (moon)
CE (e) = Celestial Equator (Earth)
SM = Synodic Month
LS = Light Speed (propagating through vacuum)
SMA (p) = Semi-Major Axis, Earth (physical)
SMI (p) = Semi-Minor Axis, Earth (physical)
OI (m) = Orbital Inclination (moon)
P = Precessional Orbital Period

Earth, Moon, Sun & Precession: Orbital-Physical Magnitudes

The ideal physical measures associated with the Earth, sun, and the moon, according to the ancients, and also with respect to the findings of this present work, are as follows:

Ideal Earth:

Tropical Year:	**360** Days
Orbital Semi-Major Axis:	**81000000** IGM
Celestial Equator:	**509142857 & 1/7** IGM
Equatorial Circumference:	**21600** IGM
Polar Circumference:	**21600** IGM
Physical Radius:	
On the Equatorial Plane:	**3436 & 4/11** IGM
Centre to Pole of planet:	**3436 & 4/11** IGM
Earth Physical Form:	**Perfect sphere**

Ideal Moon:

Physical Equator:	**6000** IGM
Celestial Equator:	**1296000** IGM
Orbital Semi-Major Axis:	**206181 & 9/11** IGM
Synodic Month:	**30** Days
Angle between Earth & Moon Orbital planes:	**0** Degrees

Ideal Sun:

Physical Radius:	**375000** IGM
Physical Equator	**2357142 & 6/7** IGM

Ideal Precession (Orbital Period):	**9331200** Solar Days, or **25920** Ideal Years

By employing the above stated laws of proportion the currently held physical and orbital magnitudes associated with the Earth, sun, moon, and precession, are as follows (NB: the key ratio of increase upon which all others rest is the precise fractional value of 1100/1079; the specific measure of increase determined to be that of the moon's own celestial equator):

Present Earth:

Tropical Year:	**365.2421840863...** Days
Orbital Semi-Major Axis:	**81801203.79971...** IGM
Celestial Equator:	**514178995.31251...** IGM
Equatorial Circumference:	**21914.531045...** IGM
Elliptical Circumference:	**21875.001265...**IGM
Physical Radius	
On the Equatorial Plane:	**3486.40266627...**IGM
Centre to Pole of planet:	**3473.82500901...**IGM
Earth Physical Form:	**Ellipsoid**

Present Moon:

Physical Equator:	**5973.445611...** IGM
Celestial Equator:	**1321223.3549582...** IGM
Orbital Semi-Major Axis:	**210194.62465245...** IGM
Synodic Month:	**29.53058551...** Days
Angle between Earth & Moon Orbital planes:	**5.1669449...** Degrees

Present Sun:

Physical Radius:	**380460.608...** IGM
Physical Equator	**2391466.681...** IGM

Present Precession (Orbital Period): **9412490.614** Solar days, or **25770.54629...**Current Years

Light Speed Values (propagation through vacuum):

Ideal:	**162000** IGM / sec or **144000** IGM / grid sec
Present:	**163927.5364** IGM / sec, or **145713.3656** IGM / grid sec

The Remaining Planets

Laws of proportion found to govern the transformation of Mercury, Venus, Mars, Ceres, Jupiter, Saturn, Uranus, Neptune and Pluto:

$$PE_{(S)} \propto V_{(SMA)}$$

$$PE^{3}_{(S)} \propto \frac{1}{V^{32}_{(OP)}}$$

$$PE^{32}_{(S)} \propto C^{3}_{(OP)}$$

$$PE^{8}_{(S)} \propto C^{21}_{(SMA)}$$

$$PE^{30}_{(S)} \propto \frac{1}{M^{19}_{(OP)}}$$

$$PE^{27}_{(S)} \propto \frac{1}{M^{17}_{(SMA)}}$$

$$PE^{14}_{(S)} \propto P^{31}_{(OP)}$$

$$PE_{(S)} \propto \frac{1}{P^{9}_{(SMA)}}$$

$$PE^{3}_{(S)} \propto \frac{1}{Ma^{7}_{(OP)}}$$

$$PE_{(S)} \propto \frac{1}{Ma^{5}_{(SMA)}}$$

$$J^{2}_{(OP)} \propto \frac{1}{S_{(OP)}}$$

$$PE^{8}_{(S)} \propto J^{47}_{(OP)}$$

$$PE^{16}_{(S)} \propto \frac{1}{S^{47}_{(OP)}}$$

$$PE_{(S)} \propto S^{32}_{(SMA)}$$

$$PE^{24}_{(S)} \propto J^{7}_{(SMA)}$$

$$PE^{78}_{(S)} \propto \frac{1}{U^{25}_{(OP)}}$$

$$PE^{70}_{(S)} \propto N^{27}_{(OP)}$$

$$PE^{11}_{(S)} \propto U^{26}_{(SMA)}$$

$$PE^{17}_{(S)} \propto \frac{1}{N^{9}_{(SMA)}}$$

Of the above, with the exception of the value PE(s), each letter or combination of letters is representative of or denotes a given planet, and is indicative of a ratio of change of either the Semi-Major Axis (SMA) or the Orbital Period (OP) of that planet. All the ratios are taken to be of the form: present magnitude / past magnitude. As detailed elsewhere, the value PE(s) is the standard ratio against which all the others are set and is the proportional increase of the physical equator of the Sun, which indeed is itself identical to that of the Earth tropical year increase. The planets are designated as follows:

V = Venus
M = Mercury
Ma = Mars
J = Jupiter
N = Neptune
C = Ceres
P = Pluto
S = Saturn
U = Uranus

The Orbital Magnitudes of the Planets: Ideal & Present

Based upon the analysis in Chapter 10, the ideal orbital values of the planets are given in summary as follows:

	Orbital Period (days)	*Semi-Major Axis* (IGM)
Mercury	90	32400000
Venus	225	58320000
Mars	691.2	125000000
Ceres	1440	225000000
Jupiter	4320	405000000
Saturn	10800	781250000
Uranus	32000	1562500000
Neptune	57600	2531250000
Pluto	90000	3240000000

By employing the above stated laws of proportion, the currently held orbital values associated with the planets are as follows (NB: the value for PE is generated by a 3 to 4 power law transforming the key ratio 1100/1079; the specific measure of increase determined to be that of the moon's own celestial equator, as given elsewhere):

	Orbital Period (days)	*Semi-Major Axis* (IGM)
Mercury	87.96890	31664554
Venus	224.69526	59169233
Mars	686.93077	124639106
Ceres	1680.08949	226242557
Jupiter	4330.64331	425579862
Saturn	10746.97942	781603024
Uranus	30588.71799	1572085945
Neptune	59799.82257	2463064490
Pluto	90589.51397	3234799794

Appendix B – Planetary Motion: The Geometry of Astronomy

In the early chapters of this work, in presenting evidence to support the validity of the proportional laws as said to apply to the Earth and moon, a brief examination of Kepler's work was given. However, it would indeed be well to consider in slightly more detail some of his key findings with regard to the laws that govern celestial motion, including the history as led up to his discoveries. To that end, what follows is a more extensive account of the actual development of astronomy over the past few millennia, culminating in a more detailed look at all of Kepler's 3 laws.

Ptolemy to Copernicus

Observation of the night sky is a most ancient activity, one that has been conducted throughout the world since time immemorial. In the history of Europe, for well over a thousand years prior to the Renaissance of the 15th century, among the most learned, a general understanding of the motion of the celestial bodies of the heavens rested primarily upon the ideas of Aristotle from the 4th century BC, and those of Ptolemy from the 2nd century AD [1]. From the former, came a well developed schemata as to the general ordering of the cosmos and of the observed bodies therein contained, upon which the latter, during his own age, established a most comprehensive mathematical model to chart the activity of the system; one that furnished a most detailed account of the movements of the then known planets against the perceived background star-field. It was an intricate system to be sure, but one that did possess a significant degree of predictive power in charting the positions of the major planets as viewed from the Earth. Coming at the time when it did, and being somewhat superior to other similar models of the period, Ptolemy's system rapidly became established as the standard model of the heavens throughout Europe. Its acceptance amongst the learned was such that it dominated astronomy throughout the region for well over a thousand years. Indeed, so ingrained did it become in the minds of men that it was not until the beginning of the 17th century that it was eventually overthrown in its entirety [2].

Of Ptolemy's decision to actually use the ideas of Aristotle concerning the ordering of the celestial phenomena of the heavens, there was one most notable consequence. Namely, that he was bound to accept as true a certain idea, quite commonly held throughout Greece during the time of Aristotle, that the Earth as a physical body is fixed in space and unmoved; the central body of the entire cosmic arrangement [3]. Indeed, many of the greatest thinkers of the age during this period were thoroughly convinced that the sun, moon, and all of the (then) known planets including the background stars, were all in orbit about a fixed, motionless Earth. Under this general scheme, each of the various types of celestial phenomena were thought to orbit the Earth in transparent spherical shells at various set distances, with the background stars held to be the farthest away. Such was the very system then that Ptolemy was led mathematically to capture during his own age, some six centuries after the height of classical Greek culture. In the end, the downfall of the Ptolemaic system was inextricably bound up with the overthrow of the idea of a fixed Earth lying at the centre of the cosmos. The first serious challenges to the system outright began at the time of the Renaissance.

One of the first astronomers to deviate from the Ptolemaic system, almost 1300 years since its founder lived, was a man named Tycho Brahe (1546 - 1601 AD). During his life Brahe carried out many observations of the planets. And, operating without even the use of a telescope, his measurements were of a level of accuracy at least 10 times more precise than anyone else had obtained previously [4]. In contrast to Ptolemy who thought that all celestial bodies orbited the Earth directly, Brahe favoured an Earth centred system wherein the known celestial bodies did not orbit directly about the Earth, but instead about the sun, which itself was then held to possess a direct orbit about the Earth. The only exception to this general scheme concerned the moon. Of this particular body, Brahe held that like the sun, it too possessed its own direct orbit about the Earth [5], but one that was much closer. Concerning the background stars, they were held to be far more distant than any of the major celestial bodies, just as under the Ptolemaic system. One can easily see then that Brahe's system was only a slight variation upon that of Ptolemy with regard to the issue of the general orbits of the planets. He still considered the Earth to be unmoved. However, a near contemporary of Brahe who also developed an alternative to the Ptolemaic system, Nicholas Copernicus (1473-1543), was indeed willing to 'set the Earth in motion'. In the system of Copernicus, the sun was placed at the centre of the cosmos, with all of the known planets, including the Earth, held to orbit the sun in circular paths of various sizes; each body engaging in uniform circular motion about the sun [6]. Indeed, circles were thought to be perfect and that it was quite natural that God would proscribe such orbits to the planets.

In comparing the various systems, a clear disagreement obviously existed between Copernicus, Brahe and Ptolemy, as to the general ordering of the planets within the heavens. Contrasting in particular the Ptolemaic system with that of Copernicus, one can also note certain key points of difference in the way that they each dealt with or accounted for the observed movements of the planets, which to an observer upon the ground looking up at the night sky, do appear most unusual. Indeed, when astronomical observations of the sky are carried out from the Earth, even by modern astronomers today, they are done so from the perspective of plotting the positions of various bodies upon the inside of a large sphere, the technical term for which is the Celestial Sphere. An observer at night will have the horizon line all around them, and will usually note the angle up from the horizon of a star or planet, and the angle from an established direction line on the ground e.g. facing north. In this way the positions of celestial bodies may be plotted and their movements recorded. When actual observations are made of the planets over extended periods of time though, a number of distinct peculiarities can be seen in their movements. Viewed from the Earth they appear to trace out periodic zigzag like patterns in the sky, constantly changing their direction of movement. They are also seen to speed up and slow down in a non-uniform manner, even to the point of coming to an apparent standstill at certain times, before moving back the way they came – what astronomers refer to as retrograde motion. One of the primary differences between the Ptolemaic and Copernican systems is how they each deal with the retrograde motion of the observed planets. Under the Ptolemaic system, this is done by the introduction of what are known as epicycles. Essentially, although the observed planets, including the sun and the moon, are said to orbit the Earth in circles at various distances, Ptolemy held that their orbits were in fact governed not by just one large circle, but by the interplay of two circles; one large, and one much smaller. The larger of the two is the deferent, whilst the smaller circle is known as the epicycle [7]. Employing the use of an epicycle then in addition to the main or primary orbital circle of a planet did give an apparent solution to the seemingly bizarre retrograde motion as observed from the Earth:

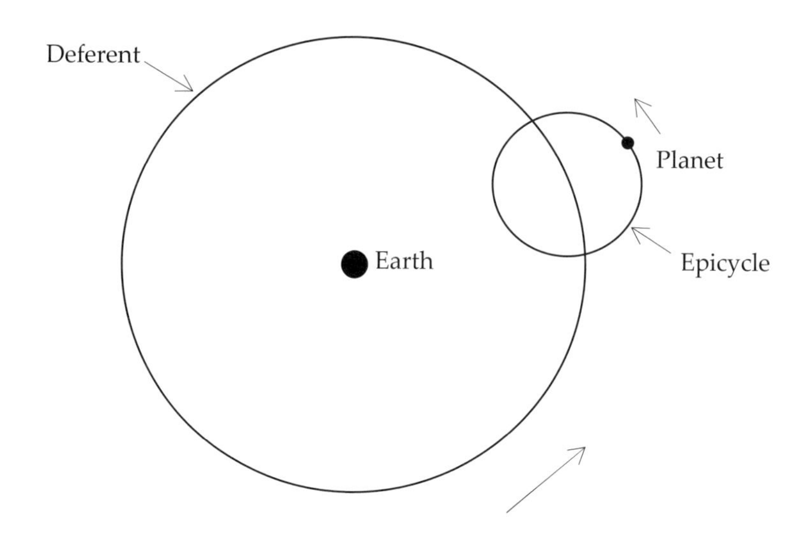

Diagram 1 (The Ptolemaic Model): If a planet is engaged in a smaller orbit (epicycle) in addition to its main orbit (deferent) about a central Earth; then at those times when the planet moves in the same direction in its smaller orbit as with its main orbit, it will appear to move quite rapidly though the star-filled sky from the view of an Earth bound observer. However, when the planet begins to turn about in its smaller orbit so as to move in the opposite direction to that of its main deferential path, the planet will seem to slow down in the sky from an observer's point of view; briefly stopping and going back the way it came, before turning once more to resume its journey through the sky at its more rapid pace.

In contrast to the Ptolemaic system, the Copernican system did not require at all the use of epicycles to account for observed retrograde motion, because under a sun-centred system retrograde motion amongst the observed planets is naturally accounted for [8]. This is due to the fact that the planets each possess different orbital periods, with those furthest away taking the longest to orbit the sun. As a result of this they are given to 'overtake' one another at various intervals of time; the very mechanics of such overtaking readily accounting for retrograde motion in a far more elegant way than under an Earth-centred system.

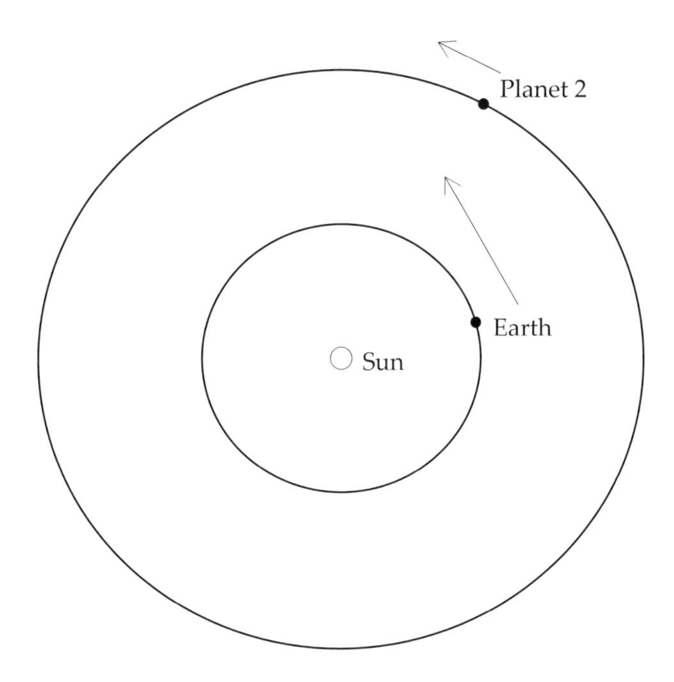

Diagram 2 (The Copernican Model): Here it can be seen that the Earth, being much closer to the sun than Planet 2, possesses a much faster orbital speed, as indicated by the different sized arrows. As a result, at some point the Earth will catch up to Planet 2; such an event occurring when there is a conjunction between both planets and the sun. From the perspective of an Earth based observer, as the conjunction is approaching, Planet 2 will appear to slow down as viewed in space against the background star field. At the exact point of the conjunction, it will briefly cease to move, before appearing to go backwards as the Earth overtakes. However, as the Earth continues in its orbit it will eventually turn such that its direction of motion is directly opposed to that of Planet 2, the most extreme point being when the sun is exactly in between both bodies. As this occurs, Planet 2 will appear to speed up and move quite rapidly through space as a result (were it possible to observe it in the absence of the light of the Sun). From the dynamics of the planetary bodies, such an ongoing recurring cycle readily accounts for apparent retrograde motion.

In the Copernican system, it can be seen then that epicycles are unnecessary as a means to explain retrograde motion. However, it is interesting to note though that Copernicus nevertheless did still retain the use of them, even within a newly formulated sun centred system [9]. His actual reason for this was because planetary observations indicated that even when the slowing down and speeding up of the observed planets due to retrograde motion was precisely accounted for, the planets still nevertheless did not seem to travel at uniform speed about the sun. Rather, the observations clearly demonstrated that they appeared to travel faster through space when closer to the sun and slower when further away from it. Indeed, this noted fact that the planets did not maintain a constant distance from the sun at all times in their orbits led Copernicus to offset his major orbital circles so that they were not precisely centred on the sun. Thus, in holding fast to his circles, and through his conviction that the speed of the planets was uniform, he was forced to retain small planetary epicyclical orbits as a subtle way to account for the continued presence of their apparent non-uniform motion about the sun. This was really nothing more though than a mathematical manipulation employed in order not to have to discard the primary aspects of his system, to allow it to better match actual observations, and also to allow him to claim that any observed non-uniform motion was not real, but illusory. A more detailed example of Copernicus' system with the inclusion of an epicycle is given as follows:

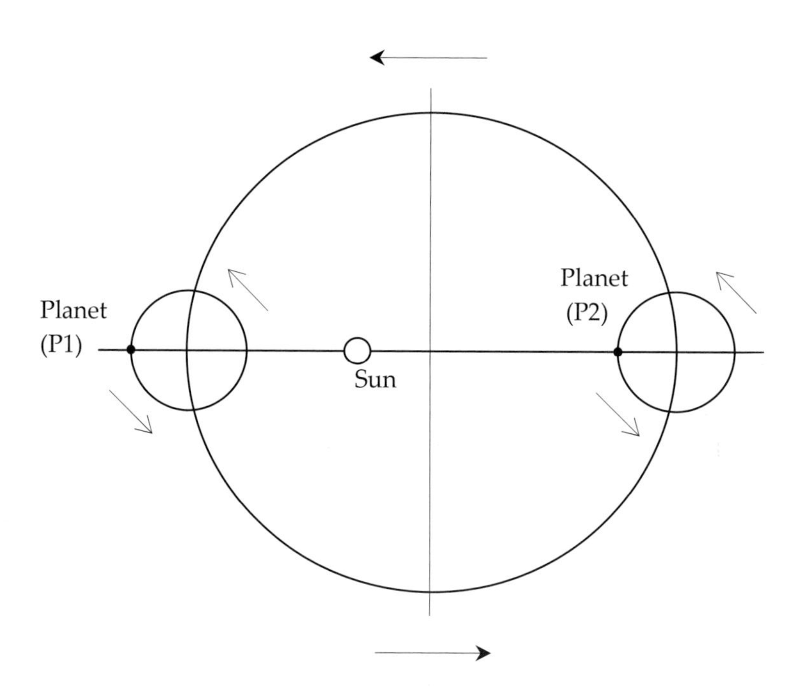

Diagram 3 (Epicycles within the Copernican Model): A planet in orbit about the sun is travelling anti-clockwise in its epicycle and also anti-clockwise in its main orbital circle; the centre of the smaller circle moving in uniform motion upon the circumference of the larger circle. When positioned at P1 the combined speeds of both 'orbital circles' are pointed in the same direction and thus are added. This necessitates that the planet is at this moment travelling at its fastest in orbit about the sun. From this initial position, as the planet moves yet further about the sun on its main circle, the uniform speed of the planet in its epicycle is such that by completing half of its main orbit about the sun, reaching point P2, it has completed one full orbit of its epicycle. At this point, the main orbit is counter to the direction of that of the epicycle, and thus the planet, at its most extreme position from the sun, is travelling at its slowest. If one were to plot the actual path of one full orbit about the sun, the planet would be found to trace out an elongated circular path as opposed to an exact circle. Such is the result of combining two uniform circular orbits in the proscribed manner.

Of Accuracy and Underlying Method

Upon the issue of system accuracy, one should note that the Copernican system actually afforded no greater level of precision than that of the Ptolemaic system [10]. Indeed, Copernicus' continued use of epicycles was in fact necessary simply to allow it to achieve the same level of accuracy as was present under the Ptolemaic system. Without them, although the general ordering of the planets about the sun would have been correct, the predictive power of the system would have been very much weaker. Upon comparison then, both systems possessed a corresponding level of accuracy; neither one being greatly superior to the other. In terms of their margin of error, both could be up to 5 degrees off the mark [11] with respect to observations. It was not simply the case though that this was attributable to inaccurate observations as such. Rather, one has to understand that the comparable level of error in both systems was due in actuality to an identical flaw *inherent* to them both. Essentially, it was the very *method* underlying their development and construction that was flawed; the choice of such method by Copernicus and Ptolemy being itself governed by their failure to understand the proper nature of the task they had set themselves. In essence, they had a certain false objective, in that what they both sought to attain in charting the course of the heavens was the means to develop nothing more than simply a *precise mathematically descriptive model* of its primary bodies; a goal, *that by necessity thus determined the very nature of their attempts to capture the course of the planets; and for whose reason, when evaluating their respective systems, whether Earth or sun centred,* ***it is impossible to know which if any is based upon truth*****.**

For all of their efforts, all that Ptolemy and Copernicus had sought to uncover was nothing more than a comprehensive series of *mathematical functions to capture the **apparent** movement* of the planets as seen against the background stars. Essentially, this is exactly akin to someone attempting to determine some future position of an object moving across their field of vision by conducting a limited number of observations of the object as it moves, so as to uncover a relationship between the spatial separation between the object at various points, and the corresponding time interval between those points i.e. how much time it takes to get from one observed point to another. In the case of an object engaged in apparent uniform motion, one would be hoping to discover a simple equation for the speed of the object: distance / time. Once obtained, the object's future position could then be calculated from any observed point given the passage of a certain amount of time. Indeed, the observer may have no actual knowledge of the true speed of the object at all, and simply measure the angle it sweeps out in the distance, such as is the case with the movement of planets upon the inside of the Celestial Sphere. Either way, the method employed is simply one wherein a mathematical function for speed is sought, whether true speed or angular speed; the aim being to 'superimpose' it upon an object perceived by the senses, and so capture its apparent movement. With this method, Ptolemy and Copernicus thus charted the *blind progress of the planets* through the sky. They thus concerned themselves with *appearance only* and of 'connecting the dots' of planetary positions to work out mathematical functions from which to determine a future position. In sum: ***They did not look into causes***. If true advancement in understanding the planets was thus to be achieved, what was needed was someone who was indeed inclined to focus upon the underlying causes that governed their activity [12].

Such a man was Johannes Kepler.

Johannes Kepler (1571-1630 AD)

What distinguished the work of Kepler from his predecessors, both ancient and contemporary, was his method. Unlike Ptolemy, Copernicus or Brahe, Kepler employed a far different method in his study of the heavens; one that allowed him to develop a far more *truthful* understanding of the motions of the planetary bodies. Whereas others simply focused upon the bodies themselves as they moved through space, Kepler stood back, and sought instead to account for their apparent or sense perceived motion set against the Celestial Sphere, through discovery of the underlying causal principles responsible for truly determining their physical action. For this reason his work differed in a true qualitative sense from that of any other. Indeed, what Kepler understood, that his colleagues throughout the ages did not, was that developing a series of mathematical functions and 'hanging them' upon a changing event in the physical world does not lead to knowledge, for it is blind to causes. Knowledge itself is attained only when one discovers the underlying causal principle behind what is observed by the senses. Moreover, the high predictive accuracy that one undoubtedly strives for, through study of such bodies as the planets, is attained almost naturally as a *secondary outcome* following on from the initial discovery of the lawful principles found to underlie their activity. With this firmly understood, Kepler was able to far surpass the work of all those who had gone before him by discovering a set of truly universal physical principles responsible for generating the non-uniform motion of the planets. Indeed, as a direct result of this he was able to falsify the very existence of epicyclical orbits; discarding them altogether. In addition to even this though, Kepler went on yet further to discover a unique set of harmonic intervals physically operative within the whole solar system, that unified in a most exacting manner all of the known planets about the sun in a way that had never been done before. Today, Kepler's primary discoveries are by convention grouped together into 3 general Laws of Planetary Motion. An understanding of them is vital, for they indeed form the very foundation upon which the discoveries of this present work rest heavily. And thus, they are hereby related, each in turn:

Law 1

'Planets orbit about the sun in an elliptical orbit, where the sun is positioned at one of the focal points of the ellipse.'

As mentioned above, it was readily apparent to the astronomers of the past through observing the planets that none of them were found to travel at uniform speed when viewed from the Earth, and that even retrograde motion itself could not fully account for this on its own; hence the continued use of epicycles by Copernicus. And yet even with their presence, a persistent margin of error of up to 5 degrees still remained when the formulations of his system were set against actual observations. Thus, it was all too evident that the movements of the planets did not conform too well to the mathematics of the epicycles that were superimposed upon them; a fact that could not be ignored. To Kepler indeed, it was something that represented a real challenge. Initially focusing upon the planet Mars, Kepler conducted an extensive examination of its orbit with the aim of developing a far better understanding of its characteristic activity than recognised under the Copernican model. To aid him in this, Kepler actually made use of the observational data gathered by his contemporary, Tycho Brahe, whom he had once worked with for a time [13]. It was following his careful examination of Mars that Kepler was eventually able to discover a series of causal principles found to govern the motion of this planet, the reality of which all but eliminated the errors present under the Copernican epicyclical system. Not only this, his work also led him to uncover the unique underlying geometry of all orbital bodies. Combined, these discoveries showed up the epicycles for what they really were: nothing but a mathematical fiction. Indeed, the strongly held belief among his contemporaries that the planets travelled in perfectly circular orbits, as intuitively they were considered divine, was also overturned.

The actual breakthrough that Kepler made was the realisation that Mars travelled not in a circular orbit, or even a combination of circular orbits, but rather in an elliptical orbit about the sun, and that the non-uniform motion of the planet itself was intimately connected to this type of geometrical figure. Thus, under the Copernican system, the combined use of *two circles* – one large and one small – for specifying its orbit, was discarded by Kepler, who replaced them with a *single ellipse* representative of its total orbital path; an ellipse that deviated just slightly from an exact circle. Furthermore, under such an orbit, Kepler determined that the sun was not positioned at the exact centre of the ellipse but rather at one of its two focal points, the other being empty. Indeed, it was this very feature of the orbit that determined the characteristic pattern of non-uniform motion which the planet engaged in. The application of these discoveries to the orbit of Mars overcame the errors inherent to the epicyclical system, reducing them to almost nothing. Moreover, these principles were found to apply to all of the known planets and not just Mars. Each appeared to possess an ellipse uniquely characteristic of its total orbit, with the sun located at one of the focal points of the ellipse. With these discoveries, Kepler was able then to break out of the restricted geometry of the circle and of uniform motion, to one that was inclusive of elliptical forms and of fundamentally non-uniform motion. Understanding the essential nature then of an ellipse shape and of how it differs from that of a simple circle is an absolute requirement if one is to fully comprehend the principles of planetary motion as discovered by Kepler.

The Nature of Circular & Elliptical Forms

When considering any shapes in nature that are bounded by curvature, it is critical to understand that such shapes are generated through the activity of rotation – this being the very underlying principle governing their formation. So it is with both the circle and the ellipse. In comparing these shapes it should be noted that the former is to the latter, as a square is to a rectangle. The circle is but a special case of an ellipse. Indeed, both shapes are generated from exactly the same essential components, namely, two centre points and a double radius. In the case of an ellipse it is quite easy to see the truth of this statement due to the fact that the two centre points are distinctly separate, as indeed are each of the two radii. It is not so readily apparent though in the case of the circle, for upon visual inspection of this shape, the two centres appear merged, as do the two radii, giving the impression that it possesses only one of each of these two components, though two are indeed present. The truth of this may be grasped when the generation of each shape through an act of rotation is considered.

The Circle

The formation of a circle begins from a *point*, mathematically considered to be 'a certain something possessed of the property of zero extension'; in essence a singularity. From the point, two lines, which by contrast do indeed possess extension – at least or rather at most – in one specific dimension, emerge to produce a double radius of a set length. The lines are then rotated one full cycle of 360 degrees upon a 2-dimensional plane of existence, the result of which sweeps out an area bounded by curvature, defining the form of the circle itself. Though it may seem that only one point and only one line (radius) are required to produce the circle, two of each is necessary, as detailed below:

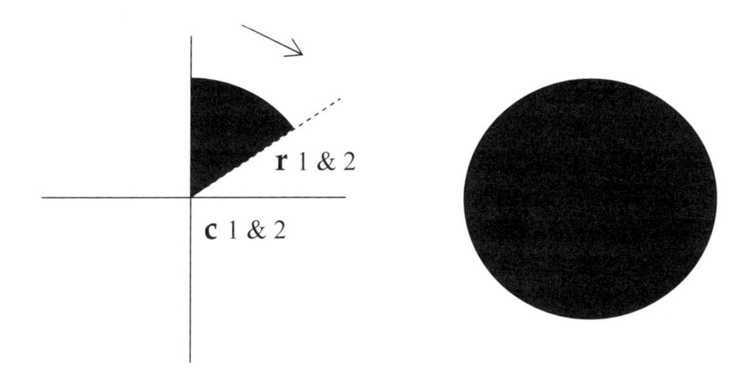

r = radius (1 & 2, merged) **c** = centre (1 & 2, merged)

*Diagram 4: On the left, it may be seen that both centre points **c1** & **c2** are merged, simultaneously existing in the same location at the exact centre of the shape about which the rotation occurs. Also, two lines of extension, each representative of the radius of the circle, **r1** & **r2**, are also merged and thus appear as one. The image on the left shows the partially swept out circle. On the right is the completed form of a circle fully swept out. Strange at it may seem, because the shape is produced from the actions of two lines, the area is actually swept out twice, though this is not obvious.*

The Ellipse

An ellipse shape has the general appearance of a somewhat squashed circle. It is of course rigorously constructed though using exactly the same components as those underlying the circle. In the case of the ellipse however, the two centre points about which each of the two radii rotate, are separated by a discreet distance. This gives rise to a certain added level of complexity in the formation of this shape that is not present in the case of the circle. Specifically, the rotational action of each radius about its centre point, ultimately generating the elliptical form, involves a non-uniform fluctuation in the length of each radius, even though the total length of both taken together remains constant throughout. This process is detailed in the diagram below:

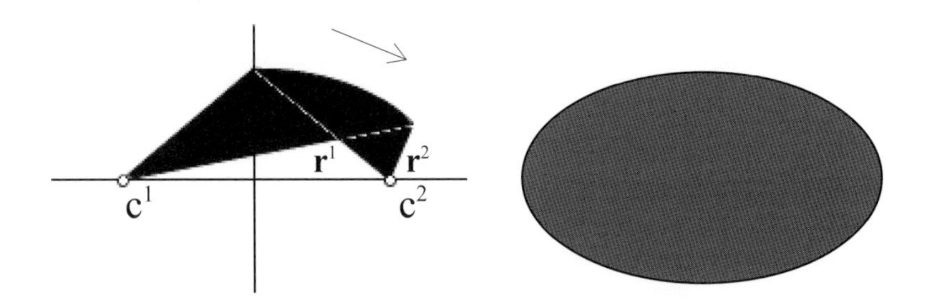

r = radius (1 & 2)   c = centre point (1 & 2)

*Diagram 5: On the left, the diagram details the positions of each of the two centre points **c1** and **c2,** each being separated by a discreet distance. From an initial position wherein each radius is of the same length, as shown above, there is a clockwise rotation of both about their respective centres. The fundamental aim is to attain the greatest possible extension of each, so that the maximum area permissible is covered in order to generate the shape. The actual perimeter of the ellipse is 'drawn' from the point of contact of each radius as they rotate. As the process unfolds, there is a transfer of* *length from each radius to the other. For one half of the formation of the shape* **r1** *increases its length whilst* **r2** *decreases. In the second half, the process is reversed with* **r1** *decreasing its length as* **r2** *increases. However, the total length of both radii when added together remains constant throughout the entire process. The complete formation of the ellipse is detailed on the right, and as may be perceived from a careful consideration of the rotational movement that generates the shape, the total area of the form is swept out twice.*

The Components & Properties of the Circle & Ellipse

Though the circle and the ellipse possess very similar properties, certain essential differences do exist between the two forms due to the added complexity of the latter. Firstly, it may be noted that a circle is symmetrical, infinitely so when cut through the centre at any angle. An ellipse however is symmetrical in two directions only, a result of the fact that the centre points are separated, whereas in the circle they are merged. By convention the centre points of both shapes are known as focal points, for they are the points about which rotational action occurs. The essential component parts of each form are detailed as follows:

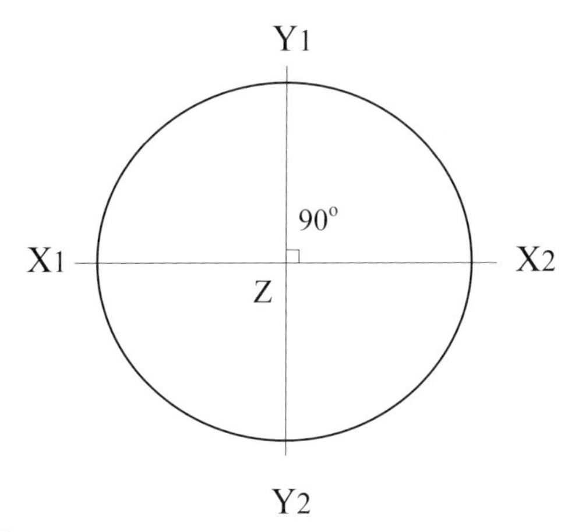

Diagram 6a: Circle

Diameter (or Major Axis) = any straight-line cut through the form passing through the exact centre of the shape, e.g. X1–X2 or Y1–Y2
Radius (or Semi-Major Axis) = Z (centre of circle) to any point on the perimeter, e.g. Z–X2
NB: Distance between focal points is always equal to 0 in the case of a circle.

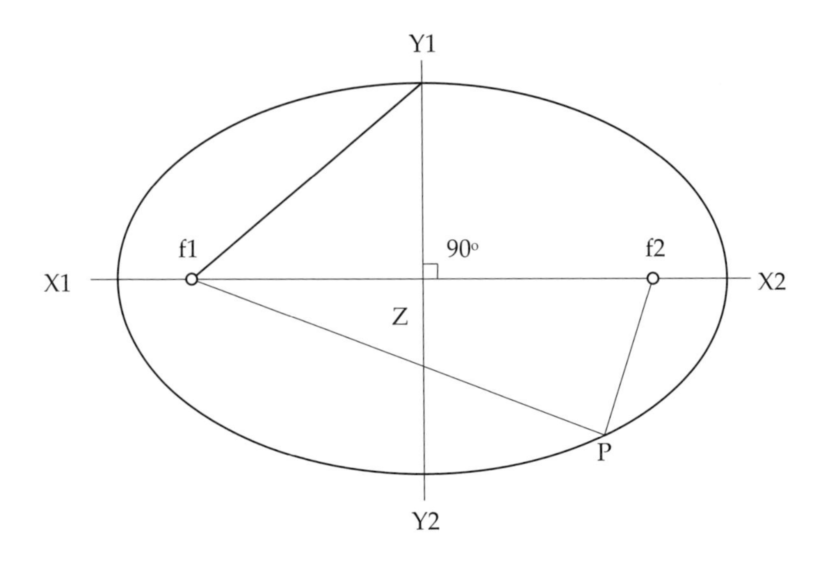

Diagram 6b: Ellipse

Major Axis = X1 – X2
Minor Axis = Y1 – Y2
Semi-Minor Axis = Z to either point Y1 or Y2
Semi-Major Axis = Z to either point X1 or X2, and also either focal point f1 or f2 to either point Y1 or Y2. E.g. f1 – Y1 as detailed
Focal point to Centre = Z to either focal point, f1 or f2

Notable Properties of the Ellipse:

1) A straight line drawn from both focal points to *any* point (e.g. P above pictured) on the perimeter of the ellipse is always equal to 2 × the semi-major axis of the ellipse, i.e. the major axis (as detailed in the above diagram *6b*).

2) *Eccentricity.* All elliptical figures possess a unique curvature, the measure of which is expressed mathematically as a number between 0 & 1 denoting the eccentricity of the shape. The eccentricity value itself is actually calculated by dividing the distance from either focal point to the exact centre of the figure, by the semi-major axis component of the form. Consequently, an eccentricity value of 0 always denotes a perfect circle as the focal points of this particular shape are merged upon the exact centre of the form itself. The distance between either of them to the centre is therefore always 0, and dividing 0 by any given radius must always produce an answer of 0. By contrast, in the case

of an ellipse, a given distance however small is always present between the exact centre of the form and either focal point. Thus, the eccentricity is always greater than 0. As a rule then the greater the deviation of the figure from a perfect circular form, the closer the eccentricity is to a value of 1, which itself denotes the condition of a completely 'squashed' figure where the focal points become merged with the perimeter of the ellipse.

NB: One may note that all of the planets within the solar system deviate only slightly from a circular orbit, and their eccentricity values are all close to 0.

The Position of the Focal Points within an Ellipse

In order to determine the location of the focal points within an ellipse a simple formula is used. The formula in question is a basic rearrangement of what is commonly known as Pythagoras' theorem, relating the various sides of a right-angled triangle. As shown in *Diagram 6b* above, because an ellipse has a two-way symmetry, the shortest distance from the exact centre to the perimeter is at 90 degrees to the longest distance from the centre to the perimeter. The focal points are always located on the latter line. As a result, to determine the distance between either focal point to the centre, the following formula is used:

$$a^2 - b^2 = c^2$$

This formula essentially relates the sides of the triangle shown above, whose three corners are designated by f1, Z and Y1.

a = f1 to Y1 = Semi-major axis
b = Z to Y1 = Semi-minor axis
c = f1 to Z = Distance from exact centre of ellipse to focal point

The Mean Point of Approach

When a planet reaches the point in its orbit wherein the semi-minor axis touches the perimeter of its ellipse, e.g. at either points Y1 or Y2 (pictured), at this given moment, the distance between the planet and the sun is the arithmetical mean of the entire orbit, which as stated previously, is the semi major axis of a planet's orbit. Set against it, are the two extreme values of approach to the sun. The point of closest approach is known as the Perihelion, whilst the point of farthest approach is called the Aphelion. In the example ellipse shown in *Diagram 6b,* were the sun positioned at f1, the other focal point being empty, then a planet would be at perihelion at point X1, and at aphelion at point X2. Mathematically, the closest approach, farthest approach, and the mean distance to the sun (semi-major axis) are related by the following formula:

$$\frac{(\text{P to Sun}) + (\text{A to Sun})}{2} = a$$

P = Perihelion, **A** = Aphelion
a = Semi-major axis

Law 2

Moving about the perimeter of its orbital ellipse, a planet sweeps-out from the focal point wherein the sun resides, equal areas in equal times.

Establishing that planets orbit the sun following an elliptical and not a circular path was a critical breakthrough for Kepler, for it led on to the realisation that the specific geometry of this shape held the solution to an understanding of the non-uniform motion of the planets, as observed from the Earth. This solution is what is encapsulated in Kepler's second law of planetary motion. Whereas the first law simply defines the character of the path taken by the planets as they orbit the sun, the second law uniquely accounts for the continuous variation in their speed throughout their journey, by linking this to the actual time required by each planet to complete one full orbit. The specific realisation that Kepler had in this regard was that the *total area* of a planet's orbital ellipse could be equated to the *total time* taken for the planet to complete one single orbit. As a consequence of this thinking, were a planet to complete but a small arc of movement about the sun upon its elliptical perimeter, a line continuously connecting the planet to the focal point wherein the sun resides, would 'sweep out' or cover a portion of its total area. The area covered by the line thus amounts to a fraction of the total area of the ellipse shape, and therefore a fraction of the total time for one complete orbit. In this way, a real connection may be perceived between the passage of time and the distance travelled by a planet as it moves through space. One simple consequence of this is that planets must increase their speed of movement as they near the sun and slow down as they move further away. This is detailed in the following diagram:

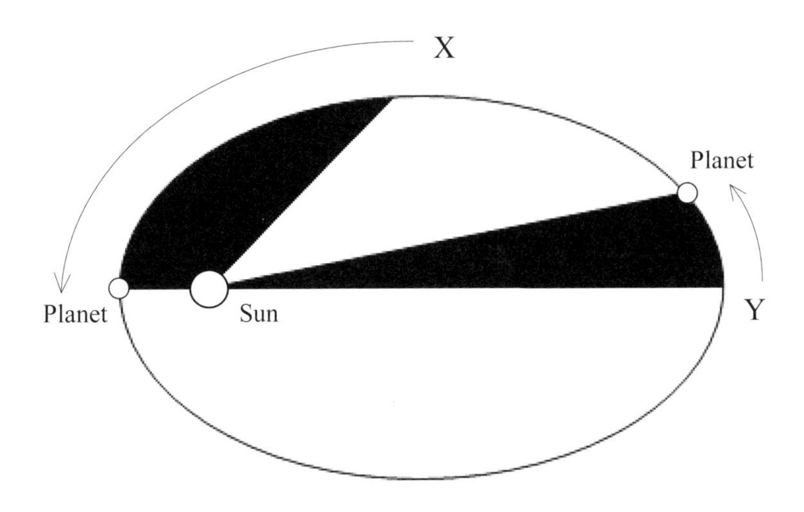

Diagram 7: As pictured, from point 'Y', the planet moves a set distance along the perimeter of the ellipse as shown by the arrow. A line from the sun (focal point) to the planet is swept out within the ellipse as the planet moves, covering an area as shown in black. This area equates to a given amount of time that passed whilst this movement took place. (The total area of the ellipse is equal to the total orbital time.) At another stage in the planet's orbit, a line from position 'X' to the closest approach to the sun is swept out. The area covered in this second example is equal to the area swept out from point 'Y', earlier. Kepler's law states that equal areas are swept out in equal times. Thus, the same amount of time passed as both sections of the orbit were completed. In one instance though, the planet covered only a short distance in its elliptical motion, when it was farthest from the sun; but when nearer to the sun it covered a far greater arc distance. The planet was therefore travelling at a significantly greater speed when it was closest to the sun, for it moved a greater arc length in the same amount of time that was taken to travel a far lesser arc length when it was at its farthest point from the sun.

The reality of Kepler's second law being operative within the solar system thus allows for a true understanding of the precise nature of the non-uniform motion of the planets. Under its direction, each planet in orbit about the sun *continuously changes* its speed as it moves through space; accelerating as it moves towards the central solar body, or decelerating as it moves away from it. At its closest point of approach (perihelion) a planet travels at its fastest speed through space, and at its farthest point (aphelion) it moves at its slowest orbital speed. Consequently, in an elliptical orbit, planets never move at a constant speed at any point during their path about the sun. Their state of motion is irreducibly one of continuous change. A most careful consideration of this point allows one to understand the true importance of the second law, and how it marked out Kepler as distinct from his contemporaries. The law of equal areas swept out in equal times *is an idea* that is *knowable* but *not observable*. Thus, what Kepler had discovered was *a principle of intention underlying orbital motion;* one that is in essence embedded within the solar system as a whole [14].

Law 3

A change to the Orbital Period of a planet, squared, is proportional to a change to its Semi-major Axis, cubed.

This law is commonly referred to as Kepler's 'Harmonic Law', and is usually expressed mathematically as follows, where **p** = Orbital Period, **a** = Semi-major Axis:

$$p^2 \propto a^3$$

Of the sum total of all the components defining the orbital ellipse of each planet, Kepler's 3rd law details a most specific relationship between the semi-major axis of each planetary body, and its orbital period, such that a precise change to either one must result in a precise change to the other, in accordance with a power relationship of 3 to 2. The semi-major axis, as already noted, is the mean distance between any given planet and the sun; the mathematical average between its distance at aphelion and at perihelion. The orbital period is the total time required by a planet to complete one full orbit about the sun, equating thus to the area of the ellipse.

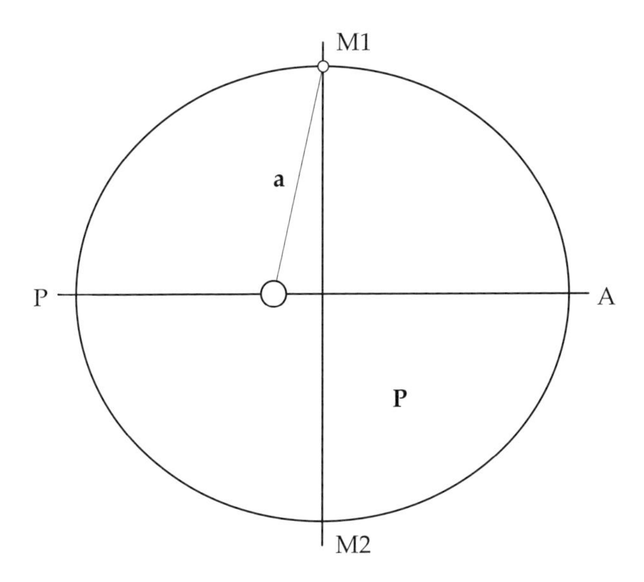

Diagram 8: In the slight elliptical orbit above, the planet is at its mean distance from the sun when at position M1 *(pictured) or* M2*. The actual mean distance or semi-major axis itself is denoted by line* **a** *connecting the sun at the focal point to the planet at* M1*. The whole area of the ellipse denoted by* **p** *equates to one complete orbital period. (Point* P *specifies the perihelion and point* A *the aphelion).*

The orbital configurations of the known planets as they stand prove the truth of Kepler's 3rd law. This is not evident however if just any one planet is examined in isolation. At least two planets must be evaluated together to demonstrate that the law is indeed valid in its expressed form. For this reason, the Law itself is usually converted into the form of a mathematical equation, allowing the details of two planetary bodies to be compared together. An evaluation of any two planets within the solar system employing the equation form of Kepler's law readily confirms its validity, enabling one thus to understand how it is capable of unifying the whole solar system around the sun, as the primary central body:

$$(p \,/\, \mathrm{p})^2 = (a \,/\, \mathrm{a})^3$$

1st Planet	2nd Planet
p = Orbital Period	p = Orbital Period
a = Semi-Major Axis	a = Semi-Major Axis

By inputting values into this equation for the semi-major axis and orbital period of any two planets within the solar system, one can see exactly how the law captures a real relationship between the two planets. An example, making use of the Earth and Mars will serve to illustrate this, by solving the harmonic equation for one unknown value, using three known values. In this instance, the orbital period of the Earth, the semi-major axis of the Earth, and the orbital period of Mars will be chosen as the three known values. The remaining unknown to be calculated shall be the semi-major axis of Mars.

According to modern sources, the actual magnitudes for each of the three chosen known values are as follows [15]:

Earth Orbital Period
= 365.2421897 Solar days (24 hours)

Earth Semi-Major Axis
= 81801193.26 Ideal Geographical Miles (1 IGM = 6000 ft)

Mars Orbital Period
= 686.9297110 Solar days (24 hours)

Substituting the above values into the stated equation, with Mars being designated planet 1, and the Earth, planet 2, one is thus able to determine the semi-major axis of Mars (a) as follows:

$$(686.9297110 / 365.2421897)^2 = (a / 81801193.26)^3$$

$$(686.9297110 / 365.2421897)^2 = a^3 / 81801193.26^3$$

$$a^3 = (686.9297110 / 365.2421897)^2 \times 81801193.26^3$$

$$a = \sqrt[3]{[(686.9297110 / 365.2421897)^2 \times 81801193.26^3]}$$

a = 124637065.17 IGM

One need only compare the calculated answer using Kepler's law, with a known value for the semi-major axis of Mars, to see that they are close [16]:

124638661.47 IGM

The actual observation of Mars does confirm then that the answer is indeed correct, and the expressed power relations of the law are physically valid. It is important to realise though that this is not something that is confined just to these two planetary bodies. Any two planets could be used in this equation and it would prove true. Kepler's law in effect unites the planets of the whole solar system, revealing that *the total time taken to complete one orbit about the sun determines the semi-major axis of a planet*. Thus, in the above example, if Mars, as a result of some outside agency, were to alter its orbital period so that it was completed in only 500 solar days instead of 686.9297110, then the semi-major axis of Mars would itself be forced to change in order to maintain a stable orbit. In such an instance Mars would be required to move nearer to the sun. The determination of its new semi-major axis would involve essentially judging the old orbit against the new. Thus, the ratio of 500 / 686.9297110 could be input into the above equation. The answer as produced for '**a**', the semi-major axis of Mars, would be that as would accompany a 500 day orbital period. Kepler's 3rd law thus embodies the idea of a principled relationship existent between two very specific planetary features: orbital period and semi-major axis.

Appendix C – The Stability of the Solar Day Examined

In general, the measure of a solar day is the time taken for the sun to appear overhead in the sky on two successive occasions e.g. noon to noon. For the most part such is determined by the rotational action of the Earth on its axis. However, orbital considerations do also play their part. Indeed, as a consequence of its elliptical orbit, there can be variations in the length of a solar day by up to 15 minutes depending upon where the Earth is in its orbit during any given year. However, these variations average out over a full year to produce a solar day whose time is almost exactly 24 hours, or 86400 seconds.

As a celestial phenomenon the solar day has been researched quite extensively over the years, with a number of studies carried out to determine if the rotational speed of the Earth on its axis has remained stable over time. Data collected in the modern period over the past 400 years or so is quite revealing. From 1623 AD to 1997 AD observational records indicate that from an exact value of 86400 seconds, the length of the solar day has fluctuated both above and below this total by a slight fraction of a second. For 41% of the time between these years the Earth has had a solar day slightly faster, and 59% of the time it has been slightly slower in its duration [1]. Yet other studies have reached further back into the past. A number of them have examined ancient records of eclipses that took place and were noted thousands of years ago. Through knowledge of the time and place of an eclipse in the distant past, researchers can determine whether the angular speed of the Earth has changed over vast lengths of time from the present era. From the period of 2000 BC to 2000 AD, an evaluation of multiple eclipses recorded during this 4000-year range indicates that the Earth has in total suffered an overall increase of 0.07 seconds to the length of a solar day. On average this is 0.0018 seconds per century [2]; an extremely small change for such a long period of time.

In consideration of the noted studies, it would appear then that the Earth is a very stable body, whose angular rotation on its axis does not vary over long periods of time by anything other than a small fraction of a second. But also that it even appears to fluctuate around the exact value of 86400 seconds. Thus, even though the Earth does show signs of long-term trends in its axial speed, such as is evidenced by the timing of ancient eclipses, this is no indication that it has always maintained such a trend. It may very well be that it is a part of a general cycle of oscillation about 86400 seconds exactly. In light of this, it would not be unreasonable at all then to suppose that the solar day has been stable stretching many years into the past. From this one could at least state that if such stability were maintained during an actual overall increase in the total orbital period of the Earth year, then the mathematics of a proportional link between this feature of the Earth and its physical equator are supported.

Appendix D – On the Irrational Nature of PI

Mathematically there is no single number called PI that may be written out as a fixed complete value. This is due to the fact that it would actually take an infinite number of calculations to achieve an answer as to its exactness, and a finite answer cannot ever be given to capture an infinite. Indeed, PI has much in common with certain other numbers of a similar nature, which also cannot be expressed as completed figures; one well known example being the square-root of 2 (1.414213562...). Grouped together, such values are generally referred to as 'irrational numbers'; a term given to describe the fact that as they unfold through calculation, being seemingly refined with ever greater precision, the series of numbers that follow on after the decimal point do not repeat themselves in any ordered or 'rational' pattern. Indeed, in ancient times, Pythagoras, one of the wisest men of Greece who studied both nature and music and the relations between them, referred to irrational numbers as 'unutterables', for they could never be spoken or 'captured' in words or writing with any sense of true completion. As a result of this, when irrational numbers such as PI or the square-root of two are employed in such as the sciences, they are only capable of being expressed to a certain practical level of exactness by specifying them to a certain number of units following the decimal point. With PI for instance, even as these words are being written, there are supercomputers in the world at this very moment that have been programmed to calculate PI continuously with ever greater refinement. And yet, even with such continual refinement, a single value representative of the ratio between the circumference of a circle and its diameter can never be achieved; only known to higher levels of accuracy and given as an approximation to an unattainable ideal. Were the ideal ever to be known to the mind of man, if such a thing were possible, it would be so only in the absence of calculation.

Appendix E – Physical Geometry Vs Mathematical Geometry

To obtain the most accurate measure of the celestial diameter of the moon orbit from knowledge of its equator, or the diameter of any circular form for that matter by way of its circumference, it is invariably assumed almost without question that one should divide the magnitude of the circumference by PI, for only this can provide the 'correct' answer. From a purely mathematical perspective there is a certain truth to this, but from a physical perspective, the choice of this operation may be wholly inappropriate, essentially giving rise to a fraudulent answer. The reason for this is due to the fact that there is a critical difference between what constitutes the domain of formal mathematical systems, and what constitutes the domain of physically generated objects and their activity.

Most students of physics, especially those concerned with astronomy, make use of co-ordinate systems that they usually 'hang' upon a particular body as a frame of reference; such as the sun for example, allowing them to specify the positions of the planets relative to it. The co-ordinate system usually favoured for this is a three-dimensional spatially extended framework comprised of three axes at 90 degrees to one another:

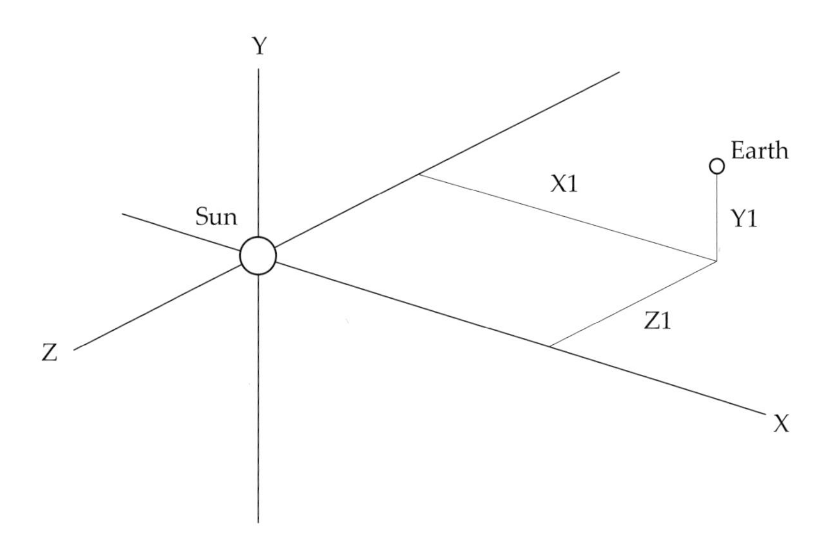

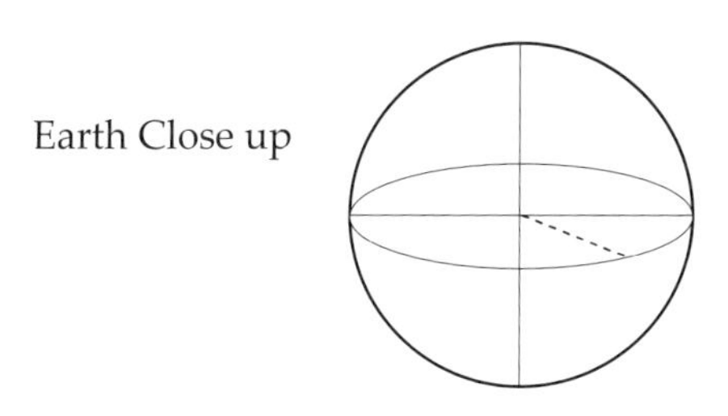

The co-ordinate system as shown above (top diagram) is set upon the position of the sun. From this, any planet, such as the Earth for example, may be assigned a position in relation to the sun with reference to each of the three axes X, Y, Z. Indeed, the Earth itself as a body is considered to actually be a component part of the established co-ordinate system which is itself built up in accordance with the system rules. As a spherical body it is held to be ***mathematically composed*** *of innumerable discreet lines that 'radiate' out from a central point in all directions (bottom diagram). The sum total of all these lines* ***form*** *the appearance of an outer surface, and in the co-ordinate system such points can also be referenced on the surface with different x, y, z values.*

The above co-ordinate system pictured is a standard Euclidean mathematical framework, named after the mathematician Euclid who wrote many treatises on mathematics in ancient times. Euclid's entire system rests upon a series of *a priori assumptions* that are created solely in the mind of the mathematician, and form the basis of all further analysis. Such assumptions are known as axioms; a term understood to mean 'self-evident' truths. Thus, Euclid speaks of concepts called line, point, and space. Within his spatially extended realm the shortest distance between two points is a straight line for example. The three-dimensional framework devised according to such 'self-evident' truths is thought to be uniform and even 'objective'. In essence, it is a blank three-dimensional 'blackboard' within which all physical objects could be placed and whose location could be specified with a series of values upon each of the 3 axes. Physical space under this system is considered to be an empty container - hence the name, space - within which 'matter-stuff' of various forms reside. Thus, by hanging this extended co-ordinate system upon a region of existence wherein reside certain 'physical entities' such as planets, many (though not all) mathematicians claim that all they are doing is *capturing what is already there without prejudice.*

In ancient times up until Kepler, the mathematical systems and laws that were devised to try to predict the future positions of the planets in the sky were essentially established in accordance with Euclidean type axioms. Mathematical / geometrical principles were formally established at the outset and superimposed upon the physical activity of the planets. In essence, the *formal laws of the system and the mathematics itself were thought primary even to the physics of the universe.* And yet critically, the laws themselves were artificially devised; chosen merely because they 'seemed right', even without any experimental evidence to back them up. Indeed, in principle, it is often forgotten that whenever the attempt is made to measure anything, a whole series of assumptions are made unwittingly as to the fundamental properties of the physical world. In the case of the Euclidian framework, space is assumed to be uniformly continuous and extended infinitely in all directions. Such things though are not actually true at all in the sense that they are physically proven. They are simply assumed to be true as axioms of a formal mathematical system.

The above considerations bear strongly upon the idea of PI as a ratio, and the geometrical system wherein it may be considered valid. Essentially, PI itself as an ideal ratio is valid only within the restricted geometry of a static form-based environment. One may consider such a domain as 2 or 3 dimensional in the sense of spatial extension; and thus consider PI in relation to either circles or spherical objects. The realm wherein PI is 'present' is thus one of *pure form in the absence of all activity*. It is for this reason that one must be careful when applying PI to a planet, such as the Earth. For example, if one would seek to determine a measure for the actual volume of the planet, the Earth would generally be examined as a static object, regardless of its motion through space i.e. its activity. In such a situation PI would doubtless be employed as part of the general mathematical formula for determining the volume of the planet to give the most accurate answer possible, which indeed could only ever be a very good approximation. Thus, within the restricted domain of purely static forms PI can have certain practical applications. However, once one departs from this domain and enters into that of physical activity, then PI loses all validity. It quite simply is not operative at all within the geometry of activity.

Ultimately, one must realise that the geometry of activity is fundamentally different to that of form. Indeed, one should not even make the assumption that they are capable *on principle* of being combined, for such is impossible. This is not to say that one cannot specify the dimensions of static volumetric forms and then simply model their apparent movement within such as a spatially extended Euclidian framework for mere 'practical' ends; as a means to model mundane movement. Indeed, such is often done when one is concerned with a 'low level' study of various phenomena for basic applications. However, upon a critical note, such a method as this which seeks to combine such fundamentally different domains as form and activity is in and of itself fatally flawed. Indeed, it is bound to almost subconsciously limit the mind of the researcher as to what is possible within nature. In addition to which, they may fail to see or discover perfectly valid underlying physical connections existent between various physical phenomena. A prime example of direct relevance to this work is that concerning the number of days contained within a tropical year. If one is content to model the Earth in terms of the axial rotation of the static planetary form in conjunction with an apparent orbit about the sun, one may then view the sum total of days per year as merely an *artificial or incidental* frequency, and not as a truly *natural-physical-**energetic*** frequency, as the arguments put forth in this present work suggest most emphatically that it is.

Appendix F – The Full Technical Details Surrounding the Evaluation of Nuclear Testing (as Presented in Chapter 15)

In order that the reader may be able to properly evaluate the analysis of the nuclear tests (or wartime bombs) presented in this work, it is necessary that precise and exacting details are given concerning the basis of the arc measures as set forth i.e. essentially, just how they were derived. As indeed, in the main text (of chapter 15), the pure latitude values and the bomb location to sun ground position arc lengths, fundamental to the whole analysis, are stated in the absence of any accompanying information as to how exactly they were arrived at. A detailed explanation is thus required in order that one may verify for themselves the truth of the presented values.

With the exception of the Trinity bomb, the primary details of all of the nuclear tests/wartime events, as evaluated in Chapter 15, are given in the following format:

Target/Shot: Yield (TNT in Tonnes):
Date: Time (GMT):
Geodetic Bomb Co-ordinates (Deg):
Lat: Long:
Geodetic Sun-Ground Position Co-ordinates (Deg):
Lat: Long:
Arc Length: Equator to Latitude of Test (feet):
Arc Length: Test Site to Sun-Ground Position (feet):

1) Of the above, the essential information concerning purely the bomb itself and the moment of detonation, consist of the following elements:

Target/Shot: Yield (TNT in Tonnes):
Date: Time (GMT):
Geodetic Bomb Co-ordinates (Deg):
Lat: Long:

The details and values cited in each of these fields for all of the nuclear tests given in this format in Chapter 15 are from the Nuclear Test Archives of Robert Johnston, which may be found on the internet at the following address:

http://www.johnstonsarchive.net/index.html

2) The Sun-Ground Position given as a geodetic set of coordinates (on the surface of the Earth) forms the next set of fields:

Geodetic Sun-Ground Position Co-ordinates (Deg):
Lat (latitude): Long (longitude):

The precise latitude and longitude values as given for the geodetic location of the sun (ground position) at the moment of a nuclear event are derived from a computer program:

Interactive Computer Ephemeris (ICE) Version 0.51
(U.S Naval Observatory Nautical Almanac Office)

Through use of the ICE software, one may input the precise time value in GMT for a given nuclear explosion (as taken from the Johnson archives) and generate a set of angular coordinates for the location of the sun with respect to the Earth, at the precise time of the event. In this instance, the ICE program is used to generate two very specific values that capture the position of the sun at the time of a nuclear event: the Greenwich Hour Angle (GHA) and the Declination of the sun; GHA being the astronomical equivalent of longitude, and declination being the astronomical equivalent of latitude. Both of these measures are of course expressed in terms of degrees, minutes, and seconds of arc.

For the purposes of nuclear analysis, the *longitudinal* values as stated for the sun-ground position of a given nuclear test are taken directly from the ICE software. However, in terms of presentation, though GHA is generally expressed by ICE in terms of 0-360 degrees (Westwards) from the Greenwich meridian, in the present work, the values have been converted into the form of 0-180 degrees, and given as either west or east of the prime Greenwich meridian. Thus, by way of example, if a given historical date of a nuclear test were to be programmed into ICE, generating a GHA for the sun of 208.1283 degrees (westwards), then a value of 151.8717 degrees eastwards would be given in the longitude field for the sun ground position (360 – 208.1283 = 151.8717):

Geodetic Sun-Ground Position Co-ordinates (Deg):

Lat: **Long: 151.8717 E**

With regard to the *latitudinal* position of the sun (ground position) it is important to note that the values as generated by the ICE software for the declination of the sun are not used directly as the ground based angular latitude of the sun. This is because astronomical values of declination are in fact geocentric and NOT geodetic angles, assuming that the Earth is a perfect sphere when it is in fact an ellipsoid. Consequently, all of the (angular) latitude values for the sun-ground position as stated for each of the nuclear events analysed, consist of the generated declination values of the ICE program, modified by a very specific mathematical equation, to convert them from geocentric to geodetic angles. The equation itself is as follows:

$$\phi = a\tan\left(\tan\phi' \,/\, (1 - f)^2\right)$$

ϕ = geodetic angle

ϕ' = geocentric angle

f = flattening measure (of a given ellipsoid)

As a further note, in the general analysis of the nuclear events in this work, there are only 2 different Earth ellipsoid models that have been used. They are the 1924 International (Hayford) model, and the WGS84 model. For all nuclear events analysed, all that were conducted between 1945 and 1966 were evaluated under the 1924 International model, and all those conducted from 1966 to the present were evaluated under the WGS84 model. The measure for flattening as used in the equation above is dependent upon the primary values of a given ellipsoid. For the two models used in this present work the flattening (f) measures are as follows:

1924 International = 1/297 WGS84 = 1/298.257223563

The reasoning behind the choice of Earth ellipsoid models is simply that the WGS84 model would not have been in use during the early period of nuclear testing. Also, the WGS66 and WGS72 models which came after the 1924 model, only differ very slightly from the more up to date WGS84 model. The difference between all three WGS models is only on the order of a few dozen feet or so, whereas all three WGS models can differ from the 1924 model in terms of latitude by several hundred feet. There was simply no reason therefore to make use of so many different models for each of the tests analysed when only 2 seemed to suffice.

To summarise then the above, in terms of specifying the latitude of the sun-ground position, all declination values generated directly by the ICE program (either north or south of the Earth equator i.e. + or –) are modified by the above stated equation. Thus, by way of example, if ICE were to give a declination value for the sun at the time of a given nuclear test as 15.99374167 degrees north – representative of a geocentric angle – then when input into the above equation (e.g. using the WGS84 model f value) it would be converted into a geodetic angle of 16.09596624 degrees north, and this is the angle that would be given in the latitude field, as follows:

Geodetic Sun-Ground Position Co-ordinates (Deg):

Lat: 16.09596624 N Long:

3) With the above points 1 and 2 thus covered, one can see how for each nuclear event, one is able to specify two sets of geodetic reference coordinates:

i) A geodetic latitude and longitude for the ground position of the bomb at the moment of detonation of the device (taken directly from the Johnston Archives).

ii) A geodetic latitude and longitude for the sun 'ground position' upon the surface of the Earth at the moment of detonation of the device (derived from the ICE program, with the above stated modifications).

With the coordinates relevant to the bomb and the sun thus in hand, an Earth ellipsoid model is selected upon which to determine the relevant arc measures which form the following fields:

Arc Length: Equator to Latitude of Test (feet):
Arc Length: Test Site to Sun-Ground Position (feet):

As already noted, all nuclear events (analysed) that were conducted between 1945 and 1966 were evaluated under the 1924 International model, and all those conducted from 1966 to the present were evaluated under the WGS84 model. In actually deriving through calculation the arc measures as given for each nuclear event, The Great Circle Calculator computer program of Ed Williams was used. This can be found on the internet at the following web address:

http://williams.best.vwh.net/gccalc.htm

Upon access to this site, one is confronted with the main page of the calculation program itself, and one is able to see a set of boxes to input two sets of latitude and longitude coordinates. In addition to this, there are two drop-down menus off to the left. One of them contains a list of various Earth ellipsoid models, and the other is for specifying the distance unit to be used for the answer, which is itself given in an 'output' box just underneath the 'input' box wherein the reference coordinates are typed in. In this present work all arc length measures are specified in feet, selected from the appropriate drop-down menu. And dependent upon the year of the test/event, either the 1924 International model or the WGS84 model is used (NB: in The Great Circle Calculator program itself the WGS84 model is listed as WGS84/NAD83/GRS80).

With the data for each test input, the units selected as feet, and the correct Earth ellipsoid model selected also, one need only press the 'Compute' button to generate an answer for the surface arc of separation between the two coordinates i.e. between the bomb position and the sun ground position. Of course, in the case of determining just the value for the latitude placement of the device at the moment of detonation, then only one value needs to be input into the program; simply the geodetic latitude value of the test as taken directly from the Johnston Nuclear Archives. The remaining other 3 boxes can be left at 0:00:00 deg/min/sec (which they are set at by default anyway).

Appendix G – The Proposed Harmonic Unified Equations of Bruce Cathie: A Critical Evaluation

For those who have read Bruce Cathie's books there will be general agreement as to their basic format and composition. Each are found to contain an introductory set of chapters so to speak, that lay out certain basic discoveries made by Cathie, which include primarily his discovery of an Earth energy grid matrix, and also a set of mathematical equations referred to by him as Harmonic Unified Equations, which he puts forward as an extension to the basic $E = mc^2$ energy equation in physics. Indeed, the first several pages in three of his most prominent books [1] are practically identical, with only minor differences. Once the basics have been covered, the bulk of the remaining chapters are concerned with certain discreet analyses of various phenomena in relation to the identified Earth energy grid, and also Cathie's Unified Equations. The emphasis is generally upon identifying if possible various mathematical and/or physical relationships between the phenomena under study, and the values built in to the grid structure and derived from his equations.

Like many no doubt who have purchased his books, this author has looked very carefully at his proposed Harmonic Unified Equations, and made a personal attempt to evaluate them in light of the analysis presented by Cathie himself. In terms of the equations themselves, Cathie states in his books that they are based upon the extended Leibniz energy equation; an expression of the total energy 'stored' in a given amount matter (see chapter 14 for full details):

$$E = mc^2$$

Energy = Mass × speed of light, *squared*

(NB: '**C**' denotes light speed in vacuum)

Following on from Cathie's early work in reconstructing the basic mathematical structure of the Earth's energy grid (which as one may recall, he found to be composed of a set of parallel lines spaced at intervals of 7.5 minutes of arc running east-west and north-south, primarily orientated to the Earth's magnetic field, including a set of interlocking great and small circles), Cathie goes on to explain how certain mathematical values found to be 'built in' to the grid, appear to hold the key to being able to modify the basic energy equation, such that one could express mass (M) purely in terms of light (C). This is achieved, according to Cathie, by the following equation:

$$m = c + \sqrt{1/c}$$

Consequently, with mass now given purely in terms of light, Cathie is able to substitute this new expression into the extended Leibniz energy equation, to produce the following:

$$E = mc^2 \quad \rightarrow \quad E = \left(c + \sqrt{1/c}\right) c^2$$

In total, this newly derived equation constitutes one of a set of three equations discovered by Cathie, which he refers to generally as Harmonic Unified Equations in his books. The particular one as given here (above right) he designates simply as Harmonic Equation 1. The other two are just slight variations on the first. Of course, the key to discovering the whole set relied essentially upon the basic equation for mass (M), and of determining just how mass could be expressed purely in terms of light speed. And so, just how exactly did Cathie arrive then at his basic mass equation? The mathematical values that Cathie identified that led him to the breakthrough are well noted in all of his books, and reproduced below, along with the labels Cathie himself assigns to them [2]:

17025 Earth mass harmonic
14389 Speed-of-light harmonic
2636 Unknown harmonic

The first of the three noted figures Cathie explains is what would be a 5 figure numeric value for the total volume of the Earth in terms of cubic nautical miles i.e. 170250000000 cubic nautical miles, in full. In presenting his figures though in the early chapters of all of his books, Cathie does not provide an exacting calculation with a full working out as to the total Earth volume. The 5 figure harmonic approximation only is provided. However that may be though, a specific value is indeed given in a later section of one of his books; notably in *The Bridge to Infinity* [3]. Here Cathie specifically states that any spherical type body will have a volume of $1.7018068 12^{11}$ cubic minutes of arc, and furthermore, that in the case of the Earth itself, 1 minute of arc (angle) is equal to 1 nautical mile. Importantly however, though this figure is oftentimes noted as a volume, it is also (especially as a numeric sequence) referred to many times by Cathie in *The Bridge to Infinity* as a "mass harmonic at the earth's surface" (see pages 109 and 172). On the matter of just how Cathie actually derives this figure, careful reconstruction would appear to reveal the answer:

1) As mentioned previously in this present work, the most advanced Earth model in use is the WGS84 model, based upon an equatorial radius of 6378.137 kilometres and an inverse flattening measure of 298.257223563. From this, one can derive a set of values for both the Earth equatorial and polar radius, as given in statute miles:

Equatorial radius
= (6378.137 × 0.6213711922) = 3963.190591704 statute miles

Polar radius
= 3963.190591704 – (3963.190591704 × (1/298.257223563))
= 3949.902764022 statute miles

2) From the above figures one may thus now proceed to determine a value for the nautical mile that would 'accompany' such an Earth:

((3963.190591704 + 3949.902764022) × PI × 5280) / 21600
= 6076.819455305 feet

3) With the nautical mile established, one can now convert both the equatorial and polar radius values into this measure:

3963.190591704 / (6076.819455305 / 5280) = 3443.519505245 nm
3949.902764022 / (6076.819455305 / 5280) = 3431.974036324 nm

4) A certain pause is required here whilst one considers the next step. Ordinarily one would proceed, in light of the fact that the geometry of the Earth is that of an ellipsoid, to use the values for both the equatorial and polar radius of the planet in the following formula, as is recognised for determining the volume of any given ellipsoid:

$$4/3 \times PI \times a \times a \times b$$

Where: a = Equatorial radius; b = Polar radius

Thus: 4/3 × PI × 3443.519505245 × 3443.519505245 × 3431.974036324

= 170465971375.741 cubic nautical miles

5) As one can see however, the answer as calculated is not really close at all to that given by Cathie in his book: 1.701806812^{11}. How can this be explained? Quite simply, Cathie does not actually use the recognized formula for determining the ellipsoid volume of the Earth. Rather, he makes use of a subtle modification of the basic formula. Instead of multiplying directly the equatorial radius squared by the polar radius i.e. a × a × b, he appears to first calculate an average radius of the two, and once generated, he then proceeds to 'cube' this value. Thus, the equation that he would appear to use to achieve this would be as follows:

$$4/3 \times PI \times ((a + b) / 2)^3$$

Thus: $4/3 \times PI \times ((3443.519505245 + 3431.974036324) / 2)^3$

= 170180681184.608 cubic nautical miles

Compare with 1.701806812^{11}

6) From the above point, it is now clear how Cathie derives a value for the volume of the Earth in cubic nautical miles, at least as is stated in his book *The Bridge to Infinity*. However, one can still see that there is a marked difference between a five figure harmonic of the value as just calculated, of 17018, and a five figure harmonic of 17025 given previously; the latter being the very basis of his unified equations. In order to account for this one must set aside *The Bridge to Infinity,* and turn to another of his books, *The Harmonic Conquest of Space,* for an answer.

In this latter publication there appears to be no reference at all to the stated mass harmonic of 1.701806812^{11}, and no mention of a nautical mile of 6076.819455305 feet, direct or implied. Rather, in this book, Cathie appears to have adopted a fixed nautical mile of exactly 6076 feet for his calculations. One can cite his use of this value in particular on page 36 [4]. Now, if one converts the radius values of the Earth as given above in statute miles (in step 3), into nautical miles of 6076 feet, and makes use of the modified ellipsoid volume formula, then one can indeed achieve an Earth volume comparable to the 5 figure value stated by Cathie:

170249546032.048 cubic nautical miles

Compare with 170250000000

With this, one can be fairly satisfied that this figure as determined is indeed very close to what Cathie must have derived himself when evaluating the volume of the Earth.

The Speed of Light Harmonic

Concerning the speed of light, Cathie states that a value of 14389 is itself representative of a 5 figure speed-of-light harmonic. A basic evaluation of this figure would indeed appear to offer a ready match with the current speed of light, given that one makes use of Cathie's preferred units of measure for speed: nautical miles per grid second. In this particular instance, the fixed value of 6076 feet, as noted above, is taken to represent 1 nautical mile:

299792.458 Kilometres / sec.

299792.458 × 0.6213711922 = 186282.397 Statute miles / sec

186282.397 / (6076 / 5280) = 161878.054 Nautical miles / sec

161878.054 × (8 / 9) = 143891.6035946 Nautical miles / grid sec

Compare with 14389

Again, one may readily perceive that the calculated value for the speed of light as given in nautical miles per grid second is a close match to the 5 figure value of Cathie.

The Unknown Harmonic

The value that Cathie initially refers to as an unknown harmonic, of 2636, was found by him to be derived from a relatively simple manipulation of the speed of light. The procedure for generating this value is contained in his basic mass equation, as follows:

1 / 143891.6035946 = 0.0000069496758324

Square-root of 0.0000069496758324 = 0.0026362237827030

Compare with 2636

The Basic Mass Equation, Employed

From the above it is quite clear that the volume of the Earth, the speed of light, and its inverse-square root derivative, do all match to a good level of accuracy the values that Cathie himself provides in terms of their basic numeric sequences, as harmonically expressed. However, the *actual* relationship between the figures is certainly not expressed in any true sense in terms of his basic mass equation, as is stated in his various books. For to reiterate, what is presented in his publications is a very precise equation of the following form:

$$m = c + \sqrt{1 / c}$$

It might interest the reader at this point to consider attempting to solve the equation and determine a value for Mass (M) by making use of a given speed of light value. In doing so, by adhering to basic mathematical conventions, one may note the following (employing the light speed value 'C' above) as an example:

1 / 143891.6035946 = 0.0000069496758324

Square-root of 0.0000069496758324 = 0.0026362237827030

143891.6035946 + 0.0026362237827030 = 143891.606230891

Therefore: M = 143891.606230891

As one can see in this example, the end result is what would appear to be simply an alternate value for light speed that is only just slightly greater than the original value used at the outset. A value representative of the Earth volume as one would hope to achieve is simply not generated. And the reason for this is quite obvious. The initial figures as Cathie presents them in his books only return the desired answer when the values of interest are modified such that the decimal points are in the 'correct' places before one solves the actual equation itself. Consequently, were one thus to use a very particular speed of light in Cathie's basic mass equation, with the intention of producing an Earth volume containing Cathie's noted 17025 sequence, the basic mass equation itself would have to be modified as follows:

$$m = c \times 1^{6} + \sqrt{1^{26} / c}$$

Using the stated light speed value as given above, this variation on Cathie's basic equation produces the following answer:

$$m = 143891603594.6 + 26362237827.0$$

$$= 170253841421.698 \text{ cubic nautical miles}$$

As one can see, in order for this answer to be achieved, one must add to the basic equation a set of additional multipliers so to speak. The speed of light value to the left of the + sign must be multiplied by 1 (×10) to the 6th power, or basically 1 million. Whereas in the case of the values to the right of the + sign, the simple division of 1 by the speed of light is inadequate, and instead one must divide a value of '1' followed by 26 zeros. This ensures that the decimal points are in the correct place prior to adding the two values together to achieve the desired Earth volume figure.

The above analysis draws attention thus to an important point, which is that the basic mass equation as stated by Cathie in his books is not as simple as it appears. And indeed, as all of his 3 noted Harmonic equations have embedded into them the basic mass equation, this consideration applies to them also. One may cite as a further example Cathie's Harmonic Equation 3 in particular, as is given on page 26 of *The Harmonic Conquest of Space*:

$$E = \sqrt{\ } [(2c + \sqrt{1/2c})(2c)^2]$$

As can be seen from the basic form of the equation, it differs only very slightly from Cathie's Harmonic Equation 1, in that all of the light speed values (C) are doubled, and E is determined ultimately by taking the square root of the final value of the resolved internal calculations.

To further emphasise then the use of built in multipliers to Cathie's Harmonic equations, one may cite Cathie's use of this equation as given on the aforementioned page of the above noted publication. In the example given, in generating a value for 'E', Cathie employs a speed of light value of 143838.1966 nm / grid second. Multiplying this initial value by 2 to produce 287676.3932, he introduces the doubled light value into his equation as follows:

$$E = \sqrt{[287676.3932 + \sqrt{(1/287676.3932)}(287676.3932)^2]}$$

From this initial line given on page 26 of *The Harmonic Conquest of Space* Cathie goes on through various workings to eventually end up with an answer of:

$$E = 26944444\ldots$$

However, if in reading through his particular workings, one were to attempt to try to obtain this very answer by starting out from the initial line of the equation as Cathie provides, one will certainly not arrive at 26944444... as a numeric sequence. And the reason, once again, is that in order to actually arrive at such a figure (as Cathie does) one must carefully and selectively make use of the correct ×10 multipliers to modify the various light speed values prior to both either adding them together, or before taking their square-root.

An Evaluation of Cathie's Unified Harmonic Equations

As a result of the above analysis, any full evaluation of the Unified Harmonic Equations as put forward by Cathie in his books would appear to centre upon 3 topic areas. They are each discussed in turn:

1) Cathie's use of a modified equation for calculating the volume of the Earth ellipsoid

From the figures as presented it is quite clear that if the standard formula for an ellipsoid volume was used, then the accuracy of the match between the relevant harmonic values would be severely compromised. However, it would certainly not be fair to dismiss the basic mass equation purely because Cathie does not use the standard formula, but a variation. Indeed, in the early chapters of this present work this author himself makes use of a modified ellipsoid formula, by insisting that 22/7 should be employed as opposed to PI. And several pages of analysis were given in the attempt to justify this unusual decision, by pointing out that, though it is an unusual decision, it is still nevertheless physically valid. Consequently, the very correspondence, in terms of the accuracy of the 3 values as evaluated harmonically by Cathie, may indeed constitute the very proof of the validity of the alternate ellipsoid formula that he uses.

Overall then, one cannot really say that Cathie's use of such a modified formula invalidates his basic mass equation. And indeed further to this, neither could one claim that the slight inaccuracies in the match between the harmonic values themselves invalidate it, for as mentioned previously in this present work, the slightest alteration made to the primary values that constitute the Earth ellipsoid model one may choose to employ, could easily eliminate any mathematical discrepancies, and nobody could truly argue that the WGS84 model is perfect. In all, one cannot claim that Cathie's Harmonic equations are invalid purely as a result of any inaccuracies of the mathematical values that he generates through their use. The error rates are just not large enough to justify such a criticism.

2) Cathie's required use of additional '×10 multipliers' to make his equations 'work'

Can one truly accept on principle that the use of so called additional multipliers, as is necessary for solving Cathie's equations, (in the way that he does) is acceptable? For indeed, though Cathie's basic equation for Mass *mathematically adds up when the appropriate multipliers are employed*, this does not necessarily imply that such harmonic manipulations are themselves *physically valid within the proposed mass equation*. One must understand however, that this is not a criticism of harmonic analysis per se; of such as dividing or multiplying successively a given number e.g. by 10 or 6 etc, to modify a given value, which itself is of the character of a defined physical unit of extension i.e. a real distance unit or arc length. As indeed, this is valid science that applies to music and matter, and to the physics of resonance. The matter of harmonic analysis in general is not the issue. Rather, it is something far more fundamental; and goes to the heart of the proposed equations. Indeed, for those who may have read the examples that Cathie provides in his books, such as the Harmonic Conquest of Space, it is not clear that he makes use of any particular ordered procedures when he accomplishes his internal harmonic manipulations to get the answers that he does.

In sum, being as the equations as are *presented* by Cathie are set out as 'pure', without any indication of the need for extraneous internal manipulations; the very fact that such are indeed required, when there does not appear to be any obvious pattern to them, is a concern, that may lead one to doubt the fundamental validity of all of Cathie's Harmonic Unified Equations as a result. At best they may be incomplete, and at worst, the values as produced by them are simply a gross abstraction of no physical consequence.

3) Is the 'M' is Cathie's Unified Equations really Mass, or is it something else entirely different?

Respecting in particular Cathie's proposed equation for mass – the fundamental basis of all of his noted equations – one is bound to wonder if the true identity of the variable 'M' as given is indeed mass. Admittedly, mass as a physical concept is hard to pin down. However, though physicists themselves may disagree somewhat as to what exactly mass is, they pretty much do all tend to agree quite well on what mass is not. And in the realm of physics, mass is not volume; a fact which is evident in light of the recognised equation for density, which holds that density = mass / volume. However that may be though, in Cathie's basic mass equation, the actual value he produces for the cubic volume of the Earth is often referred to many times in his works as a 'harmonic of mass' (e.g. *Bridge to Infinity*, p.40). But how can this be? In light of the relevant facts it does not appear that it can be. Rather, what he describes as mass would seem to be nothing more than simply volume. And indeed, that such a point is raised leads on to one very important further consideration: the issue of the equation actually balancing itself in terms of its physical units.

Under careful scrutiny it is easy to perceive that on one side of the mass equation there are two instances of speed (of light), whose physical units are both distance / time. But on the other side of the equation one has volume; an expression of distance, cubed. As a result, the overall equation just does not appear to balance in terms of its actual units: two instances of speed (albeit one slightly manipulated), equating to an instance of volume. That this would appear to be so is a strong indicator that the Unified Equations of Cathie either lack true physical validity, or that they are valid but that the variables themselves are not exactly what they appear to be.

An Overall Assessment

In view of the above it would seem then that Cathie's equations do not appear to be as strong as one might imagine. When viewed as actual physical equations, there is the distinct impression that the relations as specified are not as clear as they could be. In the basic Mass Equation for example there is doubt as to the precise identity of the 'M' variable. Is it truly mass, or is it volume, or indeed is it something entirely different, being a subtle product of speed? In consequence, there is the very real sense that Cathie's basic Mass Equation expressed in terms of pure light cannot be conceptualised as readily as when in its original form i.e. as the extended Leibniz equation. With

Cathie's alteration, it just seems unclear as to what the values as may be generated by the mass equation (including therefore all of Cathie's Unified Equations), would be representative of.

In view of the above there is thus a question mark over not only Cathie's basic Mass Equation, but also all of his Harmonic Equations. That being said however, to suggest that they are all somehow physically invalid and produce values of no consequence does not follow at all. And indeed, this present author certainly does not draw such a conclusion. But instead, he would remind the reader that Cathie has proved himself in so many ways as a result of his studies. His discoveries that pertain to the workings of nuclear weapons as are indeed of great relevance to this present work cannot be denied. More than this however, it would only be proper to defer to Cathie himself on his equations. One important reason for this will certainly be known to his readers. In a number of his books Cathie relates a visit by a mysterious stranger who came to his home to question him about his work [5]. The very same visitor is reported by Cathie to have actually asked him "point blank", after having looked at the values contained in his papers, if he had an equation that "tied everything together". Indeed, Cathie relates how after his visitor had departed he was inspired to re-examine his work to see if he could produce just such an equation. The basic Mass Equation as given is the very equation that resulted. Only Cathie himself would know the full significance of the remarks of his visitor, and just what he was looking at when he was prompted to question his host.

In addition to the above, it is only right also that one clarify a very important matter respecting the full evaluation of Cathie's equations. In actually presenting the various points of study, it is well to state that though the evaluation is a critical evaluation; in truth one could say fairly that it is also a very basic evaluation. For the points as raised, though pertinent, are quite simple. And they express just the basic concerns of (any) one who might set their mind to examining his equations. In all, the equations as are given are *physical equations;* and this present author is certainly no physicist, nor even an expert mathematician. And further to this, in *The Harmonic Conquest of Space,* one of Cathie's later books, the early chapters as lead in from his equations to his gravity and light tables are largely beyond the understanding of this present author. The significance of the information presented is not entirely clear. Indeed, this whole area of study that Cathie himself has explored i.e. the linkage between gravity and light and the periodic table of elements etc. has not really been touched upon at all in this present work. Rather an alternate methodology has been employed with a different focus.

In sum, upon the issue then of the fundamental validity of Cathie's equations, this present author does not feel himself able to pass any sort of decisive judgement at this time. Indeed, as will be quite obvious to the reader of this present work, no analyses within have been conducted at all making use of Cathie's equations. The reason for this being simply that the present author does not fully understand them. If one does not understand a set of mathematical equations, one certainly cannot use them, nor is one in any kind of position to truly judge them. In all, the presented evaluation as given is really a statement of the concerns of this present author, and in essence a list of the reasons why anybody may have difficulty in understanding the Unified Equations, as discovered by Cathie.

Appendix H – Critical Areas of Difference: Bruce Cathie & the Present Author

It goes without saying that this present work would not have been written if this author had not first come into contact with the work of Bruce Cathie. That being said however, the various analyses as given in this present work tend to differ significantly from those of Cathie, and generally approach matters from a somewhat alternative perspective. Admittedly though, many of the things examined by this current author would not have been, if they had not first been mentioned and evaluated by Cathie in his own publications. In light of this fact, it would be well therefore to explore more deeply some of the ideas and analyses set forth in Cathie's books, and to identify certain critical differences between them and the ideas and theories contained in this present work. In proceeding, it is assumed that the reader will have some knowledge of Cathie's work, as contained in the books noted below:

The Bridge to Infinity: Harmonic 371244
Adventures Unlimited Press (1997)

The Harmonic Conquest of Space
Nexus Magazine (1995)

The Energy Grid: Harmonic 695, the Pulse of the Universe.
Adventures Unlimited Press (1997)

The issues at hand as will be discussed, each in turn, relate to the following topic areas:

1) Fundamental units of measure as employed in harmonic analysis
2) The basic units of the Earth energy grid matrix
3) The speed of light and its various associations
4) The Earth magnetic field; its strength and density

On Fundamental Standards

In any detailed examination of Cathie's work, set against that of this present author, there are several key differences worthy of note, the very nature of which oftentimes results in significantly different mathematical associations and/or conclusions. In summary form, the main points would appear to be as follows:

Cathie:

1) All analyses conducted on a global scale make use of a spherical Earth model.

2) Primacy is given to angular measures swept out over a spherical Earth, making use of the basic numerical values of the sexagesimal system i.e. 360, 21600, 1296000.

3) The key distance unit held as a primary standard is taken to be the Nautical Mile, the length of which is set at 6076 British feet. The Nautical Mile unit is itself equal to 1 minute of arc.

4) No recognition appears to be given to the fact that the Earth as a physical body has in real terms suffered a general expansion.

The Present Author:

1) All analyses conducted on a global scale make use of an ellipsoid Earth model.

2) Primacy is given to actual arc length measures swept out over an ellipsoid Earth form, as opposed to angular measures.

3) The key distance unit held as a primary standard is taken to be the Ideal Geographical Mile, the length of which is set at 6000 British feet, as derived from the ideal Earth form.

4) Recognition is given to the fact that the Earth as a physical body has in real terms suffered a general expansion.

As a result of the fundamental differences noted between Cathie and this present author, there is at times a marked difference between the mathematical values as are derived from the study of various phenomena. Indeed, the differences as are evident in such cases are primarily due to the fact that Cathie would seem to be operating under the general assumption that British Imperial measures are all, in effect, too short. One may cite several pertinent examples of this fact from Cathie's works. In the *Harmonic Conquest of Space* for instance (page 35) he presents the following information:

6076 British Feet	=	1 Minute of Arc
6000 Geodetic Feet	=	1 Minute of Arc

Thus can one see how Cathie uses the minute of arc as an angular measure from which to derive a unit length composed of 6000 sub-units; referred to by him as geodetic feet. In another instance from the same publication (page 30) he analyses the distance separating the Proton and the Electron in a Hydrogen atom. The values that he presents are given as follows:

$208.5528148 \times 10^{-11}$	=	British inches
$205.9441884 \times 10^{-11}$	=	geodetic inches

If one were to divide the value in British inches by that given in geodetic inches, the resultant ratio is found to be identical to that as between 6076 / 6000:

$$208.5528148 \times 10^{-11} / 205.9441884 \times 10^{-11} = 1.0126666\ldots$$
$$6076 / 6000 = 1.01266666\ldots$$

It is quite clear then from the above that there is a major difference of opinion between Cathie and this present author, as to what would appear to constitute a fundamental standard of measure. In the case of the former, emphasis is placed upon the length of the nautical mile unit as derived from a single minute of arc angular sweep (latitudinal) over the surface of the Earth; the unit itself being split up equally into 6000 sub-units – what Cathie refers to as geodetic feet. In the case of this present author however, the primary standard is held to be a unit distance referred to as the ideal geographical mile, composed of just the same number of sub-units (6000), but with each being identical to a standard British foot; with the full unit itself based upon the physical equatorial circumference of an ideal Earth.

The Basic Units of the Earth Energy Grid Matrix

That there is a major difference of opinion as to what should rightly constitute the primary standard of distance measure between Cathie and this present author, has definite implications for the issue as to what is the true measure of separation between the basic units of the Earth energy grid. As noted in the early chapters of this work, it was related how Cathie uncovered the existence of an energy grid matrix based upon the flight patterns of UFOs. As a result of his studies he arrived at the conclusion that they appeared to move upon various 'track-lines' spaced at intervals of 7.5 minutes of arc both north-south and east-west. Critically though, to Cathie, 1 minute of arc = 1 nautical mile (6076 ft). And therefore (Re; latitude for instance), the intervals between the track lines as calculated would be 6076 × 7.5 = 45570 feet. However, in the view of this present author, such a conclusion would appear to be in error.

According to the findings of this present work, there has been a general increase to both the size of the Earth and also the length of the Earth tropical year. Of the latter it was shown that from a frequency perspective, the length of one solar day remained constant, as more days were simply added to the length of the year. Following this line of reasoning it was thus held that the noted Ideal Geographical Mile of 6000 feet, as applied to the Earth form, would itself hold true as a constant, even with the general expansion of the Earth. Consequently, upon the issue of the UFO grid matrix, the true intervals would by this reasoning be equal to 6000 × 7.5 = 45000 feet. And indeed, for confirmation that this is so (at least from a latitudinal perspective), one need only point to the secret US facility at Pine Gap in Australia, as was mentioned in Chapter 20. For there it was clearly shown that the displacement from the equator of the Earth to the centre of the main building complex is equal to almost precisely 1440 IGM, and *not* 1440 nautical miles. Indeed, the level of error is only on the order of some 200-300 feet in terms of IGM, whereas in terms of nautical miles, the level of error is some 18 (nautical) miles! This is powerful evidence in support of the notion that the true energy grid intervals are in fact based upon IGM units, and not nautical miles derived from minute of arc angular measures.

The Speed of Light

Following on from the above, one of the most important aspects of Cathie's work involves his decision to measure the speed of light in terms of nautical miles per grid second. Now, as clearly explained previously, the nautical mile is a distance unit derived from a single minute of arc angular sweep over the surface of the Earth upon the polar or elliptical circumference. Consequently, the nautical mile is therefore very much a unit measure associated with – or rather derived from – the physical dimensions *of the present Earth.*

Included in Cathie's book *The Harmonic Conquest of Space* is a basic value for the recognised light speed maximum in vacuum, expressed as 143891.3649196305 nautical miles per grid second [1]. Further to this though, and cited in a previous chapter of this present work, Cathie also demonstrates in the same book a linkage between both the speed of light and acceleration due to gravity. At the equator, Cathie shows that gravity acceleration is found to have an accompanying light speed value of 143791.3643 nm/ grid second, whilst at the poles where gravity acceleration is slightly greater, a corresponding light speed value of 143033.5767 nm/ grid second is to be had [2]. Consequently, all other latitudes between the equator and either pole are shown to possess a light speed value between the two noted extremes. Indeed, it is specifically pointed out by Cathie that depending upon one's position, the speed of light varies; a fact confirmed experimentally by Michelson, as previously noted.

When one considers the light speed values associated with gravity, including that of the recognised maximum – all expressed in terms of nautical miles per grid second – it is very easy to note that they all fall just under what would be an exacting figure of 144000 (nautical miles per grid second). As a result, it is not so hard to imagine that what is held to be the recognised maximum speed of 143891.3649 nm/grid second, is in fact not so, and that there is a still greater light speed of 144000 nm/ grid second. And this indeed, is precisely something that Cathie proposes in his work. To Cathie, a value of precisely 144000 nm/ grid second is the true maximum, and a speed that is associated specifically with the equator of the Earth [3]. But is this correct?

In light of the findings presented in this present work, it would appear that this assumption of a light speed maximum of 144000 nm/ grid second, is in error. The detailed evaluation of the Earth orbit and the determination of the proportional laws governing its transformation, indicate that whereas the numerical value of 144000 is associated with the measure of light speed, *this is only so when expressed in terms of ideal geographical miles per grid second*. Moreover, an actual maximum value of 144000 IGM/ grid second is found only to have existed (naturally) when the Earth and also the solar system as a whole, were in a most harmonious configuration in some very remote age. It is a key finding of this present work that just as all of the primary measures associated with the Earth have been transformed, so too has the speed of light. Indeed, in this present age, were light speed expressed in terms of ideal geographical miles coupled with grid seconds, its value would be 145713.3657 IGM / grid second. And given in standard seconds: 163927.5364 IGM / second.

With the above points raised it would be proper to inquire therefore as to the significance of measuring light speed in terms of nautical miles and grid seconds. Expressed in these units, does it possess any significance as such? From what has been demonstrated in the course of this work, certainly one can connect up the nautical mile unit to the key numeric sequence of 144; the proof, reproduced as below from a previous chapter. But such an association *is merely a harmonic or fractional linkage,* and does not tie in to the physical concept of speed (of light), in the view of this present author.

1 Nautical Mile = (21875.00126516 × 6000) / 21600 = 6076.38924032… feet

6076.38924032 × (8 / 7) = 6944.44484608… 1 / 6944.44484608 =

0.000143999991671…

In the view of this present author, all that the above relations reveal is that in terms of 'fractional harmonics' the nautical mile does have a certain significance; being a unit measure of some consequence. Indeed, as was demonstrated with the nuclear tests codenamed Inca and Mohawk, both found to be almost exactly 694.4444…nautical miles (of 6076.38924…) from the equator; unit intervals of nautical miles can be physically efficacious in certain situations. However, this would appear to be so only in a basic harmonic sense and not with reference to the concept of speed as such.

On the Earth's Magnetic Field Strength

In all three of Cathie's books, as previously listed, he presents an evaluation of the Earth's magnetic field. And, as a result, is able to tie in the field strength to certain harmonic intervals related to his extended grid structure, as based upon the noted combinations of great and small circles. In the course of presenting his analysis he usually begins by referencing a book called *Behind the Flying Saucers* by Frank Scully. Within this book it is detailed by Scully that the Earth has the character of a huge magnet wound up by magnetic lines of force, with two distinctive groups of such lines in existence. Coiled around the Earth, each group is found to travel in a different direction, and to possess a different density. Specifically, one group has 1257 lines to the square centimetre travelling in one direction, and in the other group 1850 lines to the square centimetre travelling in the other direction [4]. Such would seem to indicate then that overall the Earth's magnetic field is both of high density and asymmetrical in nature.

With the basic information provided by Scully, Cathie sets out to evaluate the density of the lines of force of the two noted groups. As related in *The Harmonic Conquest of Space* [5] he begins, in light of the fact of the curvature of the Earth, to propose a theoretical refinement to the values given by Scully, *specifically as would hold were the lines of force in each field counted within a unit area of one square geodetic inch*. Thus, the basis of Cathie's refinements would appear to rest upon the following values:

$1850 \times (2.54 \times 1.01266666\ldots)^2$
$= \mathbf{12239.73996}$ lines of force per square geodetic inch

$1257 \times (2.54 \times 1.01266666\ldots)^2$
$= \mathbf{8316.40710}$ lines of force per square geodetic inch

Designating the two groups Field A and Field B respectively, Cathie refines their expressed density levels to produce the following [6]:

Field A: **12245.69798** lines of force per square geodetic inch

Field B: **8317.32698** lines of force per square geodetic inch

Now, what is interesting about the refinements to the actual density of both fields is that they appear to rest upon an association to a certain value found to be built in to Cathie's extended grid structure. Indeed, to those familiar with his work, it will be known that the combinations of great circles and small circles of his extended grid establish a series of 'polar squares' as he calls them; one in each hemisphere. And, that it is upon one of the corners of these squares (the one in the southern hemisphere) that the Eltanin aerial is to be found. In his analysis of the squares themselves Cathie points out that were they to be laid out flat on a 2-dimensional surface, the side lengths of the squares would be found to be equal to 3600 minutes of arc. Consequently, the internal diagonals are determined simply by multiplying 3600 by the square root of 2. And furthermore, the distance between the centre of one of the squares and any one of its corners would be half this value:

$$3600 \times \sqrt{2} = 5091.168824$$

5091.168824 / 2 = 2545.584412… minutes of arc

From this point, Cathie further notes that the reciprocal of the above determined value, when adjusted to the appropriate harmonic level, is a match to the difference in density between the two noted fields, *as refined by Cathie himself* [7]:

$$1 / 2545.584412 = 0.0003928371006$$
$$0.0003928371007 \times 10000000 = 3928.371006$$

$$12245.69798 - 8317.32698 = 3928.371006$$

With the above thus set forth, detailing Cathie's refinements of the field density values of Scully, and of his linking the values to his extended grid structure, it would be well to evaluate such numerical associations in light of the findings of this present work. In doing so, it becomes readily apparent that one is able to offer an alternate solution to the significance of the asymmetric magnetic field values.

As already mentioned previously, it is the contention of this present author that the British Imperial system of measures is a far more exacting and important standard than the geodetic measures employed by Cathie. As a result of this, were this present author to conduct an analysis of the density of the two noted fields, it should be no surprise to find that it would be conducted in terms of British units. Thus, employing the standard British inch, instead of Cathie's preferred geodetic inch, the field strength values this present author would seek to refine, are as follows:

Field A = 1850×2.54^2
= **11935.46** lines of force per square inch

Field B = 1257×2.54^2
= **8109.6612** lines of force per square inch

Now, immediately upon viewing these figures, in light of all of the findings previously set forth in this present work, one cannot help but note that the two stated values are so tantalisingly close to being exactly 12000 and 8100 lines of force per square (British) inch. And moreover, suspect that the stated values have in all likelihood been transformed from such ideals, and in accordance with an exacting set of proportional laws. A very simple analysis, involving yet again the basic Earth year/equator ratio of increase, 1.014561622...would appear to confirm just this, with the following associations:

$1.014561622^3 = 1.044324077$

$\sqrt[8]{1.044324077} = 1.005435954$

$12000 / 1.005435954 =$ **11935.12122**

$1.014561622^3 = 1.044324077$

$\sqrt[32]{1.044324077} = 1.001356226$

$8100 \times 1.001356226 =$ **8110.985438**

From the above relations, a distinct set of physical proportional laws for the transformation of each field may be had as follows:

$$PE^3_{(e)} \propto \frac{1}{F^8_A}$$

$$PE^{3}_{(e)} \propto F^{32}_{B}$$

Where: F_A, F_B & $PE_{(e)}$ are ratios:

F_A = Magnetic Field A (present) / Magnetic Field A (past)

F_B = Magnetic Field B (present) / Magnetic Field B (past)

$PE_{(e)}$ = Physical Equator, Earth (present) /
Physical Equator, Earth (past)

As can be seen in the case of Field B, the law of proportion as given is identical to that which was found to transform the orbital period of the asteroid Ceres, from 1440 days to its present value of some 1680 days. Moreover, the direct linkage between Field A and Field B itself proves to be of still greater elegance and simplicity:

$$F_A \propto \frac{1}{F^{4}_{B}}$$

Where: F_A, F_B & are ratios:

F_A = Magnetic Field A (present) / Magnetic Field A (past)

F_B = Magnetic Field B (present) / Magnetic Field B (past)

This new association is most satisfying, considering that the transformation of the Earth physical form i.e. the relative expansion of its equatorial radius to its polar radius, is itself based upon a law of proportion containing the simple powers of 4 and 3. Thus indeed, that the transformation of the magnetic field density of the planet is based upon a 1 to 4 combination of powers is rather fitting. In view of the analysis presented of the Earth's asymmetric magnetic field, as based upon the stated proportional laws, it should be noted that the evaluation as given stands directly in conflict with that of Cathie. For if the values as refined by this present author are correct, then those of Cathie must surely be incorrect, and the tie in that he makes between the density levels of the two magnetic fields and his extended energy grid structure based upon the 3928.371006 harmonic, must also be false.

Appendix I – The Philadelphia Experiment: Bruce Cathie & the Present Author: A Comparison

In Chapter 19 it was noted that in his evaluation of the Philadelphia Experiment, as related in *The Bridge to Infinity*, Cathie presented an analysis of the co-ordinates between which the Destroyer vessel was teleported. Thoroughly in line with his grid system Cathie's analysis essentially was based upon angular measures, as one may note from his book. By contrast, the analysis as given by this present author is markedly different in respect of a number of points. And indeed, it would be well to consider the differences, especially in light of what has been said in the previous two Appendices G & H. To begin; in his evaluation of the two sets of co-ordinates, Cathie would seem to combine angular measures with a spherical Earth model, where 1 minute of arc is equal to 1 nautical mile. This point is critical.

In his review of the Delaware River latitude co-ordinate of the DE173 Destroyer vessel, Cathie notes that when converted from degrees into seconds of arc, a numeric sequence is produced that is linked in to the speed of light at the Earth's surface [1]:

Latitude: 39:56:35.77 (degrees) × 60 × 60 =
143795.77 seconds of arc

In further explanation of this value Cathie states that according to his research, the speed of light is not constant in the vicinity of the Earth. Indeed, at about one thousand miles up from the surface it is some 144000 minutes of arc per grid second, but that at sea level it is reduced to about 143795.77 [2]. With such a numeric association evident, it would appear then that Cathie considers this latter stated figure to be the defining value that actually determined the latitude placement of the experiment.

Concerning the physical distance between the two positions, Cathie once more points to the apparent significance of the *angular* separation (this time in degrees), and further links this to a 'square-root of 10' association. Once again, making use of a spherical Earth model and angular measures, he notes that the separation between the Penn's Landing site and Norfolk Naval Docks is on the order of some 189.7366596 minutes of arc. Dividing this by 60 to convert to degrees, one has 3.16227766 [3]; a figure that when squared is equal to the number 10. Of course, one would doubtless hold to the fact that such exacting values as these were not independently uncovered by Cathie, but that he has refined his figures to these measures as in principle, he does believe them to be the true values intended by the naval experiment.

In view then of the two associations as noted by Cathie, one can see a clear difference to those as given by the present author in Chapter 19. What then is the truth? With all due respect to Cathie, to the mind of this present author, the two noted associations that he gives are false. Upon the matter of the pure latitude measure, the second of arc angular displacement of 143795.77 as highlighted by Cathie, does not appear to be significant as a reduced speed of light harmonic as linked to the experiment in any way. For as one may note from Chapter 19, the actual arc length in feet is already linked in to the pure harmonic sequence of 144 via the divisor 6, without any need to consider a reduced light speed harmonic:

14511882.24 (feet) successively divided by 6:

2418647.04
403107.84
67184.64
11197.44
1866.24
311.04
51.84
8.64
1.44

With regard to the square-root of 10 association as given in degrees; again this would seem to be merely a fortuitous connection, with the 1152000 feet link appearing to be far more significant. Indeed, it is the view of this present author that Cathie *reasons out* just what the correct geodetic co-ordinates are for the Destroyer. The error that he appears to make however is primarily in choosing to use a spherical Earth model in conjunction with angular measures. And thus, with his reasoning correct as to the true positions, he makes a series of improper mathematical corrections/refinements that lead him to a set of false numerical associations. All things considered, the stated associations as given in Chapter 19 using an ellipsoid Earth model in conjunction with feet, are far more likely to be true than those as detailed by Cathie in his own work, *The Bridge to Infinity*.

Appendix J – Earthquake Prediction: A Strategy for Success

A General Overview of the Earthquake Problem

It is quite clear from the research presented in this current work that the physical science of earthquakes is very exacting, being based upon the well established movements of the celestial bodies of the immediate solar system. One would suspect therefore that it should be possible to develop some sort of computer model of the heavens that could predict with high precision a given earthquake as would strike the Earth, in terms of latitude, longitude, magnitude, and also time (GMT/Universal Time). Indeed, of such a proposed model, one would expect that it should be well able to accurately predict the future location of an earthquake to an error of only several hundred feet globally, and also be able to predict the time of impact many years in advance, to an error of just a couple of seconds of time! Provided that one were of course in possession of an exceptionally accurate astronomical model of all of the major bodies of the solar system, then theoretically such a feat would seem well within the realm of possibility.

With what may be theoretically possible stated, one must of course, consider just what would in fact be practically possible. And on this point, it would be well to note just how frequently the Earth is struck by earthquakes in the regular course of its being, and of the power of such earthquakes. According to the USGS (United States Geological Survey) in the course of a given year, the Earth is struck by almost 1.5 million low yield earthquakes. Specifically, of the range of 2.0 to 2.9, they estimate that about 1300000 earthquakes occur annually, whereas at 3.0 to 3.9, the estimate is 130000, and at 4.0 to 4.9, some 13000 [1]. Moving on though to the more powerful earthquakes, one is able to note that they are even less frequent. Of the range of 5.0 to 5.9, the USGS notes that annually, only some 1319 earthquakes strike the Earth. Of the still higher range of 6.0 to 6.9, some 134 occur. With only 17 striking the Earth per year within the range of 7.0 to 7.9 [2]. And finally, of earthquakes that rank 8.0 and higher, just one event of this magnitude will usually occur in the course of a full year [3].

In A General Theory of Earthquakes, as presented in chapter 18 of this work, it was explained how earthquakes were the result of nothing more than matter being dynamically transmitted from one or more celestial bodies to the Earth, at the moment of achieving certain propitious configurations in the heavens. With the physical matter itself carried on its way via an ordered set of hyperspatial pathways, its emergence, generally just beneath the surface of the Earth, forcing its way through into the same space as that already occupied by existing matter, produces a massive disruption to the geophysical state of the planet. And it is this very event, which constitutes an earthquake. Now, bearing this in mind, it would be well to note that there are a great multitude of celestial bodies within the immediate solar system that could all partake in a matter transfer to the Earth. The major planets are of course a given. However, one must also include the moons. And in total there are at present determined to be 169 [4]. In addition to this even one may note too some of the larger objects within the Asteroid Belt, such as Ceres. Given these facts, one is led to propose that the 1.5 million or so low yield earthquakes as would seem to regularly strike the Earth each year, are a result of the celestial patterns formed by the activity of the minor objects of the solar system. Essentially the moons (with the notable exception of the Earth's own moon) and perhaps also the largest fragments of the Asteroid Belt. In the case of high magnitude earthquakes however, especially those of 5.0 and over, one would thus expect that the activity of the major celestial bodies of the solar system is primarily responsible for their occurrence.

From a practical standpoint then, in accepting the truth of the above points, in the actual development of a computer model able to predict earthquakes to extreme accuracy into the future, one could not hope to realistically capture the precise details (of time, location and magnitude) of the vast multitude of low level earthquakes, 0 to 4.9 in strength. Indeed, one may almost consider the 1.5 million low yield hits to be no more than mere 'background noise'. And thus, it is the greater strength earthquakes that must be the main focus of attention; those of 5.0 and higher. Indeed, by restricting oneself to this band, one is concentrating upon those seismic events as truly stand out, a number of which may indeed be highly dangerous, and on occasion result in major loss of life.

The Development of an Earthquake Prediction Model

In light of the above noted considerations, in actually developing an earthquake prediction model, one may set as ones goal the creation of a model capable of correctly forecasting to extreme accuracy the future occurrence of level 5.0 magnitude earthquakes and greater; and this to an error of only some several hundred feet in terms of global placement of the epicentre, coupled with an error of several seconds of time in specifying the moment of occurrence. To those who have the capability and means to develop a model to achieve this goal, the following will be of aid; being a general description of the skills of the personnel as would be required, and also the main characteristics of the model:

Required Personnel

One would expect that developing the specified mathematical model would require the use of one or more expert mathematicians fully grounded in the intricacies of ellipsoid geometry, and also one or more very capable computer programmers not unfamiliar with the realm of astronomical modelling.

Of the Development of the Mathematical Model

1) The general character of the earthquake predication model would be that of a mathematical model of the outer surface shell of the Earth's ellipsoid form. The very centre of the Earth would be a fixed origin point of stillness, with the outer form of the planet modelled as being in motion in accordance with the speed of the axial rotation of the Earth. And thus, in essence the model would be geocentric in nature.

2) Combined with the above, one would seek to overlay the various point-like positions of the major celestial bodies of the solar system over the surface of the Earth ellipsoid shell. That is, through use of the most advanced model of the solar system, centred upon the fixed origin point of the Earth, each of the principle celestial bodies of the solar system would be modelled in terms of their noted 'Ground Positions' upon the Earth form, defined by a set of precise geodetic co-ordinates, at any given moment.

3) With a computer model of the heavens so established, one would add to it a certain capability, such that by specifying nothing more than a geodetic set of co-ordinates upon the surface of the Earth (latitude and longitude) and a Universal Time (GMT), the program would immediately produce a precise set of Ground Positions for all of the principle celestial bodies of the solar system, and in addition, calculate the surface arc lengths over the ellipsoid Earth form from each of the noted ground position co-ordinates to the prime point of interest, as originally specified.

4) With the above capability programmed, one would then need to select an authoritative earthquake database with a full list of all the past earthquake events has have been accurately recorded over the recent era, and feed the

data into the model. (And in this of course, one would select only level 5.0 earthquakes and greater). This will allow one to generate an extensive amount of raw data in the form of arc measures (expressed in British feet) as momentarily existed at the time of the noted seismic events.

5) With the data in hand one would then need to evaluate it, and in a most subtle manner. Essentially, one would need to create a further computer program to analyse the data and to search for patterns. Based upon all of what has been detailed in this present work, one would thus suggest that the analysis itself would seek to examine all of the arc length measures associated with a given earthquake to see if they are close in their numeric sequence to those of the primary sexagesimal series, included allied progressions. Not only this, but one would also examine the ratios between the noted arc measures associated with a given earthquake. Again, one would be looking to isolate certain basic ratios, whose numeric sequences are also linked to the primary sexagesimal series and its allied progressions.

6) In an evaluation of the data, one is seeking ultimately to identify a correlation between various arc measures linked to an earthquake, and its specified level of magnitude. And to discover just precisely how the various arc length values combine to produce a distinctive signature as would appear evident in the case of any high magnitude seismic event, of 5.0 or greater. One would expect also that with a physical signature isolated, that a general set of equations would emerge to characterise it.

7) Once one has the above in hand, one must return to the primary model, and program it to search for future celestial configurations as will arise relative to any given geodetic point upon the Earth, that encode the critical earthquake signatures. And thus, will one be able to state in Universal Time (GMT) to the nearest few seconds or so, that an earthquake of a specified magnitude will strike at a specific location, as identified.

The above description as given is of course only a basic outline as to how one might proceed to develop an accurate computer program as would forecast future earthquakes. A certain programming genius is required along with perhaps an intuitive insight into exactly which patterns to look for and ultimately program into the model. Indeed, the whole process as briefly sketched is far from defined in any truly exacting manner. And thus the whole venture of developing an earthquake prediction model of this type will be experimental in its essential nature.

Of Nuclear Harmonics and Earthquake Harmonics

In one final point upon the matter of earthquake prediction, and of developing a computer model to achieve it, one may wonder indeed if there is some sort of fundamental physical difference between the arc measures associated with natural earthquakes, compared to those of engineered nuclear events. At least from the sample of nuclear tests evaluated in this present work, and also the natural earthquakes considered, the arc length measures and their associated ratios do appear on the surface to be entirely interchangeable and equal in terms of their 'quality' (if that is the right word). And yet in truth this cannot be correct. For indeed, when one considers the nuclear weapons testing of the major military powers, whenever they have conducted a test, there does not occur simultaneously an earthquake at exactly the same time where their devices have been set up. What this implies is that the celestial harmonics utilised for generating an artificial explosion are not as powerful as the celestial harmonics responsible for creating truly natural earthquakes. The former when achieved or realised in the heavens, would seem only capable of momentarily 'weakening' a given locale in space and time, with the nuclear device itself upon activation producing the critical breach in that vital threshold separating this realm from that of the sub domain region; just enough so as to allow the powerful forces of that realm to flood into this one. By contrast, with a seemingly inherent ability to cause a natural breach in the aforementioned energy threshold, the celestial configurations as produce 'authentic' earthquakes, one is bound to suspect, must involve far more complex patterns in the heavens. In noting this fact, one is thus led to conclude that upon the issue of natural earthquakes per se, from the least powerful right up to the most powerful, a whole series of critical energetic thresholds must permeate the entire range. And of course one would have to actively discover them, and isolate the quality and the type of the various celestial relations as would define each range, and also the measure of their power.

Appendix K – Japan under Nuclear Attack 2007?

Japan at the Mercy of a Financial Oligarchy?

In further consideration of nuclear weapons and earthquakes, one is perhaps bound to concede the point that the seismic power of many large natural earthquakes can be actively mirrored by underground nuclear weapons testing. And that someone could if they so desired, covertly attack a country with the force of a genuine earthquake with seemingly complete deniability; masking a nuclear strike under cover of what would appear to be nothing more than a mere natural disaster. Indeed, such a strike would not necessarily require that the aggressor secretly enter the target country with a specialist team and an actual nuclear weapon, covertly dig a deep shaft at some viable energetic location, and then activate the weapon by remote at the moment of a favourable pre-determined celestial configuration. For as was revealed previously, the technology in this area as possessed by certain powers, is so far advanced, that the capability exists to generate a nuclear explosion at one location, and then actively transmit it in total to a completely different location. The celestial harmonies of the major bodies of the solar system do of course govern the whole process, both of actually generating the explosion, and determining the precise location and time of its emergence from the hyperspatial pathways as actively convey it to its target site.

It is with the above in mind, that this present author notes with some interest a news story that appeared on the internet some time after July of 2007, which suggested that Japan had been the victim of a covert attack, though use of what the author of the story referred to at the time as an 'earthquake machine', responsible for causing two high magnitude earthquakes to strike Japan in the region of Niigata, 16 July 2007. In full consideration of the story, it is well worth noting just how it emerged. To begin, the primary source was a journalist by the name of Benjamin Fulford. Originally from Canada, Mr Fulford came over to Japan when he was 19, and worked in the country as a journalist for about 20 years, being the bureau chief for Forbes Magazine from 1998 through to 2004-2005 [1]. In addition to this, in the course of his work he has had published some 15 books written in Japanese, with sales of over half a million [2]. Now, it was in the course of conducting a video taped interview with no less a figure than the former Finance Minister of Japan, one Heizo Takenaka, that Fulford obtained the admission from both Mr. Takenaka and his envoy that the reason why, according to Fulford, control over the entire financial system of Japan had been turned over to 'a group of American and European oligarchs', is that Japan had been threatened with an earthquake machine [3]. In a recount of the interview sometime later, which is indeed widely available on the internet for viewing at this time [4], Fulford further stated, that as a result of exposing some of the dealings of the oligarchs in question, that he was told by the Japanese Security Police, that because of some of the things that he had said, on places such as rense.com (on the internet), that Niigata city would be hit by an earthquake. Just two days after receiving this information, quite remarkably, Niigata was indeed struck, and not by just one quake, but two in succession only several hours apart, on the 16th of July 2007 [5]. Now indeed, considering the intriguing nature of this story, it would be well to examine very carefully the associated celestial configurations of both events, beginning with the first of them [6]:

Event: *Niigata (Japan).* Magnitude: 6.7
Date: 16 July 2007 Time: 01:13:28 (GMT)
Geodetic Epicentre Co-ordinates (Deg):
Lat: 37.584 N Long: 139.378 E
Geodetic Sun-Ground Position Co-ordinates (Deg):
Lat: 21.58317728942 N Long: 163.1333333333 E
Geodetic Moon-Ground Position Co-ordinates (Deg):
Lat: 19.28498673340 N Long: 176.245 W
Geodetic Mars-Ground Position Co-ordinates (Deg):
Lat: 15.15641173947 N Long: 91.10666666666 E
Arc Length Measures associated with the event:
Equator to Latitude of Quake (feet): = 13652641.470
Sun-Ground Position to Quake (feet): = 9484004.360
Moon-Ground Position to Quake (feet): = 15556413.106
Mars-Ground Position to Quake (feet): = 17537419.881

In an analysis of the primary arc length measures as given, one may note a series of very prominent associations, beginning with the arc length up from the equator to the latitude of the earthquake; simple multiplication of the basic value by the number 3 instantly revealing a link to a most recognisable mathematical progression:

13652641.470 × 3 = 40957924.411
40957924.411 / 8 = 5119740.551
5119740.551 / 8 = 639967.568
639967.568 / 8 = 79995.946
79995.946 / 8 = 9999.493

Ideal = **10000**

Further to this one cannot help but point to the obvious significance of the ratio between the latitude arc length and the epicentre to sun-ground position:

13652641.470 / 9484004.360 = 1.439543989

Ideal = **1.44**

Also, one may consider the arc measure of Mars, and its very own standalone link to a major variation of the primary sexagesimal set, once more via a simple initial transformation involving the prime number 11:

17537419.881 / 11 = 1594310.898

1594310.898 / 9 = 177145.6553
177145.6553 / 9 = 19682.85059
19682.85059 / 9 = 2186.983399
2186.983399 / 9 = 242.9981555
242.9981555 / 9 = 26.99979505
26.99979505 / 9 = 2.999977228

Ideal = **3**

In addition to the above, upon the issue of the most basic number of the primary sexagesimal set, i.e. 6, one can see a most striking link between the Mars arc as first modified by the number 11, and the previously noted epicentre to SGP arc length associated with the 26 December 2004 earthquake, as refined:

1594310.898 / 26572050 = 0.0599995445

Ideal = **0.06**

And finally, of the associated moon arc one can suggest a very basic and obvious connection:

15556413.106 × 9 = 140007717.959
140007717.959 / 7 = 20001102.565

Ideal = **20000000**

As a result of the above analysis, one is able to derive a complete set of values actually representative of the optimum or ideal earthquake signature. They are as follows:

Latitude arc: 40960000 / 3 = 13653333.33333… feet

Epicentre – Sun ground position:
13653333.33333… / 1.44 = 9481481.481… feet

Epicentre – Mars ground position:
(26572050 × 0.06) × 11 = 17537553 feet

Epicentre – Moon ground position:
140000000 / 9 = 15555555.555… feet

With the ideal arc length measures given one is thus able to note the differences between them and the model values (feet):

Latitude arc = 691.862
EC – Sun GP = 2522.879
EC – Mars GP = 133.119
EC – Moon GP = 857.551

Given the above, one can clearly see that a truly exception celestial configuration was achieved in the heavens at the moment of the first of the two noted earthquakes to strike Niigata; but what though of the second? Does it too reveal a similar type of harmony? Indeed it does, and most especially in terms of its latitude measure and also associated Mars arc length, just as with the first. But not only this, it is most interesting to evaluate also the arc length distance separating each of the two earthquakes [7]:

Event: *Niigata (Japan).* Magnitude: 6.8
Date: 16 July 2007 Time: 14:17:34 (GMT)
Geodetic Epicentre Co-ordinates (Deg):
Lat: 36.785 N Long: 134.85 E
Geodetic Mars-Ground Position Co-ordinates (Deg):
Lat: 15.27104663905 N Long: 105.08166666666 W
Arc Length Measures associated with the event:
Equator to Latitude of Quake (feet): = 13361716.583
Mars-Ground Position to Quake (feet): = 37724364.905
Arc Length Separation of the Niigata earthquakes (feet) = 1350672.797

Concentrating upon the above noted measures of the second Niigata earthquake, one is able to produce still further important links. One may begin with an analysis of the latitude arc; the decisive link as would lead to its refinement, appearing to rest upon a ratio whose numeric sequence is the same as the expressed measure of the full elliptical circumference of the current Earth, at 131250000 feet, and also, the already noted EC–Mars arc length of the first of the Niigata earthquakes:

17537553 / 13361716.583 = 1.3125224510
1.3125224510 / 7 = 0.18750320729
0.18750320729 / (3 × 3 × 3) = 0.0069445632329
1 / 0.0069445632329 = 143.99753684

Ideal = **144**

In addition to the above, one may note further a simple refinement of the Mars arc of the second earthquake, resting once more upon the basic prime number 11:

$$37724364.905 / 11 = 3429487.718$$

$$3429487.718 \times 9 \times 9 = 277788505.213$$
$$(1 / 277788505.213) \times 4 = 0.0000000143994439112$$

Ideal = **0.0000000144**

And finally, what is of still greater interest is what would appear to be a very intriguing connection between the very epicentres of both events. The direct arc length separation can be evaluated as follows:

$$1350672.797 \times (8 / 3) = 3601794.125$$

Thus can one see the simple fraction of 8 / 3 or 2.6666666666... at work in transforming the arc value into one of the primary sexagesimal sequence. And from this one can thus achieve the following:

$$3601794.125 \times 4 = 14407176.502$$

Ideal = **14400000**

With the above values thus analysed, one may present a summary of the refined arc length measures, as may be derived by the suggested ideals:

Latitude arc: 17537553 / 1.3125 = 13361945.142857... feet

Epicentre – Mars ground position:
= (4 / 0.0000000144) / 81) × 11 = 37722908.093... feet

Arc Length Separation of the Niigata earthquakes:
3600000 × (3 / 8) = 1350000 feet

Difference between the refined ideals and the model values (feet):

Latitude arc = 228.559
EC – Mars GP = 1456.812
Arc Length Separation = 672.797

Once more only a very small discrepancy is to be had with the noted values, such that one would have a good measure of confidence in them.

HAARP Facility Implicated in the Niigata Earthquakes

In view of the writings of T. E. Bearden as discussed in Chapter 17, it was noted that in the course of his work he did indeed name a specific facility in Russia as was the location of a nuclear howitzer, actively capable of transmitting a full nuclear explosion to a desired target location. Now indeed, in the case of the Niigata earthquake events, Benjamin Fulford, in an interview conducted by both Kerry Cassidy and Bill Ryan (of Project Camelot) sometime after the two earthquakes, did note that of his own previous interview with the Japanese finance Minister, that Mr. Takenaka himself had told him specifically that the decision to hand over control of the financial system of Japan was taken because the United States had threatened to hit Japan with HAARP if they didn't [8]. HAARP then, is the very facility named by Fulford as likely being responsible for causing the earthquakes in Niigata, 2007. Is there any evidence of this however? An examination of the facility itself may well reveal the answer.

The HAARP Complex

To those unfamiliar with the term HAARP it would be well to begin by noting that it refers to a specific technical facility located in the U.S. state of Alaska, and stands for High Frequency Active Auroral Research Program. One may also note that the very facility itself is jointly managed by both the Air Force Research Laboratory and the Office of Naval Research [9]. And as such, the program does possess obvious military ties. That there is indeed truth to this is undeniable, as the whole HAARP project was initially built to realise certain military-defence capabilities as outlined in a series of patents filed in the USA, beginning in the 1980s.

So what then does the HAARP facility consist of? Basically, an array of 180 antennas spread out over an area of some 35 acres [10], capable of being able to generate a focused beam of radio waves that may be projected into the

Earth's atmosphere; specifically the ionosphere. In conjunction with its integrated transmitters, the total radiated power capability of the array is about 3,600 kilowatts [11]. Now indeed, one must state that the basic technology behind the US HAARP device is not new. There are in fact several such facilities in various parts of the world of a similar nature, that can also beam radio waves up into the ionosphere. In general they are referred to as 'ionospheric heaters', due to the fact that their basic function of projecting radio waves into this region of the Earth's atmosphere tends to heat it up. That being said however, the HAARP facility does possess certain key features not present in other similar arrays around the world. Firstly, the primary inventor behind the HAARP facility, one Bernard Eastlund, specifically chose a phased array antenna for the complex, which allows one to precisely control just where the beam, and thus the power, is directed [12]. Also, in contrast to other heaters, which tend to just produce a 'standard beam' of a generally uniform nature, HAARP has the capability of actually focusing its own beam. It can squeeze it into a tiny area, and thus transform what is a 3.6 megawatt beam into a billion watt beam [13]. In addition to these points one may also note that compared to other ionospheric heaters, the HAARP complex is of much greater power. To give an idea of just how powerful HAARP is, one may note the words of Bernard Eastlund, the primary inventor, who indicates that were it beamed for about an hour and a half, that the energy would be equal to that of a hydrogen bomb [14].

In one final comment upon the HAARP facility, one must in all fairness state that it is not a 'top secret complex' as such, in the sense that it is unacknowledged, as for example the Area 51 facility in the US state of Nevada was for so many years. HAARP is 'on the map' so to speak, and does have its own official website on the internet [15]. Moreover, it has public relations people who have even appeared on various television documentaries to discuss the work carried out at the HAARP station. In addition to this it also has ties to various universities, and indeed quite a number of scientists do work there in an ongoing capacity conducting research. There is no doubt therefore that the HAARP complex does conduct a great deal of legitimate research into the workings of such as the ionosphere, and does contribute to a general understanding of the functioning of some of the key atmospheric layers that surround the Earth. That being said however, HAARP does have its critics. There are quite a number of serious researchers who have very carefully studied the facility and its capabilities, and have come to the conclusion that as an energetic device, it could potentially be very dangerous to the Earth. And not only that, but that the radio waves that it is capable of generating can interfere with the mental state of living organisms, including both animals and also humans.

To understand more about some of the concerns raised by the critics of the HAARP facility, it would be well to examine just how the device interacts with the Earth's atmosphere, and also some of its capabilities as first outlined in some of the patents originally filed with respect to the array.

To begin, the HAARP device was primarily built to interact with the ionosphere; an atmospheric region of critical importance to the Earth, shielding the planet from some of the more intense rays of the sun as would be damaging to life. Beginning at some 30 miles up from the surface, the ionospheric band stretches up to about 300 miles into space, covering the whole globe. A continual stream of solar radiation as impacts upon this region causes the neutral gas atoms present to be ionised; to be stripped of their electrons and so become positively charged. The whole ionospheric band is thus mostly comprised of a sea of ions and free elections, the latter of which can strongly reflect radio waves, being of great importance for long distance communications.

In beaming radio waves into the ionospheric region then, just what is it that HAARP can actually achieve? Well, according to Bernhard Eastlund, the original ideas and patents concerning the array detail capabilities such as being able to destroy missiles, to control and disrupt communications, including also modification of the weather; and in addition, to lift a part of the upper atmosphere further out into space where it was hoped that it could interfere with the flight paths of missiles [16]. Perhaps of even greater interest than these suggested capabilities though is the fact that the HAARP array, in beaming its radio waves into the upper atmosphere, can interact with a specific portion of it known as the Auroral Electrojet, which is an intense stream of electric current flowing through the lower layers of the ionosphere at high latitudes near to the Earth's polar regions. In doing so, it can generate extremely low frequency (ELF) waves that can then return back to the Earth and penetrate deep into the interior of the planet over great distances. But what can this achieve specifically? Well, essentially, through use of ELF waves, which are characterised as energy waves of between 0 and 1000 cycles per second, one can effectively probe the Earth, to detect such as mineral deposits or metals of various types. Or indeed detect the presence of cavities or hidden tunnels. In addition to this, one may also point towards the capability of communicating with submarines at sea. Such abilities as these may seem perhaps quite benign, or one may say rather legitimate. And indeed to an extent this present author would agree. However, what is also of note is that the ELF waves that HAARP is capable of generating are within the very same range as those that govern human mental processes. And it is this one fact that is greatly disturbing to many people. For quite a number of researchers, experts in the field of bio-energetics and the interaction between light waves and the human body, as a result of this capability, have directly suggested that the HAARP array could potentially be used for mass mind control over vast swathes of the globe, by producing exactly the sorts of light wave frequencies as may induce any desired emotional state, such as anger, depression, grief, happiness, elation, etc.

Truly then there are many aspects to the HAARP complex as would trouble the mind deeply. And thus it would be very wise for further investigations to be made as to its true capabilities, and to consider stringent controls to protect the populace from abuse were the suspected and suggested capabilities of the device to be as above outlined. That being said however, the immediate concern of this present work is not that of mind control, but rather whether or not HAARP is capable of causing extreme seismic disturbances.

Can HAARP Cause Earthquakes?

Perhaps the most detailed study of the HAARP array of the recent period has been the book co-authored by Dr. Nick Begich and Jeane Manning, *Angels Don't Play this Haarp*. In this work the authors present an in-depth study of some of the key patents as relate to some of the more exotic features of the HAARP array. Within this publication it is noted early on that a particular document linked to the HAARP array states that ionospheric disturbances have been both detected and ascribed to earthquakes. Indeed, the book notes a case in point; an Alaskan earthquake as occurred on 28 March 1964 [17]. Further to this, the point is also made that earthquakes do cause the ionosphere to react electromagnetically [18]. Which of course raises the question, could the reverse be true: could an ionospheric disturbance cause an earthquake? Now indeed, upon this point, one should note of the HAARP planners, that in their own literature, there was a stated intention in interacting with the ionosphere to generate a "runaway" effect [19]. This particular reference is to this present author most intriguing. And in light of it one must recall the essential functioning of a nuclear weapon. Upon implosion the very device itself is specifically designed to achieve a runaway effect, as it reaches a supercritical state.

At least from the basic assessment as given, one may suggest then that if a focused radio wave beam from HAARP were capable of disturbing the ionosphere in such a controlled fashion so as to replicate the energetic signature of a supercritical nuclear device, then theoretically, it is possible that the HAARP array could trigger a nuclear event in the form of a subsurface earthquake. However, in doing so, one would of course be forced to concede that the same limitations as apply to nuclear weapons apply also to HAARP. Essentially, and as discussed previously, just as a nuclear device is not capable of directly producing a massive explosion of its own 'power', but can only tap into external fields and energy stores to produce such an extreme physical outcome, neither could HAARP. In terms of its basic functioning then, were HAARP to be employed to cause an earthquake, the scientists behind the device would have to forecast a propitious celestial configuration relative to a specific target site, and then fire off a pulse of energy into the ionosphere directly above the site at the exact moment when the configuration reached its optimum level. Theoretically one would assume that this action would somehow stimulate the hyperspatial pathways linking up the relevant bodies to the Earth at that moment, in order to draw from them a portion of their matter, which would then be channelled to the Earth, emerging under the target site as an earthquake.

In addition to the above, and to complete the analysis of the HAARP array in terms of its capabilities, one must consider also the reach of the facility. Even though possessed of a phased array which does offer some control over the direction of the radio beam and the ability to focus it, one would still be inclined to think that HAARP could only disrupt the ionosphere generally just above the facility itself in Alaska. But is this true? A careful review of certain patents associated with HAARP would suggest that such a limitation has been overcome. And one may consider US patent number 5202689 in this regard, entitled "Lightweight Focusing Reflector for Space" [20]. This particular patent appears to make specific reference to the ability to use reflectors to "redirect a microwave power beam" [21]. A space based reflector could thus actually bounce a focused energy beam as projected up from the Earth, such as HAARP is capable of generating, and transmit it back down to another target location over a completely different part of the globe. Were one thus to have a series of such reflectors carefully positioned in space, it is not hard to imagine then that the HAARP array could be used to disrupt the ionosphere above any region of the world.

Was HAARP Responsible for the 2007 Niigata Earthquakes?

From the analysis of the celestial configurations evident at the time of both the primary and secondary Niigata earthquakes, one would have to suggest that were HAARP to have been involved in actually triggering the quakes, it would undoubtedly have been via artificial stimulation of the ionosphere over both sites at just exactly the right moment, and that an over the horizon capability would have been employed to achieve this; one would suspect making use of space reflectors. In light of this, one can then present a theoretical case to the effect that HAARP is indeed capable of being able to trigger earthquakes in much the same way as nuclear weapons, and this being so, that it would have had the potential to cause the two noted Niigata earthquakes. And yet, that being said, one may offer up an alternative scenario to account for the two earthquakes, one that in fact puts HAARP in the clear.

Japan: Victim of an Elaborate Con?

The essential link as given between the Niigata earthquakes of 2007 and HAARP as the instrument of causation, is the sole interview as recounted by Benjamin Fulford that he had with Heizo Takenaka, the former Japanese Finance Minister. Now, in studying some of the video footage of Fulford with reference to the HAARP threat, this present author is very much inclined to believe the truth of what he says. And that he has not fabricated the content of his interview with Mr. Takenaka, but has indeed faithfully reported the fact that the former Finance Minister had told him that certain US interests had threatened Japan with HAARP, and of artificial earthquakes via the device, had various economic policies not been adopted against the interests of Japan. Further to this, this present author is also inclined to believe the veracity of the statement made by Fulford that he was warned of Niigata being hit by earthquakes, and that indeed just two days later it was so struck. But does any of this mean that the earthquakes in question were artificially caused, or that HAARP specifically was responsible? Not necessarily.

In the previous Appendix (J) it was noted that there would appear to be a crucial difference between the celestial configurations whose 'energetic signatures' may be employed to trigger nuclear explosions, and those characteristic of authentic natural earthquakes, far more powerful in nature. Now indeed, in their desire to develop nuclear weapons, the various military powers upon the Earth would doubtless have studied very carefully the celestial patterns as are found to be evident at the time of natural earthquakes, as an aid to their nuclear programs. And thus, it would be very naïve to think that in doing so they would not have developed over the years an extremely accurate model capable of forecasting high magnitude seismic events to an exceptional level far into the future. Accepting this as an established fact leads to one very interesting possibility; namely that those who have acquired this knowledge would be in a position to be able to use it opportunistically to further their own interests by conning those unfamiliar with the essential science.

Officially, as far as the mainstream 'experts' are concerned, in the realm of geology, it is impossible to predict earthquakes with any degree of accuracy. And indeed, with such a major disconnect between the politicians that comprise regular governments, and the deep black world of advanced military science, it is generally the official experts of the mainstream world that regular politicians turn to for an understanding of the current state of scientific progress, for most would not know any better. Consequently, one is thus bound to wonder at the meeting that various Japanese officials had with those who, according to Fulford, attempted to blackmail the country. And in this, one might suggest just what the key objectives of the latter group would have been. Essentially, of the oligarchic interests, one would assume that they would have had three immediate goals. Firstly, to threaten Japan with artificially induced earthquakes, but in such a manner so as to insure that the Japanese representatives would doubt that they could deliver on their threats. Secondly, to deliberately provide them with a very distinct time and location as to when the attack would occur assuming non-compliance, and thirdly, to 'throw out' a plausible name i.e. HAARP, of a facility that one might imagine was capable of causing earthquakes.

In this scenario it is thus supposed that the noted oligarchic interests, with full knowledge of the highly advanced science of earthquakes, had possession of a computer model of the heavens that allowed them foreknowledge of the exact details of a twin set of natural high magnitude earthquakes as would shortly strike Japan. Now because they were not triggered by them, they certainly could not prevent their occurrence. But they could use knowledge of their occurrence opportunistically in an elaborate con game. After the meeting had taken place between both the Japanese officials and the oligarchic interests; following the Niigata earthquakes, which one would presume were not long afterwards, the Japanese would no doubt have examined very carefully the HAARP complex. And in doing so, they would – as was intended – have practically convinced themselves that they truly had been attacked. For indeed they would have consulted their 'experts', who would have told them with supreme confidence that it is impossible to predict the occurrence of any earthquake. And they would have also examined the related documents surrounding the capabilities of the HAARP array, which 'conveniently', because of the fact that the facility is not a top secret installation, would have been very easy to obtain.

In one final point with regard to the suggestion that the 2007 Niigata earthquakes were entirely natural, it would be well to note just how unusual a twin set of high magnitude earthquakes is for the country of Japan.

Consulting various records it would appear that over the past 10 years or so (as of writing) Japan has been struck by about 100 earthquakes of magnitude 6.0 and higher. Further to this though the reader might find it to be of great interest that in 2004 the Niigata region specifically suffered no less than 3 level 6.0 and above earthquakes all within the space of some 38 minutes on the 23rd of October! Now, were they all an attack? Just how would one know for certain? Indeed, should one assume that Japan is being covertly attacked on a regular basis? Unless one knows the true qualitative different between the energetic signatures of natural earthquakes, and those signatures of a far weaker nature that may be exploited via artificial stimulation to trigger nuclear reactions, it would be practically impossible to know for certain the difference between an authentic earthquake and what may well be an attack.

Two Possibilities

As a result of the above considerations, one can see that there are in fact two ways to account then for the Niigata earthquakes of 2007:

1) Japan was in fact threatened and deliberately attacked by the US using HAARP, which according to the evidence may very well have the ability to induce earthquakes via controlled ionospheric heating.

2) The earthquakes were not in fact artificially caused at all, but that a certain group of interests with advanced knowledge of the science of earthquakes had been able to successfully forecast the natural occurrence of the Niigata quakes, and with this knowledge they then proceeded to con certain Japanese officials into economic surrender.

In view of the two possible scenarios, it would only be fair to note with respect to the point made by Fulford that the threat to Japan came from the United States and European interests, that one should not be so naïve as to think that this would imply that any official governments as such would have been behind the threat. The key term used in some of Fulford's accounts referencing his interview with Mr. Takenaka is: oligarchic interests. And indeed such interests as characterised by the term oligarchic care nothing for any nation state. They do not even believe in the actual 'legitimacy' of nations at all. Representing a transnational force that would seek to use governments at will to service their own ends, they are aligned with that most fantastic tradition; that of the 'Nietzschean superman'.

Appendix L – The Echelon Matrix

The US Arm of Echelon

In Chapter 20 of this work it was revealed that various US facilities have been constructed over the course of the past few decades that would appear to have a distinct energetic association to such ancient structures as the Great Pyramid and Stonehenge. Indeed, two sites in particular were noted: Pine Gap in Australia and Menwith Hill in Great Britain. And moreover, of them both it was well stated that they are manned almost exclusively by US personnel, and as certain researchers have suggested, that both form a part of a large global surveillance system. Given this one is of course bound to suspect that the US would undoubtedly possess a similar base in its own native territory, which would itself be linked-in globally to the other two bases in Great Britain and Australia. Such indeed would appear to prove true. The critical base in question located within the United States as would appear to be quite possibly the main US arm of the Echelon system, may be identified as Buckley Air Force Base. The global co-ordinates of the base are as follows [1], with a restatement also of those of Pine Gap and Menwith Hill:

Buckley AFB:	Latitude: 39:42:06 North Longitude: 104:45:06 West
Pine Gap:	Latitude: 23:48:03.6 South Longitude: 133:44:12.75 East
Menwith Hill:	Latitude: 54:00:27.18 North Longitude: 01:41:18.42 West

In the previous analysis of the latter two noted bases it was shown that the arc length distance of separation between them was almost exactly 3 / 8 of the full equatorial circumference of the Earth. Now indeed, if Buckley AFB were itself a part of a unified system then one would expect that it would be very carefully tied in to the other two facilities. And that this would be via the key values of the basic sexagesimal system. Such does indeed prove to be true. And in fact the relations themselves as may be had are truly astounding in terms of their exacting nature:

Under the GP2007 Earth Ellipsoid Model:

Arc Distance between Buckley AFB & Pine Gap:
= 46966245.577 feet

Arc Distance between Buckley AFB & Menwith Hill:
= 23962358.251 feet

Ratio: 46966245.577 / 23962358.251
= 1.960000976732

With: 1.960000976732 / (7 × 7) = 0.040000019933

And: 0.040000019933 × 6 × 6 = 1.440000717599

Ideal = **1.44**

In terms of a suggested refinement to isolate what would have been the intended arc measures between Buckley AFB, Menwith Hill and Pine Gap, one would note the following connection:

46966245.577 / (8 × 8 × 8 × 8 × 8 × 8 × 8 × 8 × 8 × 8)
= 0.043740724751

With: 0.043740724751 / (9 × 9 × 9)
= 0.000060000994172

Ideal = **0.00006**

Reversing the mathematics from the noted ideal of 0.00006 allows one to generate a theoretical ideal for the arc of separation between Buckley AFB & Pine Gap:

46965467.38176 feet

With: 46965467.38176 / 1.96 = 23961973.15395 feet

From these two refined values then the global errors are as follows:

Buckley AFB to Pine Gap = 778 feet

Buckley AFB to Menwith Hill = 385 feet

In view of the accuracy of the above relations and their numerical significance, there is no doubt that all three noted facilities must be a part of a unified system, and that they have been sited with the same considerations as would have governed the placement of both the Great Pyramid and Stonehenge. But to go further than just the suggestion that they are involved in what may be termed 'mundane' electronic surveillance; in the view of this present author, it is quite likely that the three bases in question are covertly engaged in certain activities of a highly advanced nature that one must suspect involve hyper dimensional capabilities.

References

Introduction

[1] Russell, B. (1995) p.224 — *History of Western Philosophy*
Published by Routledge

[2] Freke, T. & Gandy, P. (2000) p.19 — *The Jesus Mysteries*
Thorsons

[3] Russell, B. (1995) p.225 Ibid. — Eratosthenes had the estimate of 7850 miles for the diameter of the Earth; a figure only fifty miles or so short.

[4] A theory by Christopher Dunn. See: *The Giza Power Plant* (1998), Bear & Company

[5] Cathie, B. (1995) p.97 — *The Harmonic Conquest of Space*
Nexus Magazine

Chapter 1

[1] Childress, D. H. (1997) p.5 — *Anti-Gravity and the World Grid* (Editor)
Adventures Unlimited Press

[2] Concerning the reality of the UFO (Unidentified Flying Objects) phenomenon:

> There are many books in existence that have examined this question thoroughly, and to date it has been well established to the satisfaction of more than enough researchers that this phenomenon is real. Thus, research has gone well beyond the simple question of its validity, and to the study of what lies behind it. Of course it is true that the vast majority of 'sightings' or encounters are of mundane objects such as relatively common atmospheric disturbances, or planets mistaken for something more than they are. However, there is that small percentage of cases that cannot be so easily explained away, and which do truly constitute genuine mysteries. Of this latter group, research indicates that these particular cases may fall into any one of the following categories:
>
> 1) Extraterrestrial craft based upon a technology well in advance of that currently held by present day humanity, piloted by other life forms not from the Earth. A variation on this is that they are not so much extraterrestrial but inter-dimensional craft, and that the beings within are not necessarily of 'our' physical reality as such, but are nevertheless able to manifest here along with their craft.
>
> 2) Man made craft based upon scientific and technological breakthroughs in the physical sciences that have been kept secret by various governments and private

companies for decades. In general, the craft that they have developed based upon such secret scientific progress is used for covert operations.

3) Phenomena not necessarily 'machine like' in form, but which do demonstrate ordered patterns of movement and/or structure that imply an underlying purpose or intelligence. Some such phenomena could be the result of active manipulation by certain groups of the Earth's energy fields through advanced technology to produce certain controlled effects, or they could perhaps even be the result of more natural but ordered Earth-energy processes that are at present little understood. An example of the latter includes ball lightning formations at various locations upon the Earth.

Usually, most credible researchers of the UFO phenomenon would consider that the 'genuine mystery' cases fall into one of the above stated categories. For those interested in knowing more, one may consult any number of works upon the subject that offer a more detailed analysis of such mysteries.

[3] Cathie, B. L. (1997) p.10 — *The Bridge to Infinity: Harmonic 371244*
Adventures Unlimited Press

[4] Cathie, B. L. (1997) p.7 — *The Energy Grid: Harmonic 695, the Pulse of the Universe*
Adventures Unlimited Press

[5] Sitchin, Z. (1978) p. 18-19. — *The 12th Planet*
Avon Books

Further note:
Specifically as to the emergence of Babylon, Sitchin gives a date of circa 1900 BC. As to its fall, 539 BC, when conquered by Cyrus the Achaemenid.

[6] Cathie, B. L. (1993) (video) — *Secret Technology and the World Energy System*

[7] Cathie, B. L. (1997) p.14 — *The Bridge to Infinity: Harmonic 371244*
Adventures Unlimited Press

Cathie notes the work of Aime Michel

[8] Cathie, B. (1995) p.11-12 — *The Harmonic Conquest of Space*
Nexus Magazine

[9] Cathie, B. L. (1997) p.25 — *The Energy Grid: Harmonic 695, the Pulse of the Universe*
Adventures Unlimited Press

Further note:
In his many published works Cathie has written extensively upon how exactly he came to discover the particular grid configuration covering the planet that seems to be associated with this object. One may consult any number of his books listed throughout this present work for a more detailed account.

[10] Cathie, B. (1995) p.11-12 *The Harmonic Conquest of Space*
Nexus Magazine

Further note: Cathie relates that the position of the aerial was given some time after its discovery in a local newspaper

[11] P. Moore (1987) p.296 (General Editor) *The Astronomy Encyclopaedia*
Mitchell Beazley Publishers

Further note: The technical term for this angle is the *Obliquity of the Ecliptic*

[12] *Microsoft Encarta 98,* from Microsoft Table entitled "*Characteristics of Earth*"

[13] http://members.aol.com/JackProot/met/spvolas.html

[14] http://members.aol.com/JackProot/met/spvolas.html

[15] http://roland.lerc.nasa.gov/dglover/dictionary//n.html

[16] Sitchin, Z. (1978) p.19 Ibid. *The 12th Planet*
Avon Books

[17] Sitchin, Z. (1978) p.21 Ibid.

[18] Sitchin, Z. (1978) p.49 Ibid.

[19] Sitchin, Z. (1990) p.214 *Genesis Revisited*
Avon Books

[20] Alford, A. F. (1998) p.180 *Gods of the New Millennium*
Hodder and Stoughton

The table of values reproduces the number sequence used by Alford.

[21] Bostock, L. & Chandler, S. (1994) p.163 *Core Maths for A-Level*
Stanley Thornes (publishers) Ltd

[22] P. Kenneth Seidelmann (Editor) (1992) p.698 *Explanatory Supplement to the Astronomical Almanac*
University Science Books

The reference given is for the Earth tropical year (Equinox to equinox)

Chapter 2

[1] Cathie, B. L. (1997) p.15-16 *The Bridge to Infinity: Harmonic 371244*
Adventures Unlimited Press

[2] Cathie, B. L. (1997) p.15 Ibid

[3] Cathie, B. L. (1997) p.16 Ibid

[4] Cathie, B. L. (1997) p.8 Ibid

Ordinarily the completion of 1 day takes 86400 seconds. Cathie instead splits up the Earth day into 97200 units, which he refers to as grid seconds. Thus, one grid second is 8/9th of a standard second, as 86400 / 97200 = 8/9.

[5] P. Kenneth Seidelmann (Editor) (1992) p.716 — *Explanatory Supplement to the Astronomical Almanac* University Science Books

Velocity of light given as 299792458 metres per second
Thus, 299792458 / 1000 = 299792.458 kilometres / second

[6] P. Kenneth Seidelmann (Editor) Ibid (1992) p.716

1 metre is given as 0.0006213711922 miles
Thus, 1 kilometre = 0.6213711922 miles

[7] Cathie, B. (1995) p.37 — *The Harmonic Conquest of Space* Nexus Magazine

[8] Cathie, B. L. (1993) (video) — *Secret Technology and the World Energy System*

Chapter 3

[1] P. Kenneth Seidelmann (Editor) (1992) p.700 — *Explanatory Supplement to the Astronomical Almanac* University Science Books

[2] Kenneth Seidelmann (Editor) 1992 Ibid

1) Mars Orbital Period is derived from the following:
Tropical Period of Mars (Julian years) = 1.88071105 (p.704)
1 Julian year = 365.25 days (p.730)

Therefore, based upon the above values, this author determines the Mars orbital period as follows:

Tropical (Orbital) Period of Mars = 1.88071105 × 365.25 =
686.9297110125 days
2) The Mean distance of Mars (from Sun) in AU given as 1.5236793419 (p.704), and 1 AU = 149597870 kilometres.

Therefore, based upon the above values, this author determines the Mean distance of Mars from the Sun as follows:

$1.5236793419 \times (149597870 \times 0.6213711922 \times 0.88) =$
$1.5236793419 \times 81801110.012 =$ **124638661.47** IGM

Chapter 4

[1] This estimate is based upon the semi-major axis value of the moon orbit divided by the equatorial radius of the Earth, as they are physically equivalent component values:

Mean distance between centre of Earth and centre of moon (Moon semi-major axis) = 384404 kilometres
Cited value by Abell, Morrison & Wolf (1987)
The Exploration of the Universe,
And referenced from Cathie B. L. (1995) p.120
The Harmonic Conquest of Space
Nexus Magazine

Earth Equatorial Radius = 6378140 metres
Cited in P. Kenneth Seidelmann (Editor). (1992) p.696
Explanatory Supplement to the Astronomical Almanac
University Science Books

Therefore: $384404 / (6378140 / 1000) = 60.26$

[2] The physical semi-major and semi-minor Earth axis values are calculated as follows:

Semi-major axis = $21914.531382 / (\mathrm{PI} \times 2) =$ **3487.8059949** IGM

Semi-minor axis = $3487.8059949 - (3487.8059949 \times (1 / 298.257223563)) =$
3476.1120418 IGM

And, the elliptical Earth circumference = $(3487.8059949 + 3476.1120418) \times \mathrm{PI} =$
21877.793744 IGM

[3] The orbital semi-major and semi-minor axis Moon values are calculated as follows:

Moon semi-major axis = 384404 kilometres
$384404 \times 0.6213711922 \times 0.88 =$ **210194.6631544** IGM

Moon semi-minor axis:
$210194.6631544 \times 0.05490 = 11539.6870071$

With:

Semi minor axis = $\sqrt{210194.6631544^2 - 11539.6870071^2}$ = **209877.6597029** IGM

Also, the values for the Celestial Equator and Elliptical
Orbit of the Moon are calculated as follows:

Elliptical Orbit = (210194.6631544 + 209877.6597029) × PI **= 1319696.123465** IGM

Celestial Equator = 210194.6631544 × PI × 2 = **1320692.019179** IGM

NB: The value given as 0.0549 is the Mean Eccentricity of the Moon orbit, as referenced from: P. Kenneth Seidelmann (Editor) (1992) p.701 Ibid

Chapter 5

[1] Graves, R. (1992), pp.21-22 — *The Greek Myths*, Penguin

[2] Graves, R. (1992), p.22 Ibid

[3] De Santillana, G & (1977), p.50 Von Dechend, H — *Hamlet's Mill*, A Nonpareil Book David R. Godine; Publisher; Boston

[4] De Santillana, G & Von Dechend, H. (1977) p.12-13 Ibid

[5] De Santillana, G & Von Dechend, H. (1977) p.26-27 Ibid

[6] De Santillana, G & Von Dechend, H. (1977) p.36 Ibid

[7] De Santillana, G & Von Dechend, H. (1977) p.37 Ibid

[8-9] Sowerby, R. (1993) p.5 — *York Notes on The Iliad, Homer*, Longman York Press

[10] Homer (1950) p. XV — *The Iliad*, Translated by E. V. Rieu, Penguin Books

[11] Sowerby, R. (1993) p.8 Ibid

[12] Sowerby, R. (1993) p.8 Ibid

[13] Publius Ovidus Naso (Ovid) — *Metamorphoses*, Translated by M. M. Innes, Penguin Books (1955) p.98

[14-16] Homer (1991) p.114 — *The Odyssey*, Translated by E. V. Rieu, Revised by D. C. H. Rieu, in Consultation with Dr. P. V. Jones, Penguin Books

[17] Ovid *Metamorphoses,* (1955) p.98-99 Ibid

[18-22] Homer, *The Odyssey,* (1991) p.115-17 Ibid

[23] De Santillana, G & Von Dechend, H. (1977) p.177 Ibid

[24] One should recall here the astronomical fact mentioned earlier that the closer a planet is to the sun, the shorter its orbital period. Thus, it is the 'overtaking' manoeuvres of the planets due to their differing orbital periods that are the direct cause of conjunctions. This then reveals the physical truth of the so called embrace of Mars and Venus from the view of the Sun, as related in the myth.

[25-39] Ovid *Metamorphoses*, (1955) Ibid

25-26: p.49 | 30: p.52 | 33-34: p.55 | 39: p.60
27: p.50 | 31: p.53 | 35: p.57
28-29: p.51 | 32: p.54 | 36-38: p.58

[40] Plato *Timaeus and Critias*
Translated by D. Lee
Penguin Books (1977), p.34

[41] Graves, R. (1992), p.193 Ibid

[42-43] Plato, *Timaeus and Critias,* (1977), p.35

[44] Plato, *Timaeus and Critias,* (1977), p.36

[45-46] Gahlin, L. (2003), p.80 *The Myths and Mythology of Ancient Egypt*
Anness Publishing Limited

[47] http://www.neferchichi.com/re.html

Internet Article: *Neferchichi's Tomb: Gods & Goddesses*

[48] Gahlin, L. (2003), p.80 Ibid

[49-50] http://www.neferchichi.com/re.html Internet Article, Ibid

[51-52] Gahlin, L. (2003), p.80 Ibid

[53] Gahlin, L. (2003), p.51 Ibid

Chapter 6

[1] http://webexhibits.org/calendars/calendar-ancient.html

Internet Article: Calendars through the ages: Other ancient Calendars

[2] *The Astronomical Almanac for the year 2003* (2001) Published by: The Stationary Office Page: D2

[3] http://webexhibits.org/calendars/calendar-ancient.html

[4] Gahlin, L. (2003), p.81 Ibid

[5] http://webexhibits.org/calendars/calendar-ancient.html

Chapter 7

[1] P. Kenneth Seidelmann (Editor) (1992) p.700 *Explanatory Supplement to the Astronomical Almanac* University Science Books

Mean Eccentricity = 0.016708617
1 astronomical unit of length = $1.49597870 \times 10^{11}$ m
Mean distance of Earth from Sun = 1.0000010178 AU

Therefore (authors' own calculation):

$1.49597870 \times 10^{11} / 1000 = 149597870$ kilometres
$149597870 \times 1.0000010178 = 149598022.2607$ km
$149598022.2607 \times 0.6213711922 \times 0.88 =$

Earth Semi-Major Axis = 81801193.269752711 IGM

[2] Greene, B. (1999), pp.32-33 *The Elegant Universe* Vintage 2000

[3] Rowell, G. & Herbert, S. (1986), p.195 *Physics: A course for GCSE* Cambridge University Press

[4-9] Nieper, Hans A. (1985), pp.19-20 *Conversion of Gravity Field Energy. Dr. Nieper's Revolution in Technology, Medicine and Society* MIT Verlag, Oldenburg

[10] Cathie, B. L. (1995), p.22 *The Harmonic Conquest of Space* Nexus Magazine

The Conversion into Feet is as follows:
$(9.78039 / 0.9144) \times 3 = 32.0878937$ feet / second squared
$(9.83217 / 0.9144) \times 3 = 32.2577755$ feet / second squared

[11-12] Cathie, B. L. (1995), p.23 Ibid

[13] Cathie, B. L. (1995), pp.38-39 Ibid

[14] See: *The Harmonic Conquest of Space* Ibid.
Primarily Chapter 1, for the relevant explanation

[15] Cathie, B. L. (1995), p.37 Ibid

[16] In the following chapter (Chapter 8) there is a far more detailed examination of the true physical size of the Earth in its ideal state; and not just with regard to its equatorial circumference but also its elliptical circumference. The arguments developed will indeed demonstrate strong support for the point made here that there really is no physical validity to a nautical mile unit of 6076 ft with respect to the ideal Earth.

Chapter 8

[1-2] *'Absolute Zero'* (title of CD ROM Article)
Contributed By: Fred Landis
Microsoft Encarta Encyclopaedia 1998

[3] *'Temperature'* (title of CD ROM Article)
Encyclopaedia Britannica Ready Reference,
Inc. © 2001 Encyclopaedia Britannica, Inc.

[4] *'Absolute Zero'* Ibid
One may note at least from a practical point of view that to date no gas has ever been experimentally reduced to absolute zero as such. Rather, this theoretical temperature ideal has merely been closely approached. As an example one may consider that the evaporation of liquid helium has achieved temperatures as low as 0.7 degrees Kelvin.

[5] http://www.nationmaster.com/encyclopedia/List-of-elements-by-melting-point

Internet Article: *Encyclopedia: List of Elements by Melting point*
The reference from the listed table on the web page is:
carbon (diamond) C 3550

[6] Encyclopaedia Britannica Deluxe Edition 2005 CD ROM
Encyclopaedia Britannica Inc. Article: *diamond*

[7-8] W. Russell & L. Russell (1957) p.81 — *Atomic Suicide*, University of Science & Philosophy

[9] W. Russell & L. Russell (1957) p.121 Ibid

[10-11] W. Russell & L. Russell (1957) p.114 Ibid

[12] W. Russell & L. Russell (1957) p.121 Ibid

[13] Sitchin, Z. (1990) p.212 — *Genesis Revisited*, Avon Books

[14] Sitchin, Z. (1978) pp.247-248 *The 12th Planet* Avon Books

[15] P. Kenneth Seidelmann (Editor) (1992) p.700 *Explanatory Supplement to the Astronomical Almanac* University Science Books

[16] Cathie, B. (1995) *The Harmonic Conquest of Space,* p.35 Ibid

Chapter 9

[1] P. Kenneth Seidelmann (Editor) (1992) p.701 *Explanatory Supplement to the Astronomical Almanac* University Science Books

[2] http://www.hermit.org/eclipse/when_stats.html
Internet Article: *Hermit Eclipse Eclipse Statistics*

[3] http://www.atributetohinduism.com/Advanced_Concepts.htm
Internet Article: *Shri 108 & Other Mysteries*

[4-5] http://www.atributetohinduism.com/Advanced_Concepts.htm Ibid

[6] P. Kenneth Seidelmann (Editor) (1992) p.700 Ibid:
Sun radius = 696000000 metres
Thus, 696000000 / 1000 = 696000 kilometres
696000 × 2 = diameter = 1392000 kilometres

[7] P. Kenneth Seidelmann (Editor) (1992) p.701 Ibid:
Moon, radius = 1738 kilometres
Thus, 1738 × 0.6213711922 × 0.88 = 950.3499561 IGM

Chapter 10

[1] In the section that follows with the analysis of the planets, the lengths of the orbital periods of the planets, including their semi-major axis values as stated at the point of introducing each planet, are all, with the exception of Ceres, derived from table 15.6 on page 704 of the *Explanatory Supplement to the Astronomical Almanac* (1992) – P. Kenneth Seidelmann (Editor), University Science Books. In each instance, the figures for the Orbital Periods of the planets are calculated by multiplying by 365.25 the values given for the planets under the column: Tropical Period (Julian Years). The semi-major axis values are determined by converting into IGM the values given in metres in table 15.6 under the heading: Mean Distance (10^{11} m). This is done by dividing each by 1000 and then multiplying the answers by 0.6213711922 and 0.88 successively.

[2] http://filer.case.edu/~sjr16/asteroid.html

[3] *The Astronomical Almanac for the year 2003* (2001) Published by: The Stationary Office

The diameter given for Ceres (Page: G2) in this publication is 932.6 kilometres. Converted to IGM this = 509.951

[4] http://www.princeton.edu/~willman/planetary_systems/Sol/Ceres

From this site the semi-major axis of Ceres is given as 2.7658 AU where 1 AU = $1.49597870691 \cdot 10^{11}$ m.
The orbital period is given as 4.59984 years, where 1 year = 365.25 days.

Thus: (149597870.691 × 2.7658) × 0.6213711922 × 0.88 = 226245511 Ideal Geographical Miles
And: 4.59984 × 365.25 = 1680.09156 days

[5] *The Astronomical Almanac for the year 2003* (2001) Published by: The Stationary Office Page: K7

The radii given for the two planets in this publication are:
Uranus = 25400 km, Neptune = 24300 km.
Converted into IGM the diameters are 27777 IGM and 26574 IGM, respectively.

[6] In addressing this issue attention is thus drawn to a whole new class of laws of proportion as yet undiscovered, that must link the transformation of the orbital periods of all of the planets to their physical size. A discussion of such laws is however beyond the scope of this present work. For the moment there must be contentment solely with the orbital laws of the planets.

Chapter 11

[1] *The Astronomical Almanac for the year 2003* (2001) Published by: The Stationary Office Page: C1

[2] Two particular groups which appear to support this view are the Binary Research Institute and the Sirius Research Group. Both have their own websites established at the following addresses:

http://www.binaryresearchinstitute.org/index.shtml
http://www.siriusresearchgroup.com/

One may note that the Sirius Research group supports the view that the star in question about which the whole solar system may be in orbit is that of Sirius itself, a star which is approximately 45,005,500,000,000 IGM distant from the sun; or just over 550000 times the distance separating the Earth and the sun (authors' own calculation).

[3] This particular model is one devised by this present author as one possible representation of the motions of the relevant bodies, put forward as a simple way to demonstrate the principle of how the Earth may complete a 360 degree orbit about the sun in a tropical year whilst the solar system itself is as a whole, a part of a binary system involving another star. Others of course may hold to slightly different models of the exact relations of the noted bodies.

[4-5] For a review of the work in this and other related areas one is directed to the most excellent work of the Binary Research Institute & Sirius Research Group. See note number [2] above for details.

[6] This is the personal judgment of the present author based upon reading various articles on the internet, including those listed on the websites of the Binary Research Institute and the Sirius Research Group, and certain other sites.

[7] Sitchin, Z. (1990) p.215 — *Genesis Revisited*, Avon Books

[8] P. Kenneth Seidelmann (Editor) (1992) p.696 — *Explanatory Supplement to the Astronomical Almanac*, University Science Books

Chapter 12

[1] Gahlin, L. (2003), p.80 — *The Myths and Mythology of Ancient Egypt*, Anness Publishing Limited

[2] Plato — *Timaeus and Critias*, Translated by D. Lee; Penguin Books (1977), p.36

[3] Plato, *Timaeus and Critias*, (1977), p. 36-37 Ibid

[4] Plato, *Timaeus and Critias*, (1977), p. 36 Ibid

Chapter 13

[1] Dunn, C. (1998) — *The Giza Power Plant*, Bear & Company

[2] Cathie, B. (1995), p.166-7 — *The Harmonic Conquest of Space*, Nexus Magazine

[3] Bauval, R. & Gilbert, A. (1997), p.39 — *The Orion Mystery*, Mandarin Paperbacks

[4] Cathie, B. L. (1997), p.59 — *The Bridge to Infinity: Harmonic 371244*, Adventures Unlimited Press

[5] Dunn, C. (1998), p.133 Ibid

[6] Bauval, R. & Gilbert, A (1997), p.40 Ibid

[7] Dunn, C. (1998), p.138 Ibid

[8] P. Kenneth Seidelmann (Editor) (1992) *Explanatory Supplement to the Astronomical Almanac* University Science Books

Polar radius of Earth cited in this publication (p.700) is: 6356755 metres
Therefore: (6356755 / 1000) × 0.6213711922 = 3949.904432873311 miles

[9] *The Astronomical Almanac for the year 2003* (2001) Published by: The Stationary Office Page: K13

[10] This arc length was calculated using the *Great Circle Calculator* of Ed Williams, which can be found at the following internet address: http://williams.best.vwh.net/gccalc.htm

[11-13] Calculated using the *Great Circle Calculator* of Ed Williams, Ibid

[14] http://www.celticnz.co.nz/US2.html

[15] http://www.metrum.org/measures/dimensions.htm.

[16] The three arc length measures of 'Bessel, Clark, and International', Calculated with *Great Circle Calculator* of Ed Williams, Ibid

[17-20] Calculated using the *Great Circle Calculator* of Ed Williams, Ibid

[21] Cathie, B. (1995), p.98 *The Harmonic Conquest of Space* Nexus Magazine

[22-24] Calculated using the *Great Circle Calculator* of Ed Williams, Ibid

Chapter 14

[1] http://nuclearweaponarchive.org/Usa/Tests/Trinity.html

[2] http://ed-thelen.org/Reunion2004-RonsPictures/P2791.JPG

[3] https://www.osti.gov/opennet/reports/rdannual.pdf (p.15)

[4] http://nuclearweaponarchive.org/Nwfaq/Nfaq8.html#nfaq8.1.2

[5] Sang, D. (1995), p.155 *Basic Physics 1 & 2* Cambridge University Press

[6] Cathie, B. L. (1997), p.172 *The Bridge to Infinity:Harmonic 371244* Adventures Unlimited Press

One may note the example that Cathie gives on p.172 of this publication

[7] Cathie, B. L. (1997) p.239 *The Energy Grid: Harmonic 695, the Pulse of the Universe* Adventures Unlimited Press

[8] http://www.johnstonsarchive.net/nuclear/tests/USA-ntests1.html

[9] Calculated using *Great Circle Calculator* of Ed Williams, Ibid
(Using the International 1924 ellipsoid model)

[10] Cathie, B. L. (1997), p.172 *The Bridge to Infinity: Harmonic 371244* Adventures Unlimited Press

Chapter 15

[1] Latitude and longitude coordinates hereby given were sourced from: http://explorer.altopix.com/map/0q8lys/Trinity_Site.htm

[2] Calculated using *Great Circle Calculator* of Ed Williams, Ibid, employing the International (1924) Earth Ellipsoid Model

[3] Calculated using *Great Circle Calculator* of Ed Williams, Ibid, employing the International (1924) Earth Ellipsoid Model

The arc length value itself is evident at the detonation of the
Trinity test at Universal Time: July 16 1945, 11:29:21.0,
with a Sun Ground Position of: 21.52399973677 N, 9.1333333333 E

[4] For a full breakdown of sources for the information on the details of the bombs as presented throughout Chapter 15 in the following format, see Appendix F:

Target/Shot: Yield (TNT in Tonnes):
Date: Time (GMT):
Geodetic Bomb Co-ordinates (Deg): Lat: Long:
Geodetic Sun-Ground Position Co-ordinates (Deg):
Lat: Long:
Arc Length: Equator to Latitude of Test (feet):
Arc Length: Test Site to Sun-Ground Position (feet):

[5] http://www.johnstonsarchive.net/nuclear/tests/wrjp205a.html

Chapter 17

[1] http://www.seismo.unr.edu/ftp/pub/louie/class/100/magnitude.html

[2] http://ludb.clui.org/ex/i/AK3128/

[3] Of *Cannikin*, the primary details of the bomb are taken from:

http://www.johnstonsarchive.net/nuclear/tests/USA-ntests2.html

This includes the Yield, Date, Time and the Geodetic Bomb Co-ordinates of the test. The Geodetic Sun Ground Position Co-ordinates as calculated were determined in accordance with the methodology detailed in Appendix F, as also used to evaluate various nuclear weapons tests.

All arc length measures calculated using Ed William's Great Circle Calculator: http://williams.best.vwh.net/gccalc.htm

[4-5] http://webmap.ga.gov.au/imf-natural_hazards/imf.jsp?site=natural_hazards_earthquake

[6] Of the *Cannikin Aftershock*, the basic details, including the Date, Time, and the geodetic epicentre co-ordinates of the event, are from the Australian Government 'Natural Hazards' database:

http://webmap.ga.gov.au/imf-natural_hazards/imf.jsp?site=natural_hazards_earthquake
The Geodetic Sun Ground Position Co-ordinates as calculated were determined in accordance with the methodology detailed in Appendix F, as also used to evaluate various nuclear weapons tests.

All arc length measures were calculated using Ed William's Great Circle Calculator: http://williams.best.vwh.net/gccalc.htm

[7-8] Thomas E. Bearden (2002) p.230 — *Excalibur Briefing* ADAS Press

[9-10] Thomas E. Bearden (2002) p.234 — *Excalibur Briefing* ADAS Press

Chapter 18

[1] http://www.johnstonsarchive.net/nuclear/tests/1961USSR-1.html

[2] http://webmap.ga.gov.au/imf-natural_hazards/imf.jsp?site= natural_hazards_earthquake

[3] http://ublib.buffalo.edu/libraries/asl/guides/indian-ocean-disaster.html

[4-5] Of the two *Ocean Events*, the basic details, including the Date, Time, and the geodetic epicentre co-ordinates of the events are from the Australian Government 'Natural Hazards' database:

http://webmap.ga.gov.au/imf-natural_hazards/imf.jsp?site=
natural_hazards_earthquake
The Geodetic Sun, Moon and Venus Ground Position Co-ordinates as calculated were determined in accordance with the methodology detailed in Appendix F, as also used to evaluate various nuclear weapons tests.

All arc length measures were calculated using Ed William's Great Circle Calculator: http://williams.best.vwh.net/gccalc.htm

[6] Although there have only been two earthquakes evaluated in the course of this present work, several additional earthquakes of some prominence have also been examined by this author, which indeed reveal a variety of decisive associations quite similar to the 2004 events as have been presented. However, due to the fact though that a comprehensive evaluation of many other earthquake events would make Chapter 18 inordinately long, and unnecessarily so, only the 2004 earthquake examples have been selected for presentation in this work. Suffice it to say, that though the sample of earthquakes as reviewed is not nearly as large as that of the nuclear weapons tests that have been considered, the associations of both are indicative of the same underlying physical science at work. That said; it is the intention of this present author to post further details of additional earthquake events, including detailed analysis, upon the website of this author: http://www.ancient-world-mysteries.com

[7] G. P. Verbrugghe & J. M. Wickersham (2001) p.66 — *Berossos and Manetho: Native Traditions in Ancient Mesopotamia and Egypt* University of Michigan Press

Chapter 19

[1] Berlitz, C. & Moore, W. (2004) p.115 — *The Philadelphia Experiment: Project Invisibility* Souvenir Press

[2] Childress, D. H. (1999) p.62-63 (Editor) — *Anti-Gravity and the Unified Field* Adventures Unlimited Press

The specific chapter within this publication that is referenced is called *The Vortex Arena*, by John Walker

[3] Childress, D. H. (1999) p.60, ibid (*The Vortex Arena*, J. Walker)

[4] Childress, D. H. (1997) p.115 (Editor) — *Anti-Gravity and the World Grid* Adventures Unlimited Press

The specific chapter within this publication that is referenced is called *The Philadelphia Experiment*, by Harry Osoff.

[5-6] Berlitz, C. & Moore, W. (2004) p.31 — *The Philadelphia Experiment: Project Invisibility* Souvenir Press

[7] Childress, D. H. (1997) p.119, ibid (*The Philadelphia Experiment* by, Harry Osoff)

[8] Charles Berlitz, co-author along with William Moore of:

The Philadelphia Experiment: Project Invisibility (2004), Souvenir Press,

[9-10] Cathie, B. L. (1997) p.102 — *The Bridge to Infinity: Harmonic 371244* Adventures Unlimited Press

[11] For a full account of Cathie's reasoning and analysis one should consult his book:

The Bridge to Infinity: Harmonic 371244, pages 96-106.

[12] Cathie, B. L. (1997) *The Bridge to Infinity:* p.104, ibid

[13-14] Arc length calculated using Ed Williams' *Great Circle Calculator*: http://williams.best.vwh.net/gccalc.htm

[15] Berlitz, C. & Moore, W. (2004) p.106 — *The Philadelphia Experiment: Project Invisibility* Souvenir Press

[16] Childress, D. H. (1999) p.180 (Editor) — *Anti-Gravity and the Unified Field* Adventures Unlimited Press

[17] To elaborate further upon the very issue of hyperspatial travel as suspected between covert facilities on different celestial bodies, one may note an additional point raised by Bearden with regard to nuclear howitzers. Upon the issue of the transmission of nuclear explosions through hyperspace, Bearden notes that tuners, carefully placed at a desired target site, can be utilised to actually 'receive' a transmitted explosion (Thomas E. Bearden, 2002 p.230, *Excalibur Briefing*, ibid). Of course such tuners are certainly not by any means a necessity, for as with the Cannikin evaluation, it is quite clear that nuclear aftershock events do not *require* the use of such artificial receivers. That said; they may in fact be the key component that does allow for stable hyperspatial travel over and above relying solely upon the orbital positions of the major celestial bodies.

One is thus led to theorise then that were one to transmit some object in a stable manner to another celestial body (i.e. not as an explosion) then one would *initially* be bound by the physical dynamics of the celestial bodies, in actually engineering the first transmission. However, if one were to send through some sort of special tuner or beacon that would shortly upon arrival self activate, then one may be able to use it essentially to establish a link-up between the two locations in accordance with *a unique path of least resistance*. Once such a connection was set up, then it may well be possible to artificially establish communication between two sites without further need to rely upon the dynamics of the celestial spheres. Following this line of development, one may thus suggest that it is a distinct possibility that stable portals have already been established that allow for passage between worlds at any time, and not necessarily only at those times when the planets *naturally* allow for movement.

Chapter 20

[1] http://en.wikipedia.org/wiki/Pine_Gap

[2] http://www.thewatcherfiles.com/pinegap.htm

[3] Arc length calculated using Ed Williams' *Great Circle Calculator*: http://williams.best.vwh.net/gccalc.htm

[4] http://www.euronet.nl/~rembert/echelon/mwh.htm

[5-8] Calculated using Ed Williams' *Great Circle Calculator*, ibid

[9] http://nssdc.gsfc.nasa.gov/planetary/viking.html

[10] P. Moore (1987) p.447 (General Editor) — *The Astronomy Encyclopaedia* Mitchell Beazley Publishers

[11] Cathie, B. (1995) p.126 — *The Harmonic Conquest of Space* Nexus Magazine

[12] http://www.mt.net/~watcher/monument.html

[13-15] http://www.msss.com/education/facepage/face_discussion.html

[16] P. Kenneth Seidelmann (Editor) (1992), p.697 — *Explanatory Supplement to the Astronomical Almanac* University Science Books

The value of the equatorial radius of Mars in this publication is given as 3397.2 Kilometres. In Ideal Geographical Miles:
$3397.2 \times 0.6213711922 \times 0.88 = 1857.611548444$ IGM

[17] Kenneth Seidelmann (Editor) 1992 Ibid, p.706

The value of flattening for Mars is given as 0.00647630
Therefore, inverse flattening is calculated as follows:
$1 / 0.00647630 = 154.40915337461204\ldots$

[18] Circumference calculated using Ed Williams' *Great Circle Calculator*: http://williams.best.vwh.net/gccalc.htm

[19-20] Calculated with Ed Williams' *Great Circle Calculator*, ibid.

Appendix B

[1] http://www.stetson.edu/~efriedma/periodictable/html/Pm.html

Internet Article: *Claudius Ptolemy*
Primarily, Ptolemy's actual description of the universe is based upon the Earth centred theories of Aristotle.

[2] On the Longevity of the Ptolemaic system:

The reasons for this are no doubt many and varied, but are bound to include such as the lack of strong communications networks between the learned men of the age, and also too much reverence being given to both Ptolemy and Aristotle. However, one must not forget to mention also the fact that during these times alternative thinking upon such matters was forcefully discouraged quite frequently by those who had sizable ruling power over the continent. Thus indeed, the Roman Church, a most active power in Europe during the middle-ages and a known enemy of truth, was so strongly intent upon upholding the Ptolemaic Earth-centred system for dogmatic reasons that they persecuted, sometimes to the point of death, those who thought to develop and publicise other ideas. Nevertheless, even in spite of such strong opposition, as is evident from history, the ancient Ptolemaic system did eventually meet its end, though its death was a slow process and did take several successive generations of great and courageous thinkers.

[3] http://www.321books.co.uk/encyclopedia/astronomy/ancient/greek-astronomy.htm

Internet Article: *Greek astronomy*

[4] http://www.astronomynotes.com/history/s4.htm

Internet Article: *History and philosophy of western astronomy: Renaissance*

[5] P. Moore (General Editor) (1987) p.84 Ibid

[6] P. Moore (1987) p.106 Ibid

[7] P. Moore (1987) p.114 Ibid

[8-11] http://astro.wsu.edu/allen/courses/astr150/Notes/week7.html

Internet Article: *ASTR 150 Course Pages*

[12] Science of Kepler & Fermat. EIRVI – 2001-12 (Video)

[13] P. Moore (General Editor) (1987) p.222 Ibid

[14] http://larouchein2004.net/pages/writings/2002/020125 enddelusionch1.htm

Internet Article: *Economics: At The End Of A Delusion*
Lyndon H. LaRouche, Jr. (Jan 12, 2002)

[15] P. Kenneth Seidelmann (Editor) (1992) *Explanatory Supplement to the Astronomical Almanac* University Science Books

The values used for the example calculation are referenced and derived from the above noted publication as follows:

1) The Earth Tropical year = **365.2421897** days (p.698)

2) The Earth semi-major axis is taken to be the product of the following:

1 astronomical unit of length = 149597870000 metres (p.700),

Mean distance of the Earth from Sun = 1.0000010178 AU (p.700)

Therefore: (149597870000 / 1000) × 1.0000010178 = 149598022.26071 kilometres

Converted into Ideal Geographical miles:
149598022.26071 × 0.6213711922 × 0.88 = **81801193.26975** IGM

3) Mars Orbital Period is derived from the following:
Tropical Period of Mars (Julian years) = 1.88071105 (p.704)
1 Julian year = 365.25 days (p.730)
Therefore, based upon the above values, this author determines the Mars orbital period as follows:
Tropical (Orbital) Period of Mars = 1.88071105 × 365.25 = **686.9297110125** days

[16] Kenneth Seidelmann (Editor) 1992 Ibid

Mean distance of Mars (from Sun) in AU given as 1.5236793419 (p.704)

Based upon the above values, this author determines the Mean distance of Mars from the Sun as follows:

1.5236793419 × (149597870 × 0.6213711922 × 0.88) = **124638661.47** IGM

Appendix C

[1] http://www.creation-answers.com/slowing.htm

The Slowing Spin of Earth: Is Earth's Rotation Slowing Down Throughout Time?

[2] http://www.creation-answers.com/slowing.htm. Ibid

Appendix G

[1] *The Bridge to Infinity: Harmonic 371244,* Adventures Unlimited Press (1997)
The Harmonic Conquest of Space, Nexus Magazine (1995)
The Energy Grid: Harmonic 695, the Pulse of the Universe. Adventures Unlimited Press (1997)

[2] Cathie, B. L. (1997) p.23 *The Bridge to Infinity: Harmonic 371244* Adventures Unlimited Press

[3] Cathie, B. L. (1997) p.80 ibid, *The Bridge to Infinity*

[4] Cathie, B. (1995) p.36 *The Harmonic Conquest of Space* Nexus Magazine

[5] Including: *The Harmonic Conquest of Space, The Bridge to Infinity: Harmonic 371244* and also *The Energy Grid,* all of which have been cited numerous times within this current work.

Appendix H

[1] Cathie, B. (1995) p.37 ibid, *The Harmonic Conquest of Space*

[2] Cathie, B. (1995) p.39 ibid, *The Harmonic Conquest of Space*

[3] Cathie, B. (1995) p.37 ibid, *The Harmonic Conquest of Space*

[4] Cathie, B. L. (1997) p.81-82 *The Energy Grid: Harmonic 695, the Pulse of the Universe* Adventures Unlimited Press

[5-6] Cathie, B. (1995) p.20 ibid, *The Harmonic Conquest of Space*

[7] Cathie, B. (1995) p.21-22 ibid, *The Harmonic Conquest of Space*

Appendix I

[1] Cathie, B. L. (1997) p.104 *The Bridge to Infinity: Harmonic 371244* Adventures Unlimited Press

[2] Cathie, B. L. (1997) p.102 ibid, *The Bridge to Infinity*

[3] Cathie, B. L. (1997) p.104 ibid, *The Bridge to Infinity*

Appendix J

[1-3] http://earthquake.usgs.gov/learning/faq.php?categoryID=11&faqID=69

[4] http://www.universetoday.com/guide-to-space/the-solar-system/how-many-moons-are-in-the-solar-system/

Appendix K

[1-2] http://www.rense.com/general77/fulf.htm

[3-5] http://www.youtube.com/watch?v=0VX0JvpW5q0

[6-7] Of the two Niigata earthquakes, the basic details, including the Date, Time, Magnitude, and geodetic epicentre co-ordinates of the events, are taken from the Australian Government 'Natural Hazards' database:

> http://webmap.ga.gov.au/imf-natural_hazards/imf.jsp?site=natural_hazards_earthquake
>
> The Geodetic Sun, Moon and Mars Ground Position Co-ordinates as calculated were determined in accordance with the methodology detailed in Appendix F, as also used to evaluate various nuclear weapons tests.
>
> All arc length measures were calculated using Ed William's Great Circle Calculator: http://williams.best.vwh.net/gccalc.htm

[8] http://www.projectcamelot.org/benjamin_fulford_interview_ transcript_2.html

[9-11] http://www.haarp.alaska.edu/haarp/faq.html

[12-14] http://www.realufos.net/2008/06/haarp-ionisphernic-warfare-weapon-that.html

[15] http://www.haarp.alaska.edu/

[16] http://www.realufos.net/2008/06/haarp-ionisphernic-warfare-weapon-that.html

[17] Dr. N. Begich & J. Manning (1995), p.26 — *Angels Don't Play this Haarp: Advances in Tesla Technology* Earthpulse Press

[18] Dr. N. Begich & J. Manning (1995), p.80 ibid.

[19] Dr. N. Begich & J. Manning (1995), p.81 ibid.

[20-21] Dr. N. Begich & J. Manning (1995), p.103 ibid.

Appendix L

[1] http://en.wikipedia.org/wiki/Buckley_Air_Force_Base

Latest Research & Updates

To keep up to date with the latest research on the topics as presented in this book by this author, please visit my official website, from which you may also order more copies of *The Lost Age of High Knowledge*:

http://www.ancient-world-mysteries.com